FUNDAMENTALS of KINEMATICS — and — DYNAMICS of MACHINES and MECHANISMS

FUNDAMENTALS of KINEMATICS — and — DYNAMICS of MACHINES and MECHANISMS

Oleg Vinogradov

The software mentioned in this book is now available for download on our Web site at: http://www.crcpress.com/e_products/downloads/default.asp

CRC Press is an imprint of the
Taylor & Francis Group, an **informa** business

CRC Press
Taylor & Francis Group
6000 Broken Sound Parkway NW, Suite 300
Boca Raton, FL 33487-2742

First issued in paperback 2019

ISBN-13: 978-0-8493-0257-2 (hbk)
ISBN-13: 978-0-367-39832-3 (pbk)
Library of Congress Card Number 00-025151

Library of Congress Cataloging-in-Publication Data

Vinogradov, Oleg (Oleg G.)
Fundamentals of kinematics and dynamic of machines and mechanisms / by Oleg Vinogradov.
p. cm.
Includes bibliographical references and index.
ISBN 0-8493-0257-9 (alk. paper)
1. Machinery, Kinematics of. 2. Machinery, Dynamics of. I. Title.
TJ175.V56 2000
621.8'11—dc21
00-025151
CIP

Visit the Taylor & Francis Web site at
http://www.taylorandfrancis.com

and the CRC Press Web site at
http://www.crcpress.com

Preface

The topic of *Kinematics and Dynamics of Machines and Mechanisms* is one of the core subjects in the Mechanical Engineering curriculum, as well as one of the traditional subjects, dating back to the last century. The teaching of this subject has, until recently, followed the well-established topics, which, in a nutshell, were some general properties, and then analytical and graphical methods of position, velocity, and acceleration analysis of simple mechanisms. In the last decade, computer technology and new software tools have started making an impact on how the subject of *kinematics and dynamics of machines and mechanisms* can be taught.

I have taught *kinematics and dynamics of machines and mechanisms* for many years and have always felt that concepts and numerical examples illustrating them did not allow students to develop a perception of a mechanism as a whole and an understanding of it as an integral part of the design process. A laboratory with a variety of mechanisms might have alleviated some of my concerns. However, such a laboratory, besides being limited to a few mechanisms, mainly serves as a demonstration tool rather than as a design tool, since it would be very time-consuming to measure such fundamental properties as position, velocity, and acceleration at any point of the mechanism. It would be even more difficult to measure forces, internal and external. There is yet one more consideration. With class sizes as they are, the experience of a student becomes a group experience, limited in scope and lacking in the excitement of an individual "discovery."

A few years ago I started using Mathematica in my research, and it became clear to me that this software can be used as a tool to study mechanisms. It gives a student a chance to perform symbolic analysis, to plot the results, and, what is most important, to animate the motion. The student thus is able to "play" with the mechanism parameters and see their effect immediately. The idea was not only to develop an understanding of basic principles and techniques but, more importantly, to open a new dimension in this understanding by appealing to the student's visual perception and intuition. I noticed also that it gives students a sense of pride to be able to do something on their own and have it "work." I have seen many times how their eyes brighten when they see their mechanism in motion for the first time. In general, such software as Mathematica allows students to study complex mechanisms without the limitations imposed by either a physical laboratory or a calculator. All of this prompted me to write this book.

The subject of this book is kinematics and dynamics of machines and mechanisms, and Mathematica is used only as a tool. In my view it would be detrimental to the subject of the book if too much attention were placed on a tool that is incidental to the subject itself. So in the book the two, the subject and the tool, are presented separately. Specifically, the Appendix shows how Mathematica can be used to illuminate the subject of each chapter. All the material in the Appendix is available on CD-ROM in the form of interactive Mathematica notebooks. The Problems and Exercises at the end of each chapter are also split into two types: Problems, that do not require the use of Mathematica and emphasize understanding concepts, and Exercises, based on Mathematica, that require students to perform analysis of mechanisms. The second type I call projects, since they require homework and a report.

In my opinion, the use of a symbolic language such as Mathematica should not prevent a student from developing analytical skills in the subject. With this in mind, I provide a consistent analytical approach to the study of simple and complex (chain-type) mechanisms. The student should be able to derive solutions in a closed form for positions, velocities, accelerations, and forces. Mathematica allows one to input these results for plotting and animation. As an option, students can perform calculations for a specific mechanism position using analytical solutions.

In my class, the numerical part of the course is moved to the computer laboratory. It is done in the form of projects and assumes complete analysis, parametric study, and animation. There are two to three projects during the term, which gives students sufficient exposure to numerical aspects of mechanism analysis and design. This procedure then allows the instructor to concentrate in quizzes and exams on understanding of the subject by asking students to answer conceptual-type questions without the students' spending time on calculations. Thus, instructors can cover more material in their tests.

A few basic Mathematica files (programs) are available on the CD-ROM. The intention is to provide students with the foundation needed to solve other problems without spending too much time studying the tool itself. For example, the programs for simple slider-crank and four-bar mechanisms allow students to study a complex mechanism combining them. I must emphasize, however, that the available programs cannot substitute for the Mathematica book by S. Wolfram (see Bibliography).

Specifically, the following programs written in Mathematica are on CD-ROM:

- introduction.nb outlines basic features of Mathematica with examples of numerical calculations, symbolic solutions, plotting, and animation.
- howTo.nb gives a set of Mathematica answers to specific questions arising in solutions of kinematics and dynamics problems, such as, for example, how to write an ordinary differential equation, how to solve it in symbols, how to plot the results, how to draw a line, etc.
- sliderCrank.nb provides a complete solution of the kinematics and dynamics of the offset slider-crank mechanism, with plots and animation.
- fourBar.nb provides a complete solution of the kinematics and dynamics of the four-bar mechanism, with plots and animation.
- CamHarmonic.nb provides analysis of a harmonic cam with oscillating and offset followers.
- 2DOFfree.nb provides analysis of the free vibrations of a two-degree-of-freedom system with damping.
- 2DOFforced.nb provides analysis of the forced vibrations of a two-degree-of-freedom system with damping.

All the above programs contain textual explanations and executable commands (Mathematica allows mingling of text and executable commands in one file).

An inevitable question arises when any software tool is introduced: How do students learn Mathematica? Knowing Mathematica to a sufficient degree is, of course, prerequisite to this course. Some universities introduce students to Mathematica in their calculus courses. In my third-year course in kinematics I introduce Mathematica in the first few weeks in the computer laboratory by giving students the interactive programs Introduction to Mathematica and How-To in Mathematica. The latter answers specific questions relevant to the course material. In addition, I make programs dealing with two basic mechanisms, *slider-crank mechanism* and *four-bar linkage*, available to students. Students use these two programs as starting points for studying more complex mechanisms.

All of the problems listed in this book as assignments were assigned to students as projects over the last 3 years since I began teaching this course in a new format. The students' reaction to this new learning environment helped me design this book. And for that I am thankful to all of them. My specific thanks go to my former third-year student Mr. Yannai Romer Segal who developed all of the graphics for this book. I also appreciate the support provided by the technical personnel in our department, Mr. B. Ferguson, Mr. D. Forre, and Mr. N. Vogt, at various stages of this project. My sincere thanks to the Killam Foundation for awarding me a fellowship to write this book.

Oleg Vinogradov
Calgary, 2000

About the Author

Oleg Vinogradov, Professor of Mechanical Engineering at the University of Calgary, has been involved in the design and analysis of machines in industrial, research institute, and university settings for more then 35 years. He has B.Sc. degrees in Mechanical Engineering and Applied Mechanics and a Ph.D. in Mechanical Engineering. Dr. Vinogradov has published *Introduction to Mechanical Reliability: A Designer's Approach* (Hemisphere), and has written more than 100 papers on a range of topics, including structural and rotor dynamics, vibrations, and machine components design.

Table of Contents

1 Introduction

1.1 THE SUBJECT OF KINEMATICS AND DYNAMICS OF MACHINES

This subject is a continuation of statics and dynamics, which is taken by students in their freshman or sophomore years. In kinematics and dynamics of machines and mechanisms, however, the emphasis shifts from studying general concepts with illustrative examples to developing methods and performing analyses of real designs. This shift in emphasis is important, since it entails dealing with complex objects and utilizing different tools to analyze these objects.

The objective of *kinematics* is to develop *various means of transforming motion* to achieve a specific kind needed in applications. For example, an object is to be moved from point A to point B along some path. The first question in solving this problem is usually: What kind of a mechanism (if any) can be used to perform this function? And the second question is: How does one design such a mechanism?

The objective of *dynamics* is analysis of the behavior of a given machine or mechanism when subjected to dynamic forces. For the above example, when the mechanism is already known, then external forces are applied and its motion is studied. The determination of forces induced in machine components by the motion is part of this analysis.

As a subject, the kinematics and dynamics of machines and mechanisms is disconnected from other subjects (except statics and dynamics) in the Mechanical Engineering curriculum. This absence of links to other subjects may create the false impression that there are no constraints, apart from the kinematic ones, imposed on the design of mechanisms. Look again at the problem of moving an object from A to B. In designing a mechanism, the size, shape, and weight of the object all constitute input into the design process. All of these will affect the size of the mechanism. There are other considerations as well, such as, for example, what the allowable speed of approaching point B should be. The outcome of this inquiry may affect either the configuration or the type of the mechanism. Within the subject of kinematics and dynamics of machines and mechanisms such requirements cannot be justifiably formulated; they can, however, be posed as a learning exercise.

1.2 KINEMATICS AND DYNAMICS AS PART OF THE DESIGN PROCESS

The role of kinematics is to ensure the functionality of the mechanism, while the role of dynamics is to verify the acceptability of induced forces in parts. The functionality and induced forces are subject to various constraints (specifications) imposed on the design. Look at the example of a cam operating a valve (Figure 1.1).

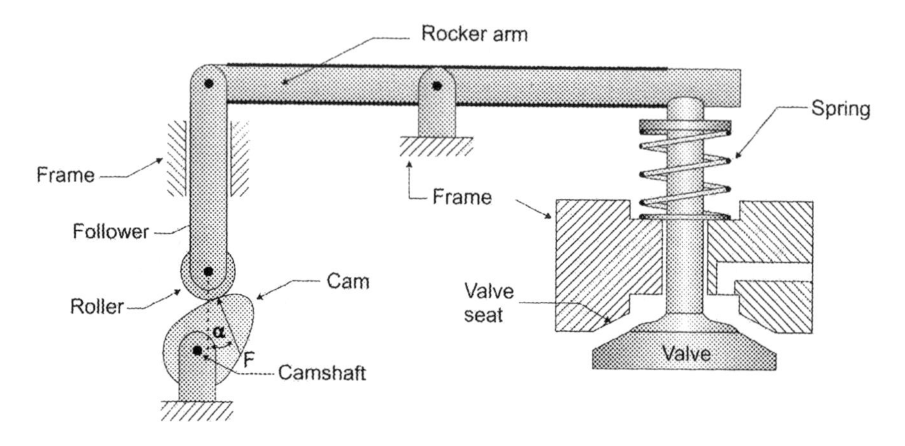

FIGURE 1.1 A schematic diagram of cam operating a valve.

The *design process* starts with meeting the *functional requirements* of the product. The basic one in this case is the proper *opening, dwelling,* and *closing* of the valve as a function of *time*. To achieve this objective, a corresponding cam profile producing the needed follower motion should be found. The rocker arm, being a lever, serves as a displacement amplifier/reducer. The timing of opening, dwelling, and closing is controlled by the speed of the camshaft. The function of the spring is to keep the roller always in contact with the cam. To meet this requirement the inertial forces developed during the follower–valve system motion should be known, since the spring force must be larger than these forces at any time. Thus, it follows that the determination of component accelerations needed to find inertial forces is important for the choice of the proper spring stiffness.

Kinematical analysis allows one to satisfy the functional requirements for valve displacements. Dynamic analysis allows one to find forces in the system as a function of time. These forces are needed to continue the design process. The *design process* continues with meeting the *constraints requirements*, which in this case are:

1. Sizes of all parts;
2. Sealing between the valve and its seat;
3. Lubrication;
4. Selection of materials;
5. Manufacturing and maintenance;
6. Safety;
7. Assembly, etc.

The forces transmitted through the system during cam rotation allow one to determine the proper sizes of components, and thus to find the overall assembly dimension. The spring force affects the reliability of the valve sealing. If any of the requirements cannot be met with the given assembly design, then another set of

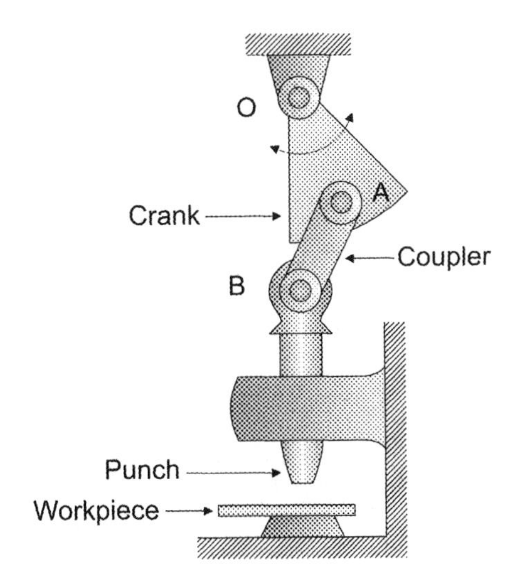

FIGURE 1.2 Punch mechanism.

parameters should be chosen, and the kinematic and dynamic analysis repeated for the new version.

Thus, kinematic and dynamic analysis is an *integral part* of the machine design process, which means it *uses input* from this process and *produces output* for its continuation.

1.3 IS IT A MACHINE, A MECHANISM, OR A STRUCTURE?

The term *machine* is usually applied to a complete product. A *car* is a machine, as is a *tractor*, a *combine*, an *earthmoving machine,* etc. At the same time, each of these machines may have some devices performing specific functions, like a windshield wiper in a car, which are called *mechanisms.* The schematic diagram of the assembly shown in Figure 1.1 is another example of a mechanism. In Figure 1.2 a punch mechanism is shown. In spite of the fact that it shows a complete product, it, nevertheless, is called a mechanism. An internal combustion engine is called neither a machine nor a mechanism. It is clear that there is a historically established terminology and it may not be consistent. What is important, as far as the subject of kinematics and dynamics is concerned, is that the identification of something as a machine or a mechanism has no bearing on the analysis to be done. And thus in the following, the term *machine* or *mechanism* in application to a specific device will be used according to the established custom.

The distinction between the *machine/mechanism* and the *structure* is more fundamental. The former must have moving parts, since it transforms motion, produces work, or transforms energy. The latter does not have moving parts; its function is purely structural, i.e., to maintain its form and shape under given external loads, like a bridge, a building, or an antenna mast. However, an example of a folding

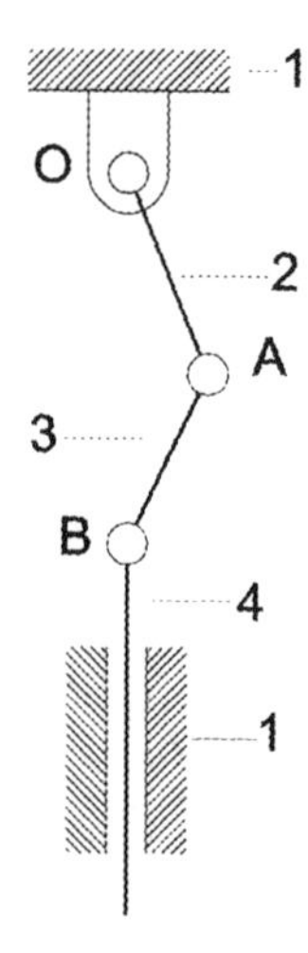

FIGURE 1.3 A skeleton representing the punch mechanism.

chair, or a solar antenna, may be confusing. Before the folding chair can be used as a chair, it must be *unfolded*. The transformation from a folded to an unfolded state is the transformation of motion. Thus, the folding chair meets two definitions: it is a *mechanism* during unfolding and a *structure* when unfolding is completed. Again, the terminology should not affect the understanding of the substance of the matter.

1.4 EXAMPLES OF MECHANISMS; TERMINOLOGY

The punch mechanism shown in Figure 1.2 is a schematic representation of a device to punch holes in a workpiece when the oscillating *crank* through the *coupler* moves the punch up and down. The function of this mechanism is to transform a small force/torque applied to the crank into a large punching force. The specific shape of the crank, the coupler, and the punch does not affect this function. This function depends only on locations of points O, A, and B. If this is the case, then the lines connecting these points can represent this mechanism. Such a representation, shown in Figure 1.3, is called a *skeleton* representation of the mechanism. The power is supplied to crank 2, while punch 4 is performing the needed function.

In Figure 1.3, the lines connecting points O, A, and B are called *links* and they are connected to each other by *joints*. Links are assumed to be rigid. Revolute joints connect link 2 to link 3 and to the frame (at point O). A revolute joint is a pin, and it allows rotation in a plane of one link with respect to another. A revolute joint also connects the two links 3 and 4. Link 4 is allowed to slide with respect to the frame, and this connection between the frame and the link is called a *prismatic joint*. The motion is transferred from link 2, which is called the *input link*, to link 4, which is called the *output link*. Sometimes the input link is called a *driver*, and the output link the *follower*.

Another example of a mechanism is the windshield wiper mechanism shown in Figure 1.4. The motion is transferred from the crank driven by a motor through the

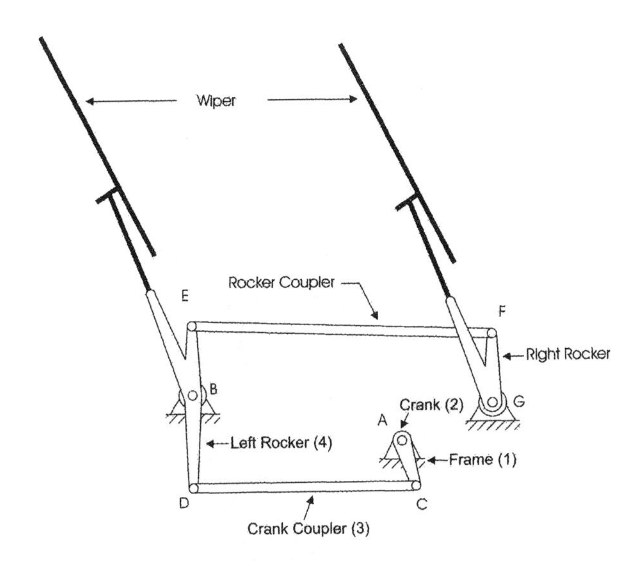

FIGURE 1.4 A windshield wiper mechanism.

coupler to the left rocker. The two rockers, left and right, are connected by the rocker coupler, which synchronizes their motion. The mechanism comprising links 1 (frame), 2 (crank), 3 (coupler), and 4 (rocker) is called a *four-bar mechanism.* In this example, revolute joints connect all links.

A *kinematic chain* is an interconnected system of links in which not a single link is *fixed.* Such a chain becomes a *mechanism* when one of the links in the chain is fixed. The fixed link is called a *frame* or, sometimes, a *base link.* In Figure 1.3 link 1 is a frame. A *planar mechanism* is one in which all points move in parallel planes.

A joint between two links restricts the *relative motion* between these links, thus imposing a *constraining condition* on the mechanism motion. The type of constraining condition determines the number of degrees of freedom (DOF) a mechanism has. If the constraining condition allows only one DOF between the two links, the corresponding joint is called a lower-pair joint. The examples are a revolute joint between links 2 and 3 and a prismatic joint between links 4 and 1 in Figure 1.3. If the constraint allows two DOF between the two links, the corresponding joint is called a high-pair joint. An example of a high-pair joint is a connection between the cam and the roller in Figure 1.1, if, in addition to rolling, sliding between the two links takes place.

A dump truck mechanism is shown in Figure 1.5, and its skeleton diagram in Figure 1.6. This is an example of a *compound mechanism* comprising two simple ones: the first, links 1–2–3, is called the *slider-crank mechanism* and the second, links 1–3–5–6, is called the *four-bar linkage.* The two mechanisms work in sequence (or they are *functionally in series*): the input is the displacement of the piston in the hydraulic cylinder, and the output is the tipping of the dump bed.

All the previous examples involved only links with two connections to other links. Such links are called *binary links.* In the example of Figure 1.6, in addition to binary links, there is link 2, which is connected to three links: 1 (frame), 3, and 5. Such a link is called a *ternary link.* It is possible to have links with more than three connections.

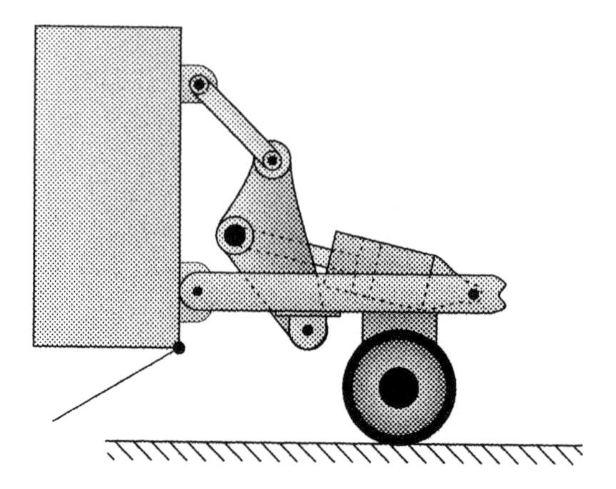

FIGURE 1.5 Dump truck mechanism.

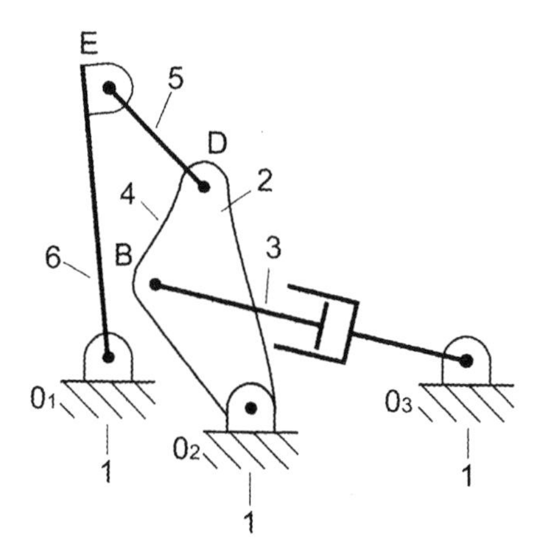

FIGURE 1.6 Skeleton of the dump truck mechanism.

1.5 MOBILITY OF MECHANISMS

The *mobility* of a mechanism is its *number of degrees of freedom.* This translates into a number of independent input motions leading to a single follower motion.

A single unconstrained link (Figure 1.7a) has three DOF in planar motion: two translational and one rotational. Thus, two disconnected links (Figure 1.7b) will have six DOF. If the two links are welded together (Figure 1.7c), they form a single link having three DOF. A revolute joint in place of welding (Figure 1.7d) allows a motion of one link relative to another, which means that this joint introduces an additional (to the case of welded links) DOF. Thus, the two links connected by a revolute joint have four DOF. One can say that by connecting the two previously disconnected links by a revolute joint, two DOF are eliminated. Similar considerations are valid for a prismatic joint.

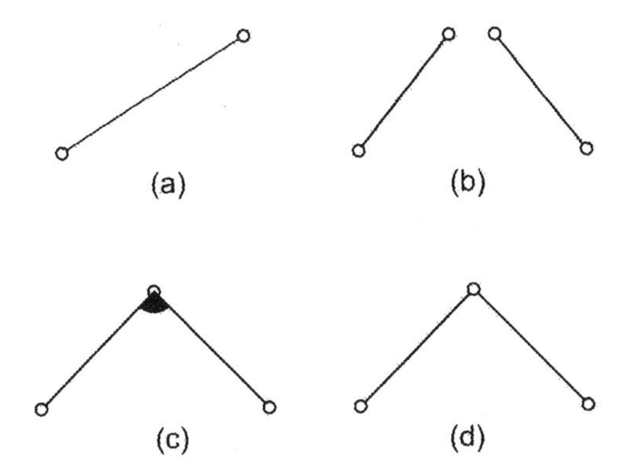

FIGURE 1.7 Various configurations of links with two revolute joints.

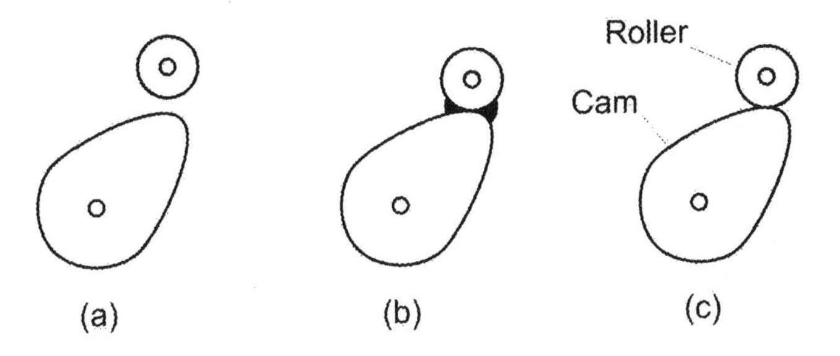

FIGURE 1.8 Various configurations of two links with a high-pair joint.

Since the revolute and prismatic joints make up all low-pair joints in planar mechanisms, the above results can be expressed as a rule: *a low-pair joint reduces the mobility of a mechanism by two DOF.*

For a high-pair joint the situation is different. In Figure 1.8 a roller and a cam are shown in various configurations. If the two are not in contact (Figure 1.8a), the system has six DOF. If the two are welded (Figure 1.8b), the system has three DOF. If the roller is not welded, then two relative motions between the cam and the roller are possible: rolling and sliding. Thus, in addition to the three DOF for a welded system, another two are added if a relative motion becomes possible. In other words, if disconnected, the system will have six DOF; if connected by a high-pair joint, it will have five DOF. This can be stated as a rule: *a high-pair joint reduces the mobility of a mechanism by one DOF.*

These results are generalized in the following formula, which is called *Kutzbach's criterion* of mobility

$$m = 3(n-1) - 2j_1 - j_2 \tag{1.1}$$

where n is the number of links, j_1 is the number of low-pair joints, and j_2 is the number of high-pair joints. Note that 1 is subtracted from n in the above equation to take into account that the mobility of the frame is zero.

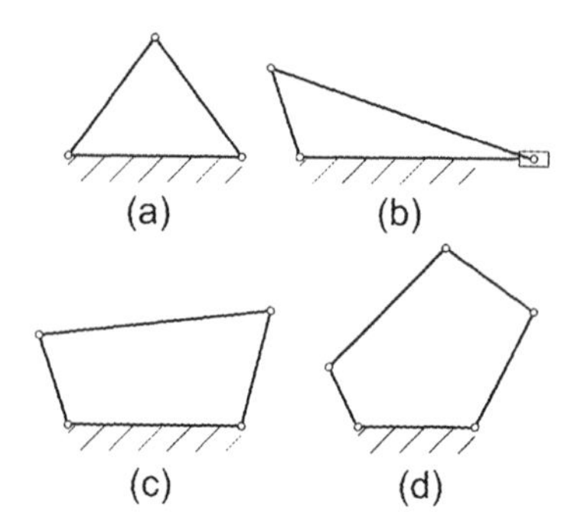

FIGURE 1.9 Mobility of various configurations of connected links: (a) $n = 3, j_1 = 3, j_2 = 0, m = 0$; (b) $n = 4, j_1 = 4, j_2 = 0, m = 1$; (c) $n = 4, j_1 = 4, j_2 = 0, m = 1$; (d) $n = 5, j_1 = 5, j_2 = 0, m = 2$.

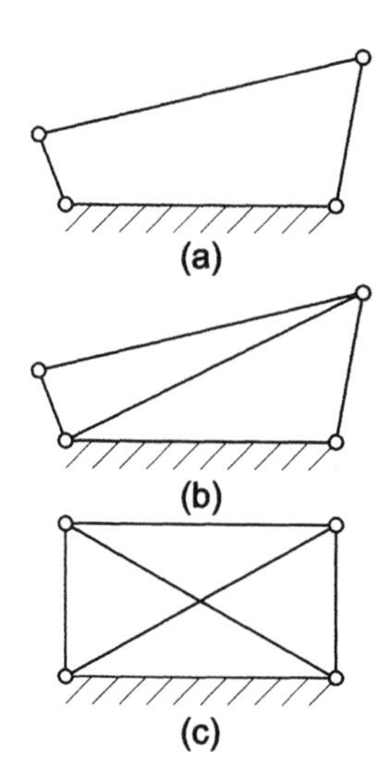

FIGURE 1.10 Effect of additional links on mobility: (a) $m = 1$, (b) $m = 0$, (c) $m = -1$.

In Figure 1.9 the mobility of various configurations of connected links is calculated. All joints are low-pair ones. Note that the mobility of the links in Figure 1.9a is zero, which means that this system of links is not a mechanism, but a structure. At the same time, the system of interconnected links in Figure 1.9d has mobility 2, which means that any two links can be used as input links (drivers) in this mechanism.

Look at the effect of an additional link on the mobility. This is shown in Figure 1.10, where a four-bar mechanism (Figure 1.10a) is transformed into a structure having zero mobility (Figure 1.10b) by adding one link, and then into a structure having negative mobility (Figure 1.10c) by adding one more link. The latter is called an *overconstrained* structure.

In Figure 1.11 two simple mechanisms are shown. Since slippage is the only relative motion between the cam and the follower in Figure 1.11a, then this interface is equivalent to a prismatic low-pair joint, so that this mechanism has mobility 1.

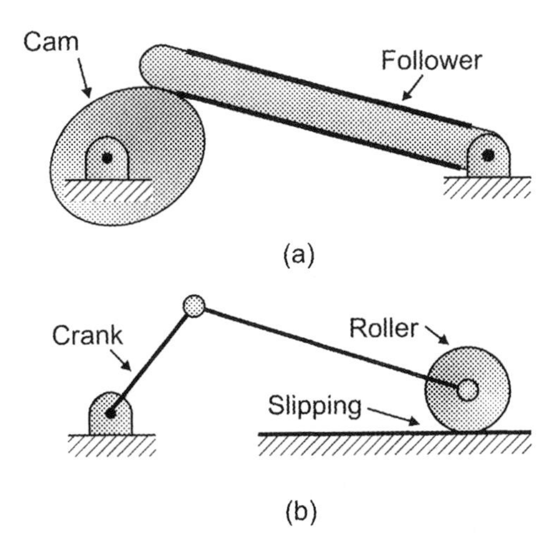

FIGURE 1.11 Mechanisms involving slippage only (a), and slippage and rolling (b).

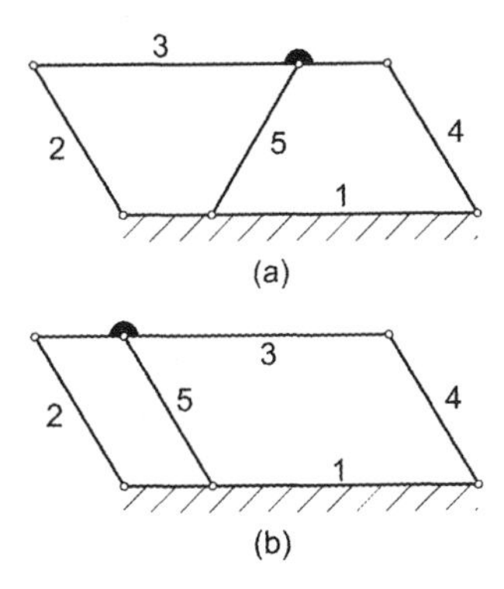

FIGURE 1.12 Example of violation of Kutzbach's criterion: (a) $n = 5, j_1 = 6, j_2 = 0, m = 0$; (b) $n = 5, j_1 = 6, j_2 = 0, m = 0$.

On the other hand, if both slippage and rolling are taking place between the roller and the frame in Figure 1.11b, then this interface is equivalent to a high-pair joint. Then the corresponding mobility of this mechanism is 2.

Kutzbach's formula for mechanism mobility does not take into account the specific geometry of the mechanism, only the connectivity of links and the type of connections (constraints). The following examples show that Kutzbach's criterion can be violated due to the nonuniqueness of geometry for a given connectivity of links (Figure 1.12). If links 2, 5, and 4 are as shown in Figure 1.12a, the mobility is zero. If, however, the above links are parallel, then according to Kutzbach's criterion the mobility is still zero, whereas motion is now possible.

It has been shown that in compound mechanisms (see Figure 1.6) there are links with more than two joints. Kutzbach's criterion is applicable to such mechanisms provided that a proper account of links and joints is made. Consider a simple compound mechanism shown in Figure 1.13, which is a sequence of two four-bar

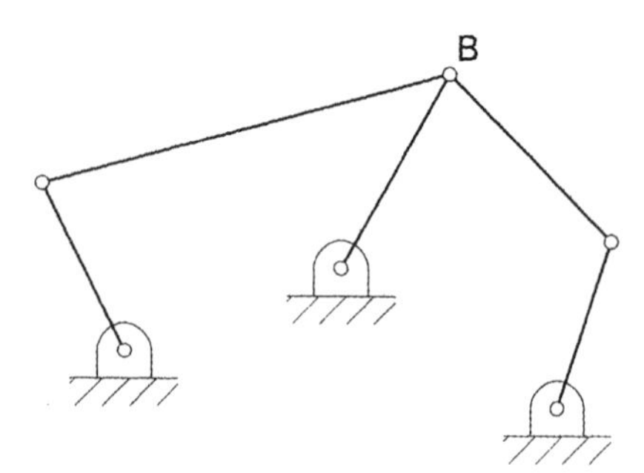

FIGURE 1.13 An example of a compound mechanism with coaxial joints at *B*.

mechanisms. In this mechanism, joint *B* represents two connections between three links. A system of three links rigidly coupled at *B* would have three DOF. If one connection were made revolute, the system would have four DOF. If another one were made revolute, it would have five DOF. Thus, if the system of three disconnected links has nine DOF, their connection by two revolute joints reduces it to five DOF. According to Kutzbach's formula $m = 3 \times 3 - 2 \times 2 = 5$. In other words, it should be taken into account that there are, in fact, *two* revolute joints at *B*. The axes of these two joints may not necessarily coincide, as in the example of Figure 1.6.

1.6 KINEMATIC INVERSION

Recall that a kinematic chain becomes a mechanism when one of the links in the chain becomes a frame. The process of choosing different links in the chain as frames is known as *kinematic inversion*. In this way, for an n-link chain n different mechanisms can be obtained. An example of a four-link slider-crank chain (Figure 1.14) shows how different mechanisms are obtained by fixing different links functionally. By fixing the cylinder (link 1) and joint *A* of the crank (link 2), an internal combustion engine is obtained (Figure 1.14a). By fixing link 2 and by pivoting link 1 at point *A*, a rotary engine used in early aircraft or a quick-return mechanism is obtained (Figure 1.14b). By fixing revolute joint *C* on the piston (link 4) and joint *B* of link 2, a steam engine or a crank-shaper mechanism is obtained (Figure 1.14c). By fixing the piston (link 4), a farm hand pump is obtained (Figure 1.14d).

1.7 GRASHOF'S LAW FOR A FOUR-BAR LINKAGE

As is clear, the motion of links in a system must satisfy the constraints imposed by their connections. However, even for the same chain, and thus the same constraints, different motion transformations can be obtained. This is demonstrated in Figure 1.15, where the motions in the inversions of the four-bar linkage are shown. In Figure 1.15, s identifies the smallest link, l is the longest link, and p, q are two other links.

From a practical point of view, it is of interest to know if for a given chain at least one of the links will be able to make a complete revolution. In this case, a motor can drive such a link. The answer to this question is given by Grashof's law, which states

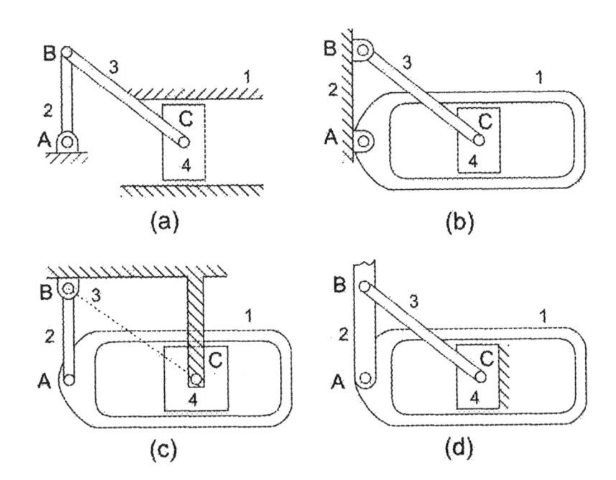

FIGURE 1.14 Four inversions of the slider-crank chain: (a) an internal combustion engine, (b) rotary engine used in early aircraft, quick-return mechanism, (c) steam engine, crank-shaper mechanism, (d) farm hand pump.

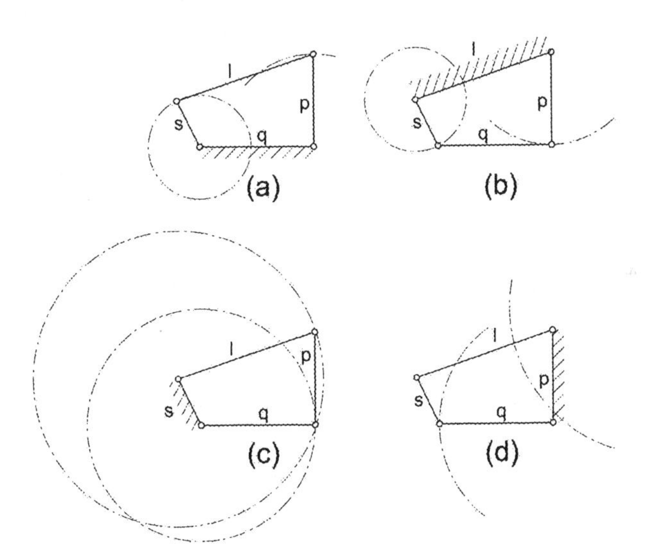

FIGURE 1.15 Inversions of the four-bar linkage: (a) and (b) crank-rocker mechanisms, (c) double-crank mechanism, (d) double-rocker mechanism.

that for a four-bar linkage, if the sum of the shortest and longest links is not greater than the sum of the remaining two links, at least one of the links will be revolving. For the notations in Figure 1.15 Grashof's law (condition) is expressed in the form:

$$s + l \leq p + q \tag{1.2}$$

Since in Figure 1.15 Grashof's law is satisfied, in each of the inversions there is at least one revolving link: in Figure 1.15a and b it is the shortest link s; in Figure 1.15c there are two revolving links, l and q; and in Figure 1.15d the revolving link is again the shortest link s.

PROBLEMS

1. What is the difference between a linkage and a mechanism?
2. What is the difference between a mechanism and a structure?
3. Assume that a linkage has N DOF. If one of the links is made a frame, how will it affect the number of DOF of the mechanism?
4. How many DOF would three links connected by revolute joints at point B (Figure P1.1) have? Prove.

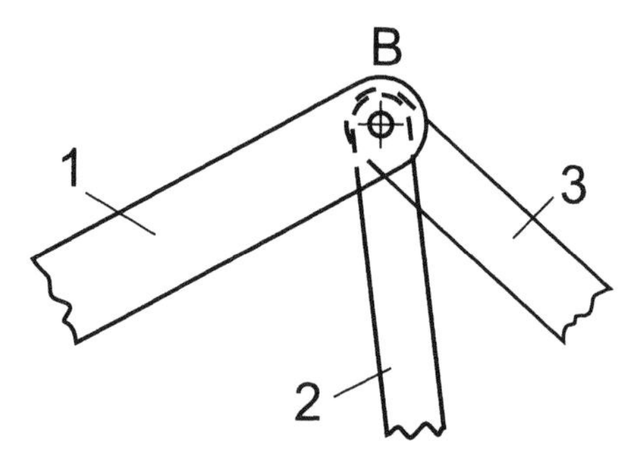

FIGURE P1.1

5. A fork joint connects two links (Figure P1.2). What is the number of DOF of this system? Prove.

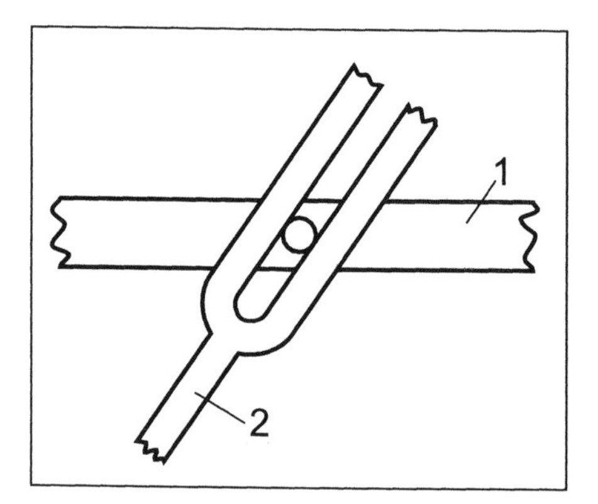

FIGURE P1.2

6. An adjustable slider drive mechanism consists of a crank-slider with an adjustable pivot, which can be moved up and down (see Figure P1.3).
   a. How many bodies (links) can be identified in this mechanism?
   b. Identify the type (and corresponding number) of all kinematic joints.
   c. What is the function of this mechanism and how will it be affected by moving the pivot point up and down?

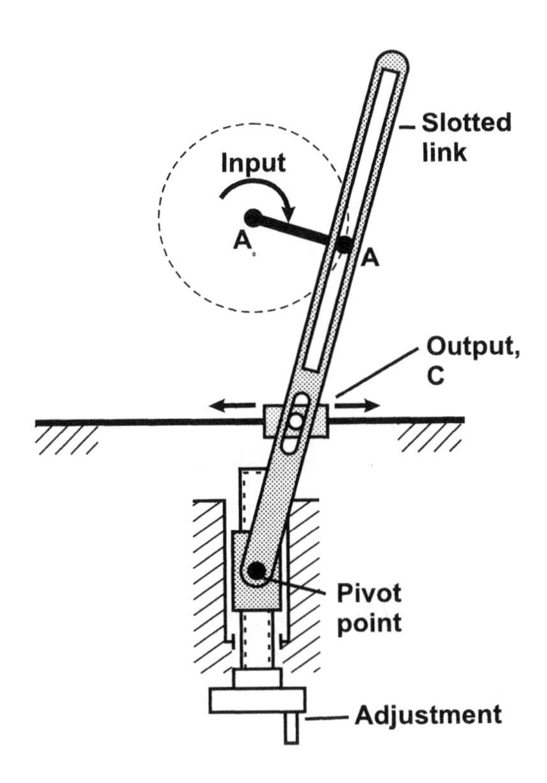

FIGURE P1.3

7. In Figure P.1.4 a pair of locking toggle pliers is shown.
   a. Identify the type of linkage (four-bar, slider-crank, etc.).
   b. What link is used as a driving link?
   c. What is the function of this mechanism?
   d. How does the adjusting screw affect this function?

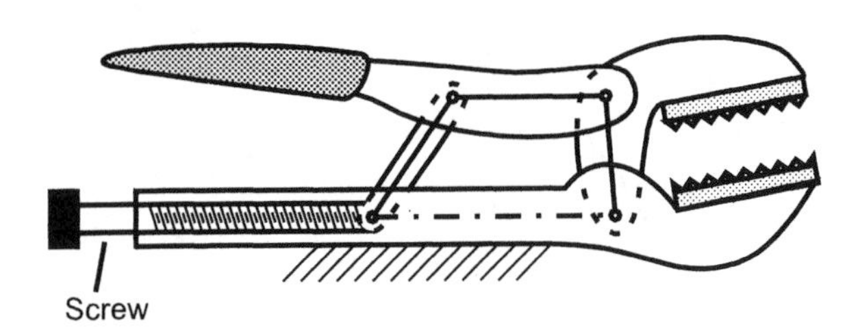

FIGURE P1.4

8. A constant velocity four-bar slider mechanism is shown in Figure P1.5.
   a. How many bodies (links) can be identified in this mechanism?
   b. Identify the type (and corresponding number) of all kinematic joints.
   c. Identify the frame and the number of joints on it.

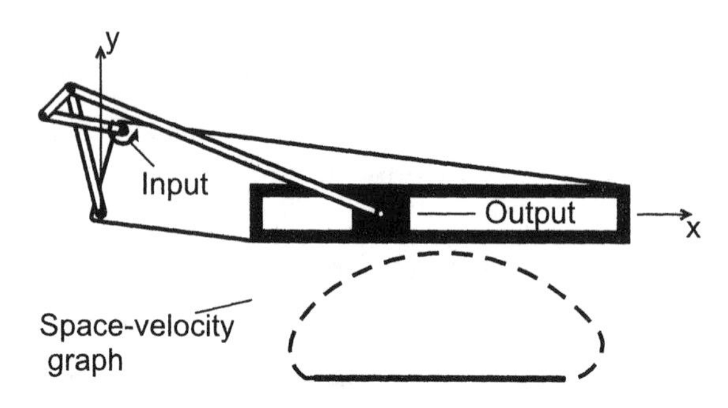

FIGURE P1.5 Constant-velocity mechanism.

9. How many driving links are in the dump truck mechanism shown in Figure P1.6?
10. What are the mobilities of mechanisms shown in Figures P1.3 through P1.5?
11. What are the mobilities of mechanisms shown in Figures P1.6 and P1.7?

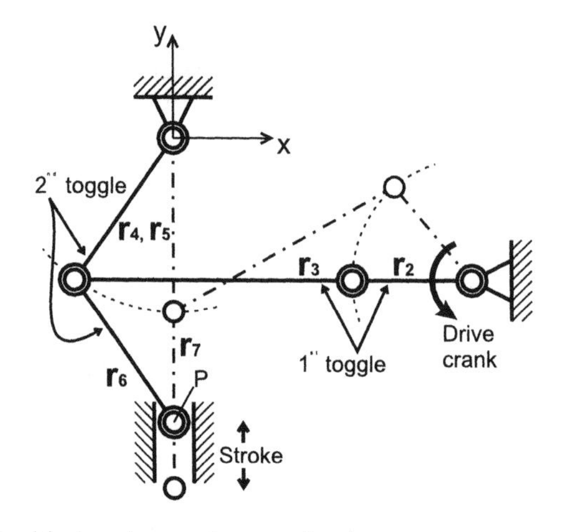

FIGURE P1.6 Double-toggle puncher mechanism.

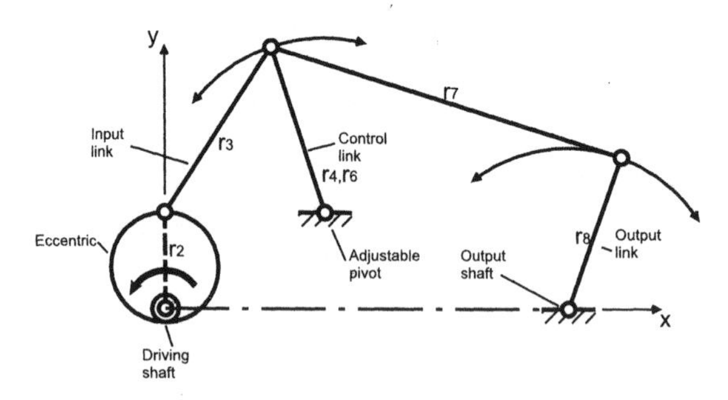

FIGURE P1.7 Variable-stroke drive.

12. Identify the motion transformation taking place in a windshield wiper mechanism (Figure 1.4). What must the relationship between the links in this mechanism be to perform its function?

2 Kinematic Analysis of Mechanisms

2.1 INTRODUCTION

There are various methods of performing kinematic analysis of mechanisms, including *graphical, analytical,* and *numerical.* The choice of a method depends on the problem at hand and on available computational means. A Bibliography given at the end of this book provides references to textbooks in which various methods of analysis are discussed. In this book the emphasis is placed on studying mechanisms rather than methods of analysis. Thus, the presentation is limited to one method, which is sufficient for simple and for many compound mechanisms. This method is known as the *loop-closure equation* method. It is presented here in vector notation.

Coordinate Systems

One should differentiate between the *global (inertial, absolute)* and *local (moving)* coordinate systems. Figure 2.1 shows a point P on a body referenced in global (x,y) and local (x_1,y_1) coordinate systems. The local coordinate system is embedded in the body and thus moves with it in a global system. For consistency, the *right-hand coordinate system* is used throughout this book for both local and global coordinate systems. Recall that the coordinate system is called *right-hand* if the rotation of the x-axis toward the y-axis is counterclockwise when viewed from the tip of the z-axis. Thus, in Figure 2.1 the z-axis is directed toward the reader.

A vector $\mathbf{r}$ has two components in the (x,y) plane: r_x and r_y (Figure 2.2). Note that the bold font identifies vectors. The following two notations for a vector in a component form will be used:

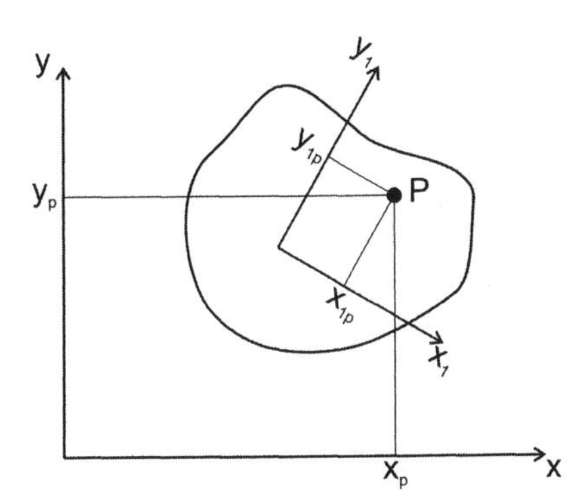

FIGURE 2.1 Global (x,y) and local (x_1,y_1) coordinate systems.

$$\mathbf{r} = \begin{bmatrix} r_x \\ r_y \end{bmatrix} = (r_x, r_y)^T \tag{2.1}$$

or

$$\mathbf{r} = |\mathbf{r}| \begin{bmatrix} \cos\alpha \\ \sin\alpha \end{bmatrix} = |\mathbf{r}|(\cos\alpha, \sin\alpha)^T \tag{2.2}$$

where r_x, r_y are x- and y-components of the vector, $|\mathbf{r}|$ is the vector magnitude and is equal to $|\mathbf{r}| = (r_x^2 + r_y^2)^{1/2}$, $\cos\alpha = r_x / |\mathbf{r}|$, $\sin\alpha = r_y /|\mathbf{r}|$, and T is the transposition sign. Denote in the following $|\mathbf{r}| = r$. A unit vector $\mathbf{u} = \mathbf{r}/r = (\cos\alpha, \sin\alpha)^T$ gives the direction of $\mathbf{r}$.

It is important for the sake of consistency and to formalize the analysis to adopt a convention for the positive angles characterizing vector directions. It is taken that the positive angle α is always directed *counterclockwise* and is measured from the positive direction of the x-axis (see Figure 2.2).

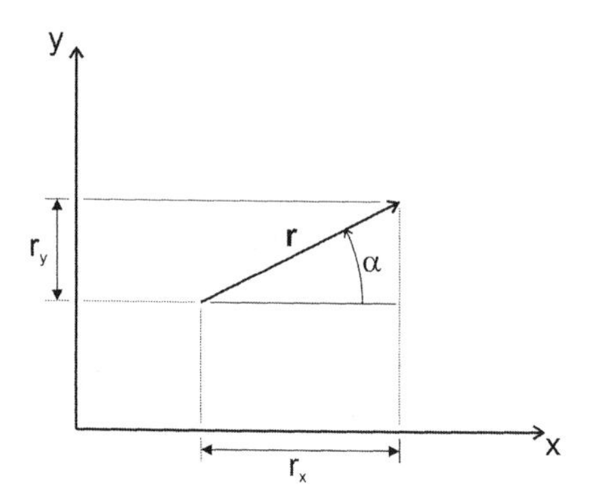

FIGURE 2.2 A vector in a plane.

2.2 VECTOR ALGEBRA AND ANALYSIS

An addition/subtraction of two (or more) vectors is a vector whose elements are found by addition/subtraction of the corresponding x- and y-components of the original vectors. If $\mathbf{a} = (a_x, a_y)^T$ and $\mathbf{b} = (b_x, b_y)^T$, then

$$\mathbf{a} + \mathbf{b} = (a_x + b_x, a_y + b_y)^T \tag{2.3}$$

The scalar (or dot) product of two vectors is a scalar, which is found by multiplication of the corresponding x- and y-components of two vectors and then the summation of results. For the two vectors $\mathbf{a}$ and $\mathbf{b}$, their scalar product is

$$\mathbf{d} = \mathbf{a}^T\mathbf{b} = (a_x, a_y) \begin{bmatrix} b_x \\ b_y \end{bmatrix} = a_x b_x + a_y b_y \tag{2.4}$$

If vectors **a** and **b** are given in the form **a** = a (cos α, sin α)^T and **b** = b (cos β, sin β)^T, then the result of their scalar product takes the form

$$\mathbf{d} = \mathbf{a}^T\mathbf{b} = ab(\cos\alpha\cos\beta + \sin\alpha\sin\beta) = ab\cos(\alpha - \beta) \tag{2.5}$$

Note that $\mathbf{a}^T\mathbf{b} = \mathbf{b}^T\mathbf{a}$. This property of the scalar multiplication is called the *commutative law.*

The result of the cross-product of two vectors is a vector perpendicular to the plane in which the original two vectors are lying. Thus, if the two vectors are lying in the (x,y) plane, then their product will have a z-direction. To find this product the two vectors must be described as three-dimensional objects. Thus, for the two vectors $\mathbf{a} = (a_x, a_y, 0)^T$ and $\mathbf{b} = (b_x, b_y, 0)^T$ their cross-product $\mathbf{u} = \mathbf{a}^T \times \mathbf{b}$ can be found by finding the three determinants of the matrix associated with the three components of the vector **u**.

$$\mathbf{u} = \begin{bmatrix} \mathbf{i} & \mathbf{j} & \mathbf{k} \\ a_x & a_y & 0 \\ b_x & b_y & 0 \end{bmatrix} \tag{2.6}$$

In Equation 2.6 **i**, **j**, and **k** are the unit vectors directed along the x-, y-, and z-axis, respectively. The components of the vector **u** are the second-order determinants associated with the unit vectors. Thus, the vector **u** is equal to

$$\mathbf{u} = (a_y 0 - b_y 0)\,\mathbf{i} - (a_x 0 - b_x 0)\,\mathbf{j} + (a_x b_y - b_x a_y)\,\mathbf{k} = (a_x b_y - b_x a_y)\,\mathbf{k} \tag{2.7}$$

It is seen that the magnitude of the vector |**u**| is

$$|\mathbf{u}| = u = (a_x b_y - b_x a_y) \tag{2.8}$$

and it is directed along the z-axis. Note that the cross-product is not commutative, i.e., $\mathbf{a}^T \times \mathbf{b} = -\mathbf{b}^T \times \mathbf{a}$.

If vectors **a** and **b** are given in the form of Equation 2.2, then in Equation 2.8 $a_x = a\cos\alpha$, $a_y = a\sin\alpha$, $b_x = b\cos\beta$, and $b_y = b\sin\beta$, and the expression for the magnitude of **u** is reduced to

$$\mathbf{u} = ab\,(\cos\alpha\sin\beta - \cos\beta\sin\alpha) = ab\sin(\beta - \alpha) \tag{2.9}$$

If vectors **a** and **b** are normal, then their scalar product is zero. It follows from Equation 2.5, since in this case $\alpha - \beta = \pi/2, 3\pi/2$. If vectors **a** and **b** are collinear, then their scalar product is $\pm ab$, since in this case $\alpha - \beta = 0, \pi$.

Vector differentiation is accomplished by differentiating each vector component. For example, if $\mathbf{a} = a(t)\,[\cos\alpha(t), \sin\alpha(\mathrm{t})]^T$, then

$$\frac{d\mathbf{a}(t)}{dt} = \frac{da(t)}{dt}[\cos\alpha(t), \sin\alpha(t)]^T + a(\mathrm{t})[-\sin\alpha(t), \cos\alpha(t)]^T\frac{d\alpha(t)}{dt} \tag{2.10}$$

Note that the latter equation can be written in the form

$$\frac{d\mathbf{a}(t)}{dt} = \frac{da(t)}{dt}[\cos\alpha(t), \sin\alpha(t)]^T + a(t)\left[\cos\left(\alpha(t)+\frac{\pi}{2}\right), \sin\left(\alpha(t)+\frac{\pi}{2}\right)\right]^T \frac{d\alpha(t)}{dt} \quad (2.11)$$

One can see that the differentiated vector comprises two components: the first has the direction of the original vector, while the second component is being rotated with respect to the first by $\pi/2$ in the positive direction.

2.3 POSITION ANALYSIS

2.3.1 Kinematic Requirements in Design

Kinematic considerations are part of machine design specifications. Although the two examples discussed here do not adequately represent the thousands of mechanisms in applications, they should help to develop a general perception of such requirements.

Figure 2.3 shows an earthmoving machine, which has a front-mounted bucket and a linkage that loads material into the bucket through forward motion of the machine and then lifts, transports, and discharges this material. Hydraulic cylinder 5 lifts arm 4, while hydraulic cylinder 6 controls, through links 10 (bellcrank) and 7, the position of the bucket. Thus, the final displacement of the bucket is controlled by two mechanisms: one is the mechanism for lifting the arm, and the other is for the rotation of the bucket. The first is an inversion of the slider-crank mechanism (see Figure 1.14b). The second is a four-bar linkage, in which link 10 is a crank, link 7 is a coupler, and the bucket is a follower link. In designing this machine, if the positions of the bucket are given, then the designer has to find such dimensions of the links that allow attaining the given positions of the bucket. Since the hydraulic cylinders have a limited stroke, the rotations of both the arm and the bucket are limited to some specific angles. Thus, in order for a bucket to be in a needed position, the two motions, of arm 4 and of crank 10, must be synchronized.

The process of finding the mechanism parameters given the needed output is called *kinematic synthesis*. If, however, the mechanism parameters are known, then the objective is to find the motion of the output link. This process of finding the output motion given the mechanism parameters is called *kinematic analysis*. In the case of the example in Figure 2.3, if the dimensions of all links were known, then the objective would be to find the displacements of the hydraulic cylinders such that the bucket is in proper position. In other words, by performing kinematic analysis the relationship between the displacements of the pistons in cylinders and the position of the bucket will be established.

Another example is an application of a slider-crank mechanism in an internal combustion engine (Figure 2.4). The motion from piston 4 is transferred through connecting rod 3 to crank 2, which rotates the crankshaft. This is a diesel engine, which means that one cycle, combustion–exhaust–intake–compression, comprises two complete crank revolutions. If all components of the cycle are equal in duration, then each will take 180°. In other words, the stroke of the engine, the piston motion

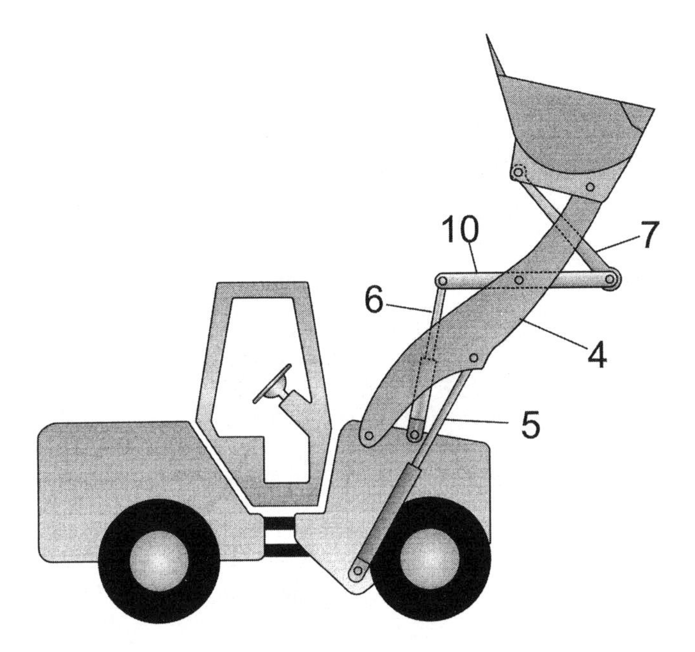

FIGURE 2.3 A loader.

from its maximum upper position to its lowest position, should correspond to 180° of the crank rotation. This is a *kinematic requirement* of the mechanism. Another kinematic requirement is that the swing of the connecting rod be such that it does not interfere with the walls of the cylinder. There are other design requirements affecting kinematic and dynamic analysis: link dimensions and shape.

2.3.2 The Process of Kinematic Analysis

Kinematics is the study of motion without consideration of what causes the motion. In other words, the input motion is assumed to be known and the objective is to find the transformation of this motion. Kinematic analysis comprises the following steps:

- Make a skeletal representation of the real mechanism.
- Find its mobility.
- Choose a coordinate system.
- Identify all links by numbers.
- Identify all angles characterizing link positions.
- Write a loop-closure equation.
- Identify input and output variables.
- Solve the loop-closure equation.
- Check the results by numerical analysis.

A *skeletal* representation completely describes the kinematics of the mechanism; i.e., it allows one to find the trajectories, velocities, and accelerations of any point on a skeleton. As long as this information is available, the trajectory of any point

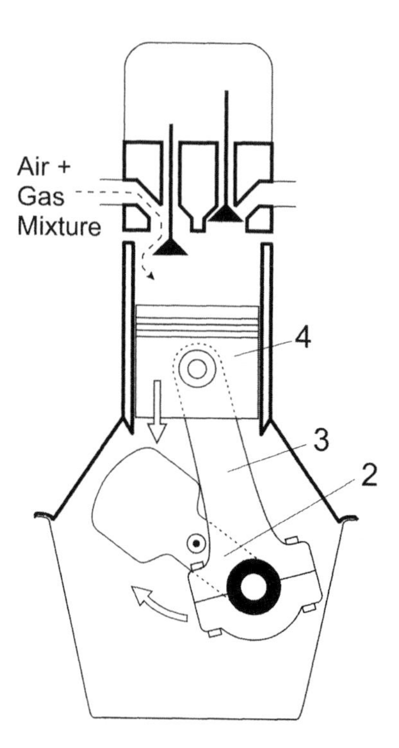

FIGURE 2.4 Schematic diagram of an internal combustion engine.

not located on a skeleton is easily found. Thus, the analysis of mechanisms is reduced to the analysis of their skeletons. The skeleton is represented by a system of connected *links*. The position of each link is identified by an angle in the chosen coordinate system.

2.3.3 Kinematic Analysis of the Slider-Crank Mechanism

Before proceeding with the general formulations for position analysis, it is worth considering a simple slider-crank mechanism used in the engine. In Figure 2.5a, a skeleton diagram of the slider-crank mechanism is shown. The coordinate system is such that the x-axis coincides with the slider axis. The crank is link 2, the connecting rod is link 3, the piston is link 4, and, by convention, the frame is link 1.

With each link one can associate a vector. The magnitude of this vector is the length of the link, whereas the direction of this vector is along the link, but otherwise it is arbitrary. If one starts from point O and continues through points A and B, one will come back again to O. The system of vectors thus makes a loop (Figure 2.5b). The mathematical representation of this requirement is that the sum of all vectors in the loop must be equal to zero, i.e.,

$$\sum_{i=1}^{3} \mathbf{r}_i = \mathbf{r}_1 + \mathbf{r}_2 + \mathbf{r}_3 = 0 \tag{2.12}$$

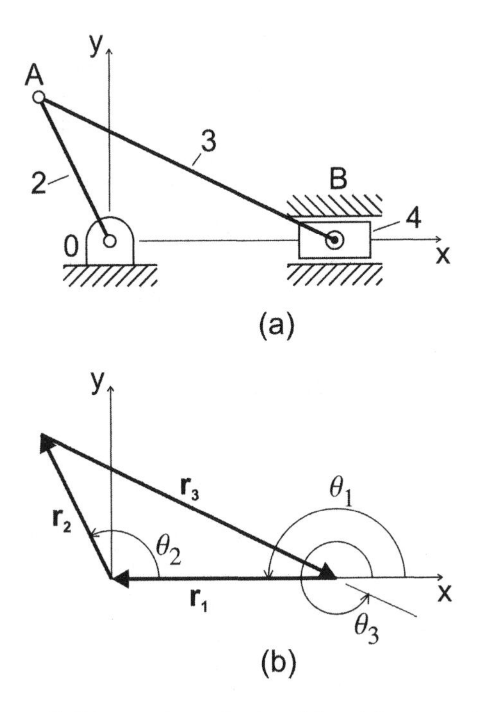

FIGURE 2.5 Skeleton of slider-crank mechanism (a) and its vector representation (b).

Recall that the connection between any two links imposes a *constraint* on links motion. The fundamental property of the loop-closure equation is that it takes into account all the constraints imposed on links motion.

Since each vector can be represented in the form:

$$\mathbf{r}_i = r_i(\cos\theta_i, \sin\theta_i)^T \tag{2.13}$$

Equation 2.12 becomes

$$r_1(\cos\theta_1, \sin\theta_1)^T + r_2(\cos\theta_2, \sin\theta_2)^T + r_3(\cos\theta_3, \sin\theta_3)^T = 0 \tag{2.14}$$

Now the problem is reduced to solving Equation 2.14. There are many ways to solve this equation (the reader should consult the books listed in the Bibliography). Presented here is only one approach, which can be applied to any planar mechanism and allows obtaining symbolic solutions in the simplest form. It is important to solve the equations in symbols because the symbolic form of the solution gives explicit relationships between the variables, which, in turn, enhances understanding of the problem and the subject matter. It should be noted that Mathematica could solve Equation 2.14 in symbols. Although the result it gives is not in the simplest form, it can be used for numerical analysis.

Equation 2.14 is a vector equation comprising two scalar ones, which means that it can be solved for only two unknowns. Since the number of parameters in

Equation 2.14 (vector magnitudes and their corresponding direction angles) exceeds two, it is necessary to identify what is given, and what is to be found.

If the mechanism shown in Figure 2.5 is an engine, then it can be assumed that the motion of the piston is given, that is, $r_1 = r_1(t)$ is known at any moment in time. The magnitudes of vectors $\boldsymbol{r}_2$ and $\boldsymbol{r}_3$ are also given, as well as the direction of the vector $\mathbf{r}_1$. Thus, one is left with two unknowns: angles θ_2 and θ_3.

If, on the other hand, the mechanism in Figure 2.5 is a compressor, then the rotation of the crankshaft is known, that is, $\theta_2 = \theta_2(t)$ is given as a function of time. Since the magnitudes of the vectors $\mathbf{r}_2$ and $\mathbf{r}_3$ are known, as well as the direction of the vector $\mathbf{r}_1$, then the two unknowns are r_1 and θ_3.

The following considers some generic loop-closure equations and their symbolic solutions. Five various cases are presented, which should be sufficient to analyze any simple planar mechanism.

2.3.4 Solutions of Loop-Closure Equations

Since any simple (as opposed to compound) planar mechanism can be described by a loop-closure equation, then a generic equation can be solved for various possible combinations of known parameters and unknown variables. For a mechanism with N links, such an equation has the form:

$$\sum_{i=1}^{N} r_i(\cos\theta_i, \sin\theta_i)^T = 0 \tag{2.15}$$

There are only a few possible cases arising from this loop-closure equation in different applications.

First Case

A *vector* is the unknown; i.e., its *magnitude* and its *direction* are to be found. If one moves all the known vectors in Equation 2.15 to the right-hand side, then this equation takes the form:

$$r_j(\cos\theta_j, \sin\theta_j)^T = b(\cos\alpha, \sin\alpha)^T \tag{2.16}$$

where $\mathbf{b}$ is the vector equal to the sum of all vectors except the unknown vector $\mathbf{r}_j$

$$\mathbf{b} = (b_x, b_y)^T = -\sum_{i=1, i\neq j}^{N} r_i(\cos\theta_i, \sin\theta_i)^T \tag{2.17}$$

In Equation 2.17

$$b_x = -\sum_{i=1, i\neq j}^{N} r_i \cos\theta_i \tag{2.18}$$

and

$$b_y = -\sum_{i=1, i \neq j}^{N} r_i \sin\theta_i \tag{2.19}$$

Thus b, cos α, and sin α in Equation 2.16 are equal to

$$b = \sqrt{b_x^2 + b_y^2} > 0 \tag{2.20}$$

$$\cos\alpha = \frac{b_x}{b} \tag{2.21}$$

and

$$\sin\alpha = \frac{b_y}{b} \tag{2.22}$$

The angle α must be known explicitly to solve for r_j and θ_j in Equation 2.16. Denote the principal solution belonging to the $x > 0$ and $y > 0$ quadrant as

$$\alpha^* = \arcsin\left|\frac{b_y}{b}\right| \tag{2.23}$$

Then the solution for α depends on the signs of cos α and sin α in Equations 2.21 and 2.22.

$$\alpha = \begin{cases} \alpha^* & \text{if } \cos\alpha > 0 \text{ and } \sin\alpha > 0 \\ \pi - \alpha^* & \text{if } \cos\alpha < 0 \text{ and } \sin\alpha > 0 \\ \pi + \alpha^* & \text{if } \cos\alpha < 0 \text{ and } \sin\alpha < 0 \\ 2\pi - \alpha^* & \text{if } \cos\alpha > 0 \text{ and } \sin\alpha < 0 \end{cases} \tag{2.24}$$

Note, that in Mathematica the ArcTan[cos α, sin α] function finds the proper quadrant.

Now, from the vector Equation 2.16 two scalar equations follow:

$$r_j \cos\theta_j = b \cos\alpha \tag{2.25}$$

and

$$r_j \sin\theta_j = b \sin\alpha \tag{2.26}$$

Since r_j and b are both positive, it follows from the above equations that the angles α and θ_j must be equal. Thus, the solution in this case is

$$r_j = b,\ \theta_j = \alpha \tag{2.27}$$

Second Case

In this case the *magnitude of one vector* and the *direction of another vector* are to be found. Again all the known vectors are moved to the right-hand side, so that the vector equation (Equation 2.15) takes the form:

$$r_i(\cos\theta_i,\ \sin\theta_i)^T + r_j(\cos\theta_j,\ \sin\theta_j)^T = b\ (\cos\alpha,\ \sin\alpha)^T \tag{2.28}$$

Assume that the two unknowns are r_i and θ_j. First, premultiply Equation 2.28 from the left by a unit vector perpendicular to the vector $\mathbf{r}_i$, namely, by the vector $\mathbf{u}_1 = (-\sin\ \theta_i,\ \cos\ \theta_i)^T$, and then by a unit vector parallel to $\mathbf{r}_i$, namely, by the vector $\mathbf{u}_2 = (\cos\ \theta_i,\ \sin\ \theta_i)^T$. The result of the first operation is

$$r_j \sin(\theta_j - \theta_i) = b\ \sin(\alpha - \theta_i) \tag{2.29}$$

and the result of the second operation is

$$r_i + r_j \cos(\theta_j - \theta_i) = b\ \cos(\alpha - \theta_i) \tag{2.30}$$

The system of Equation 2.28 has been transformed into a new system of Equations 2.29 and 2.30 having a new variable $(\theta_j - \theta_i)$. Now this new system can be further simplified by moving r_i in Equation 2.30 to the right-hand side, squaring both Equation 2.29 and the transformed Equation 2.30, and then adding them. The result is a quadratic equation for the r_i, the solution of which is

$$r_i = b\ \cos(\alpha - \theta_i) \pm \sqrt{r_j^2 - b^2 \sin^2(\alpha - \theta_i)} \tag{2.31}$$

If both signs in the above equation give positive values for r_i, then it means that there are two physically admissible mechanism configurations.

As soon as r_i is known, the signs of $\sin(\theta_j - \theta_i)$ and $\cos\ (\theta_j - \theta_i)$ are found from Equations 2.29 and 2.30. Again, this allows one to find the unique solution for the angle θ_j

$$\theta_j = \begin{cases} \theta_j^* & \text{if } \cos(\theta_j - \theta_i) > 0 \text{ and } \sin(\theta_j - \theta_i) \ge 0 \\ \pi - \theta_j^* & \text{if } \cos(\theta_j - \theta_i) < 0 \text{ and } \sin(\theta_j - \theta_i) \ge 0 \\ \pi + \theta_j^* & \text{if } \cos(\theta_j - \theta_i) < 0 \text{ and } \sin(\theta_j - \theta_i) \le 0 \\ 2\pi - \theta_j^* & \text{if } \cos(\theta_j - \theta_i) > 0 \text{ and } \sin(\theta_j - \theta_i) \le 0 \end{cases} \tag{2.32}$$

where the angle θ_j^* in this case is equal to the principal value given by

$$\theta_j^* = \theta_i + \arcsin|b \sin(\alpha - \theta_i)/r_j| \tag{2.33}$$

Third Case

In this case the *magnitudes of two vectors* are to be found. As before, all the known vectors are moved to the right-hand side, and the vector equation has the form of Equation 2.28, except that in this case r_i and r_j are the two unknowns. The solution is unique because the inverse trigonometric functions are not involved in this case. As before, eliminate one of the unknowns by premultiplying Equation 2.28 from the left by a unit vector perpendicular to the vector $\mathbf{r}_i$, and then eliminating the second unknown by premultiplying Equation 2.28 by a unit vector perpendicular to the vector $\mathbf{r}_j$. The first result gives Equation 2.29, from which a formula for r_j follows:

$$r_j = b\frac{\sin(\alpha - \theta_i)}{\sin(\theta_j - \theta_i)} \tag{2.34}$$

Similarly, the formula for r_i is

$$r_i = b\frac{\sin(\alpha - \theta_j)}{\sin(\theta_i - \theta_j)} \tag{2.35}$$

Fourth Case

This case involves *two unknown angles* in Equation 2.28, namely, θ_i and θ_j. The solution strategy in this case must be different from the previous cases because in this case multiplying Equation 2.28 by the corresponding unit vectors cannot eliminate the two unknown angles. Instead, a new unit vector $\mathbf{u}_b = (\cos\alpha, \sin\alpha)^T$ is used to transform the variables θ_i and θ_j into new variables $\alpha - \theta_i$ and $\alpha - \theta_j$. This is achieved by premultiplying Equation 2.28 first by the unit vector perpendicular to vector **b**, and then by a unit vector parallel to vector **b**. The result is the following two equations, respectively:

$$r_i \sin(\alpha - \theta_i) + r_j \sin(\alpha - \theta_j) = 0 \tag{2.36}$$

and

$$r_i \cos(\alpha - \theta_i) + r_j \cos(\alpha - \theta_j) = b \tag{2.37}$$

There are still two unknowns in Equations 2.36 and 2.37, namely, $\alpha - \theta_i$ and $\alpha - \theta_j$. However, this system is solvable. First, one can find $\sin(\alpha - \theta_i)$ from Equation 2.36, and then, by using the trigonometric identity ($\sin^2(\alpha - \theta_i) + \cos^2(\alpha - \theta_i) = 1$), reduce the second equation (Equation 2.37) to the following:

$$r_i\sqrt{1-\left(\frac{r_j}{r_i}\right)^2 \sin^2(\alpha - \theta_j)} = -r_j\cos(\alpha - \theta_j) + b \tag{2.38}$$

By squaring both sides of Equation 2.38 and simplifying, the following equation for the unknown $\alpha - \theta_j$ is obtained:

$$\cos(\alpha - \theta_j) = A \tag{2.39}$$

where it is denoted

$$A = \frac{b^2 - r_i^2 + r_j^2}{2br_j} \tag{2.40}$$

In principle, θ_j can be found from Equation 2.39, and then θ_i from either of Equations 2.36 or 2.37. However, since the inverse function in Equation 2.39 is not unique, this procedure is not the most efficient. Instead, one can substitute Equation 2.39 into Equations 2.36 and 2.37 (using again the trigonometric identity), and solve for $\cos(\alpha - \theta_i)$ and $\sin(\alpha - \theta_i)$. The result is

$$\cos(\alpha - \theta_i) = B \tag{2.41}$$

and

$$\sin(\alpha - \theta_i) = C \tag{2.42}$$

where

$$B = \frac{b - r_j A}{r_i}$$

and

$$C = \pm\frac{r_j}{r_i}\sqrt{1 - A^2}$$

In the above equations, A and B are unique numbers, whereas C may either be positive or negative. However, for any chosen C, the angle θ_i is determined uniquely, depending on the signs of B and C (similar to Equation 2.32). Since there are two options for the sign of C, there will be two solutions for θ_i.

For any found θ_i, the angle θ_j can also be found uniquely from the loop-closure Equations 2.36 and 2.37.

$$\cos(\alpha - \theta_j) = (b - r_i \cos(\alpha - \theta_i))/r_j \tag{2.43}$$

and

$$\sin(\alpha - \theta_j) = -r_i \sin(\alpha - \theta_i)/r_j \tag{2.44}$$

In summary, for a + C a set of solution angles $(\theta_i, \theta_j)^1$ is found, and for a – C another set of solutions $(\theta_i, \theta_j)^2$ is found. Since both sets are based on the solution of the loop-closure equation, they are physically admissible. In practical terms it means that a mechanism with given links allows two physical configurations.

Fifth Case

In this case the *magnitude* of one vector, the *direction* of another vector, and the *directions* of two other vectors, functionally related to the direction of the second vector, are to be found. The loop-closure equation, after the known vectors are moved to the right-hand side, has the form:

$$r_i (\cos\theta_i, \sin\theta_i)^T + r_j (\cos(\theta_i - \gamma), \sin(\theta_i - \gamma))^T + r_k(\cos(\theta_i - \beta), \sin(\theta_i - \beta))^T = b(\cos\alpha, \sin\alpha)^T \tag{2.45}$$

where θ_i and r_j are the two unknowns, and it is seen that $\theta_j = \theta_i - \gamma$ and $\theta_k = \theta_i - \beta$.

Premultiply Equation 2.45 from the left by a unit vector perpendicular to the vector $\mathbf{r_j}$, namely, by the vector $\mathbf{u_1} = (-\sin(\theta_i - \gamma), \cos(\theta_i - \gamma))^T$. The result is

$$r_i \sin\gamma + r_k \sin(\gamma - \beta) = b\sin(\alpha - \theta_i + \gamma) \tag{2.46}$$

Now premultiply Equation 2.45 from the left by a unit vector parallel to the vector $\mathbf{r_j}$, namely, by the vector $\mathbf{u_2} = (\cos(\theta_i - \gamma), \sin(\theta_i - \gamma))^T$. The result is

$$r_i \cos\gamma + r_j + r_k \cos(\gamma - \beta) = b\cos(\alpha - \theta_i + \gamma) \tag{2.47}$$

The strategy is to find r_j first. To achieve this, square both sides in Equations 2.46 and 2.47 and add the two. The result is

$$r_i^2 + r_j^2 + r_k^2 + 2\, r_i\, r_k \cos\beta + 2\, r_j\, r_i \cos\gamma + 2\, r_j\, r_k \cos(\gamma - \beta) + 2\, r_i\, r_k \cos\gamma \cos(\gamma - \beta) = b^2 \tag{2.48}$$

The latter equation is a quadratic one with respect to r_j

$$r_j^2 + c r_j + d = 0 \tag{2.49}$$

where $c = 2r_i \cos\gamma + 2\, r_k \cos(\gamma - \beta)$ and $d = r_i^2 + r_k^2 + 2r_i\, r_k \cos\beta - b^2$.

Equation 2.49 has two roots:

$$r_{j_{1,2}} = -\frac{c}{2} \pm \sqrt{\frac{c^2}{4} - d} \tag{2.50}$$

It is seen that for the solution of the original system Equation 2.45 to exist there must be a positive root in Equation 2.50. If such a root does exist, it defines the unknown magnitude r_j. The other unknown, angle θ_i, is found from the system of Equations 2.46 and 2.47, which has multiple solutions. However, in this case a unique solution can be found. Denote $\zeta = \alpha - \theta_i + \gamma$, and $\zeta^* = |\alpha - \theta_i + \gamma|$. From Equations 2.46 and 2.47 it follows that

$$\sin \zeta = (r_i \sin \gamma + r_k \sin(\gamma - \beta))/b = A \tag{2.51}$$

and

$$\cos \zeta = (r_i \cos \gamma + r_j + r_k \cos(\gamma - \beta))/b = B \tag{2.52}$$

Since A and B are known constants for the already-found r_j, then angle ζ is uniquely found from the following conditions:

$$\zeta = \begin{cases} \zeta^* & \text{if } A > 0 \text{ and } B > 0 \\ \pi - \zeta^* & \text{if } A > 0 \text{ and } B < 0 \\ \pi + \zeta^* & \text{if } A < 0 \text{ and } B < 0 \\ 2\pi - \zeta^* & \text{if } A < 0 \text{ and } B > 0 \end{cases} \tag{2.53}$$

Having found ζ, the unknown angle $\theta_i = \alpha - \zeta + \gamma$ can be determined.

2.3.5 Applications to Simple Mechanisms

Slider-Crank Inversions

One can apply the solutions found in Section 2.3.4 to some inversions of the slider-crank mechanism shown in Figure 1.14.

- Case of Figure 1.14a
  Assuming that the crank 2 is the driver, the loop-closure equation is

$$r_1 (\cos \theta_1, \sin \theta_1)^T + r_3 (\cos \theta_3, \sin \theta_3)^T = -r_2 (\cos \theta_2, \sin \theta_2)^T \tag{2.54}$$

where r_1 and θ_3 are the unknowns, and thus the equation falls into the second case category. Note that r_1 is given by Equation 2.31, and θ_3 by Equation 2.32.

A position analysis of this mechanism was done using Mathematica. Snapshots of the motion at four positions are shown in Figure 2.6 for the following input data:

$$\frac{r_3}{r_1} = 4, \ \theta_1 = \pi$$

The change of the angle of rotation of the connecting rod (link 3) during one cycle of crank rotation is shown in Figure 2.7. Note that when $\theta_2 = 0, \pi, 2\pi$, the connecting rod coincides with the x-axis. The maximum of the angle θ_3 allows one to check for possible interference with the cylinder walls.

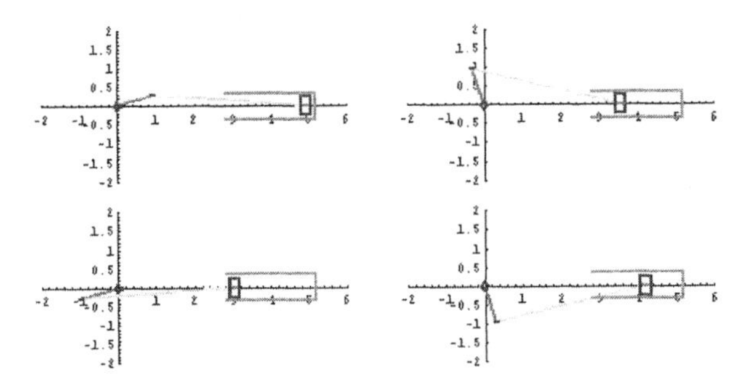

FIGURE 2.6 A slider-crank mechanism in four positions during crank rotation.

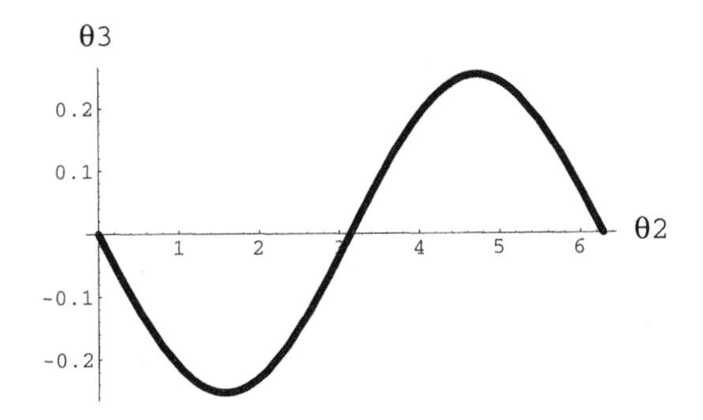

FIGURE 2.7 Angle θ_3 as a function of crank angle θ_2.

- Case of Figure 1.14a
  Assuming that piston 4 is the driver, the loop-closure equation is

$$r_2\,(\cos\,\theta_2,\,\sin\,\theta_2)^T + r_3\,(\cos\,\theta_3,\,\sin\,\theta_3)^T = -r_l\,(\cos\,\theta_1,\,\sin\,\theta_1)^T \qquad (2.55)$$

where θ_2 and θ_3 are unknowns, and thus the equation falls into the fourth case category. This case represents the actuator mechanism. The stroke of the piston is limited to a less-than-half-circle rotation of link 2. Snapshots of this system in four positions are shown in Figure 2.8 in which the following data were used:

$$\frac{r_3}{r_1} = 4, \;\; \theta_1 = \frac{5\pi}{4}$$

The angular position of link 2 as a function of piston displacement is shown in Figure 2.9, and the angular position of the connecting rod as a function of piston displacement is shown in Figure 2.10.

- Case of Figure 1.14b
  Assuming that link 3 is the driver, the loop-closure equation is

$$r_l(\cos\,\theta_1,\,\sin\,\theta_1)^T = -r_2(\cos\,\theta_2,\,\sin\,\theta_2)^T - r_3(\cos\,\theta_3,\,\sin\,\theta_3)^T \qquad (2.56)$$

where r_1 and θ_1 are unknowns, and thus the equation falls into the first case category.

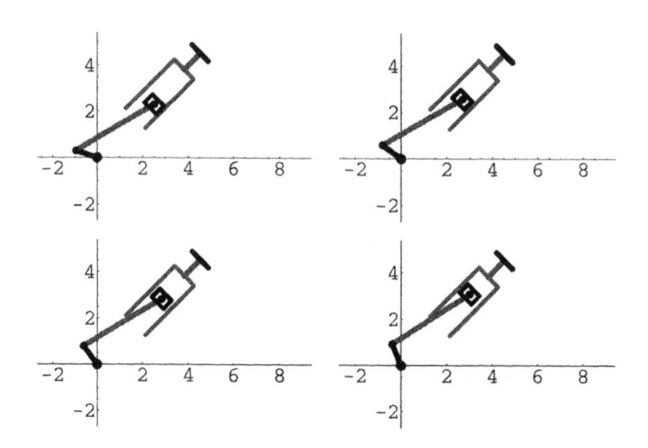

FIGURE 2.8 Four positions of the actuator.

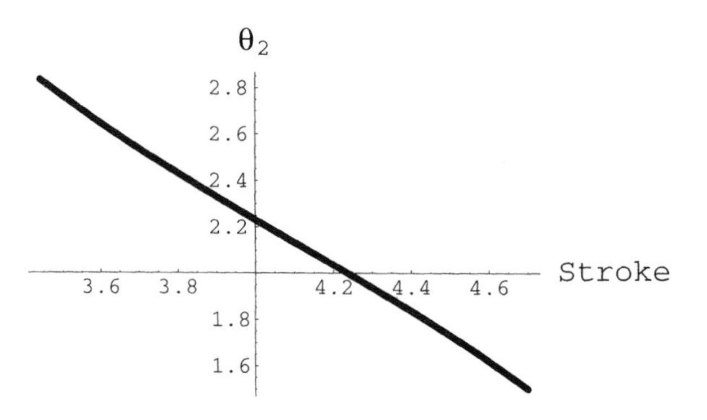

FIGURE 2.9 Angular position of link 2 vs. piston displacement.

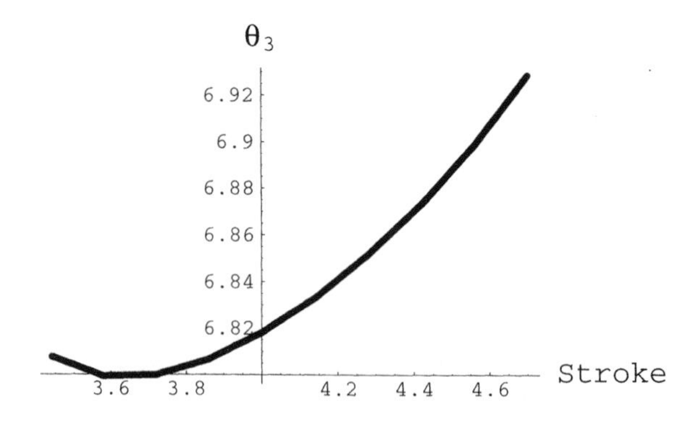

FIGURE 2.10 Angular position of the connecting rod vs. piston displacement.

Four-Bar Mechanism

For the case of Figure 1.4, crank 2 is the driver, and the loop-closure equation is

$$r_3(\cos\theta_3, \sin\theta_3)^T + r_4(\cos\theta_4, \sin\theta_4)^T = -r_l(\cos\theta_1, \sin\theta_1)^T - r_2(\cos\theta_2, \sin\theta_2)^T \tag{2.57}$$

where θ_3 and θ_4 are unknowns, and thus this equation falls into the fourth case category. In this case the vector $\mathbf{b} = (b_x, b_y)^T = b(\cos\alpha, \sin\alpha)^T$ is defined by

$$b_x = -r_1\cos\theta_1 - r_2\cos\theta_2 \text{ and } b_y = -r_1\sin\theta_1 - r_2\sin\theta_2 \tag{2.58}$$

$$b = \sqrt{b_x^2 + b_y^2} \tag{2.59}$$

and angle α by

$$\cos\alpha = \frac{b_x}{b} \text{ and } \sin\alpha = \frac{b_y}{b} \tag{2.60}$$

The following data were used to simulate the four-bar linkage:

$$\frac{r_1}{r_2} = 4, \quad \frac{r_3}{r_2} = 7, \quad \frac{r_4}{r_2} = 3, \text{ and } \theta_1 = \pi$$

In Figure 2.11 a four-bar mechanism is shown in six positions during crank r_2 rotation. As is seen, the crank does not make full circle; it rotates from $\pi/2$ to $3\pi/2$. Thus, this four-bar linkage is a triple rocker. This is confirmed by plotting the angles of rotation for the coupler (link 3) (Figure 2.12) and follower (link 4) (Figure 2.13).

In the example shown in Figure 2.11 the driving link 2, as well as two other links, was rocking. What if one wants to make this link revolving, i.e., to be a crank? This can be achieved if the relationships between the links in this mechanism are changed in such a way that they satisfy Grashof's criteria for having at least one revolving link.

Five-Bar Mechanism

An example of a five-bar linkage is shown in Figure 2.14a and the corresponding loop of vectors in Figure 2.14b. It is given that vectors $\mathbf{r}_3$ and $\mathbf{r}_5$ are perpendicular to the vector $\mathbf{r}_4$, i.e., $\theta_4 = \theta_3 - \pi/2$ and $\theta_5 = \theta_3 - \pi$.

Assuming that crank 2 is the input link, then there are two unknowns in this system: θ_3 and r_4. The loop-closure equation is

$$r_3(\cos\theta_3, \sin\theta_3)^T + r_4(\cos\theta_4, \sin\theta_4)^T + r_5(\cos\theta_5, \sin\theta_5)^T = \mathbf{b} \tag{2.61}$$

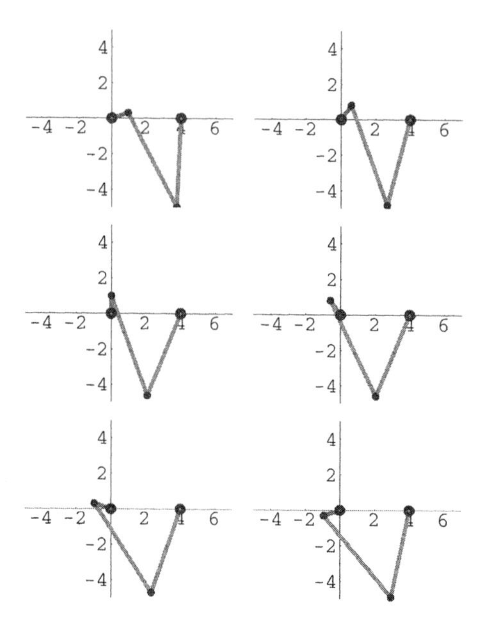

FIGURE 2.11 A four-bar linkage in six positions during link 2 rocking.

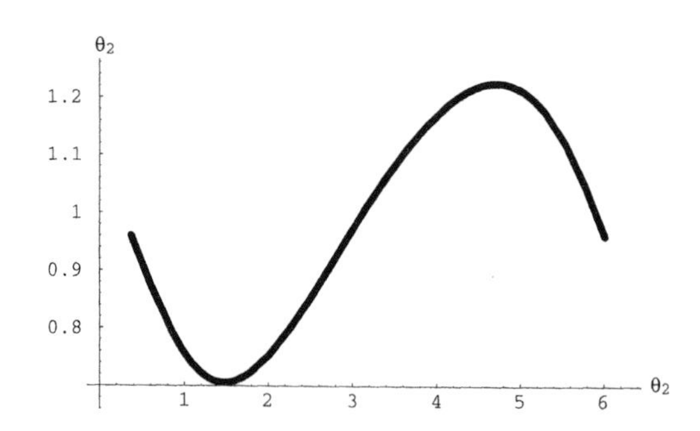

FIGURE 2.12 Angular positions of the coupler during link 2 rocking.

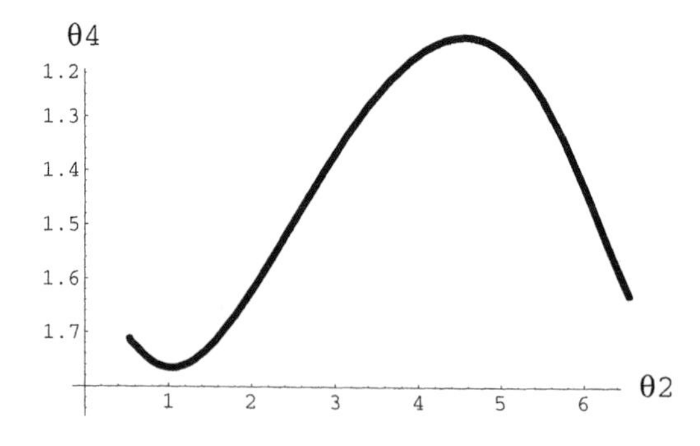

FIGURE 2.13 Angular positions of the follower during link 2 rocking.

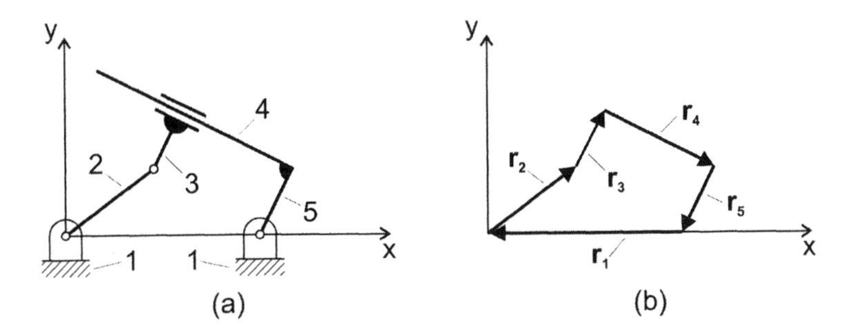

FIGURE 2.14 Five-bar mechanism (a) and corresponding vector diagram (b).

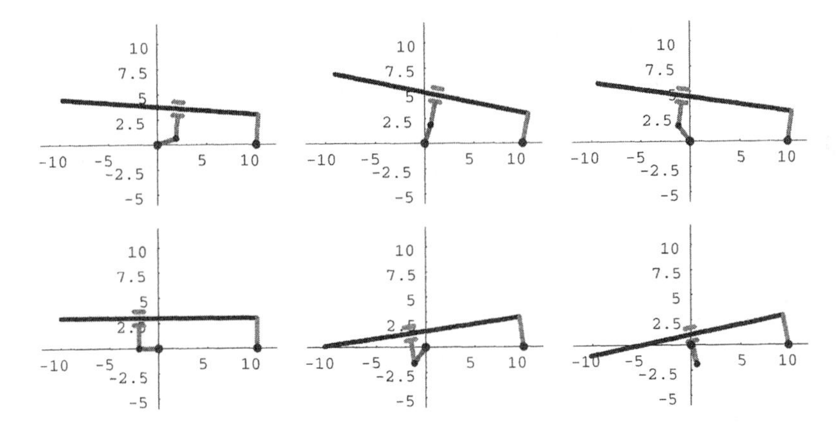

FIGURE 2.15 Five-bar mechanism in six positions during crank rotation.

where

$$\mathbf{b} = -r_1(\cos\theta_1, \sin\theta_1)^T - r_2(\cos\theta_2, \sin\theta_2)^T \tag{2.62}$$

Taking into account that θ_4 and θ_5 are functions of θ_3, Equation 2.61 becomes

$$r_3(\cos\theta_3, \sin\theta_3)^T + r_4(\sin\theta_3, -\cos\theta_3)^T + r_5(-\cos\theta_3, -\sin\theta_3)^T = \mathbf{b} \tag{2.63}$$

The vectors $\mathbf{r}_3$ and $\mathbf{r}_5$ can be summed so that Equation 2.63 takes the form

$$r_4(\sin\theta_3, -\cos\theta_3)^T + (r_3 - r_5)(\cos\theta_3, \sin\theta_3)^T = \mathbf{b} \tag{2.64}$$

Thus it is seen that this equation falls into the fifth case category. In Figure 2.15, snapshots of the animation are shown, where the following data were used:

$$\frac{r_1}{r_2} = 5, \quad \frac{r_3}{r_2} = \frac{r_5}{r_2} = 1.5, \quad \theta_1 = \pi, \quad \gamma = \frac{\pi}{2}, \quad \text{and} \quad \beta = \pi$$

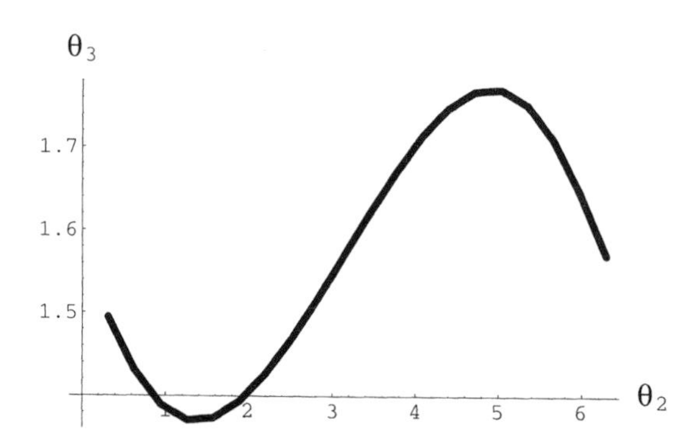

FIGURE 2.16 Angle θ_3 vs. crank angle.

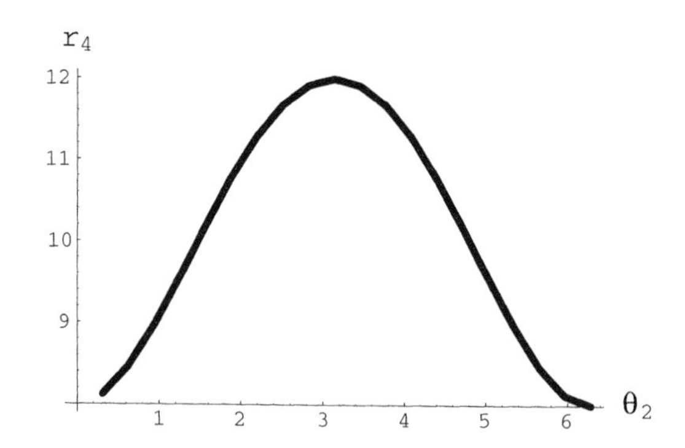

FIGURE 2.17 Slider position vs. crank angle.

In Figures 2.16 and 2.17 the variation of the angle θ_3 and the position of the slider r_4 during one cycle of crank rotation are shown.

Scotch Yoke Mechanism

In Figure 2.18a, a variation of the Scotch yoke mechanism is shown, in which crank 2 moves slider 3 up and down.

The position analysis is simple in this case. Here, however, a loop-closure approach will be applied for the sake of universality. The vertical position of slider 3 is defined by point *P*. The vector diagram is as shown in Figure 2.18b, and the corresponding loop-closure equation is

$$r_3(\cos\theta_3, \sin\theta_3)^T + r_4(\cos\theta_4, \sin\theta_4)^T = \\ -r_1(\cos\theta_1, \sin\theta_1)^T - r_2(\cos\theta_2, \sin\theta_2)^T \tag{2.65}$$

The two unknowns are r_3 and r_4, and Equation 2.65 thus falls into the third case category. Snapshots of the animation are shown in Figure 2.19 using the following data:

$$\frac{r_1}{r_2} = 2, \quad \theta_1 = \pi, \quad \theta_3 = 0, \quad \text{and} \quad \theta_4 = \frac{3\pi}{2}$$

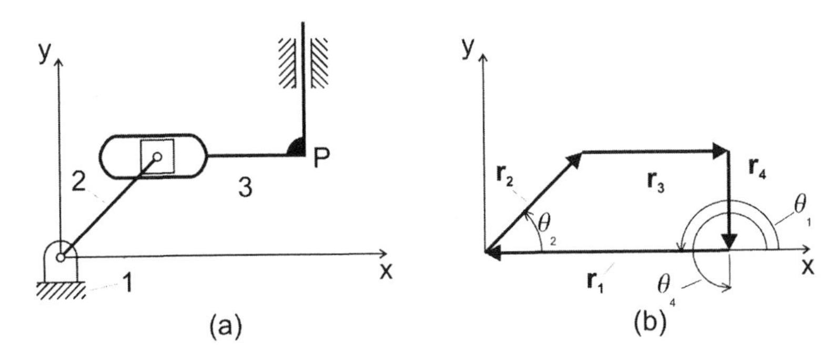

FIGURE 2.18 An offset Scotch yoke mechanism (a) and corresponding vector diagram (b).

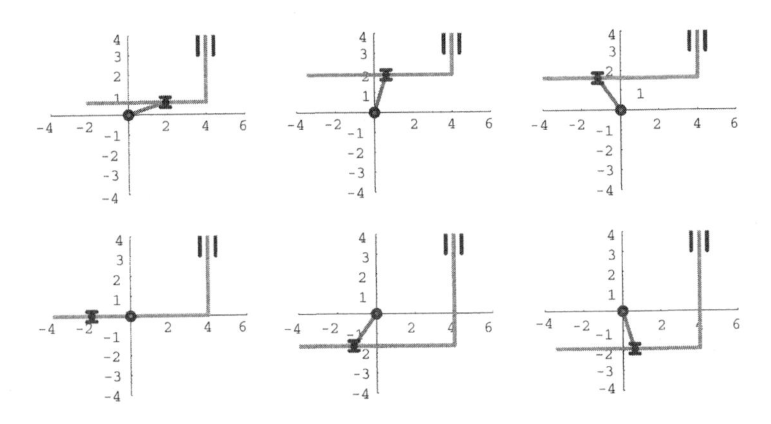

FIGURE 2.19 Six snapshots of the Scotch yoke mechanism.

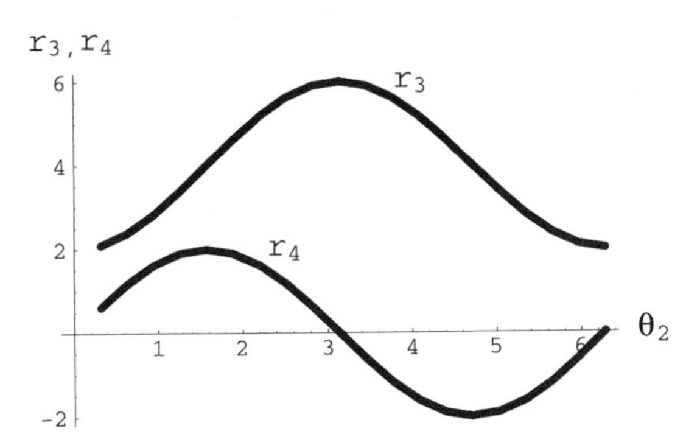

FIGURE 2.20 Displacements r_3 and r_4 vs. crank angle θ_2.

In Figure 2.20 the displacements of two sliders are shown during one cycle. It is seen that r_3 has the frequency of the crank, while r_4 is double the crank's frequency. The maximums are defined by the geometry of the mechanism.

2.3.6 Applications to Compound Mechanisms

Loader

For the loader shown in Figure 2.3, the motion of the bucket is the result of two inputs: cylinder (actuator) 5 rotating lift arm 4, and cylinder 6 rotating the bucket through bellcrank 10. This compound mechanism comprises three simple ones represented by the links (1, 2, 3), (11, 8, 9, 10), and (12, 13, 14, 15). These three mechanisms form three loops: loop 1, loop 2, loop 3, respectively. The mechanisms and the vector loops are shown in Figures 2.21 through 2.23.

For each mechanism a loop-closure equation can be written:

Loop 1

$$r_2(\cos\theta_2, \sin\theta_2)^T + r_3(\cos\theta_3, \sin\theta_3)^T = -r_1(\cos\theta_1, \sin\theta_1)^T \tag{2.66}$$

Loop 2

$$\begin{aligned} &r_8(\cos\theta_8, \sin\theta_8)^T + r_9(\cos\theta_9, \sin\theta_9)^T = \\ &-r_{11}(\cos\theta_{11}, \sin\theta_{11})^T + r_{10}(\cos\theta_2, \sin\theta_2)^T \end{aligned} \tag{2.67}$$

Loop 3

$$\begin{aligned} &r_{13}(\cos\theta_{13}, sin\ \theta_{13})^T + r_{14}(\cos\theta_{14}, sin\ \theta_{14})^T = \\ &-r_{12}(\cos\theta_9, sin\ \theta_9)^T + r_{15}(\cos\theta_2, sin\ \theta_2)^T \end{aligned} \tag{2.68}$$

In Equation 2.66 there are two unknowns: θ_2 and θ_3, which means that this equation can be solved and thus it is independent of other equations (see graph in Figure 2.24). One of the solutions of Equation 2.66, namely θ_2, becomes an input into loop 2, while the unknowns in loop 2 are angles θ_8 and θ_9. In loop 3, solutions from the two previous loops, θ_2 and θ_9, are used as inputs, while the unknowns are θ_{13} and θ_{14}.

Equations 2.66 though 2.68 all belong to the fourth case of equations considered above. Indeed, in all of them the unknowns are two angles: θ_2 and θ_3 in Equation 2.66, θ_8 and θ_9 in Equation 2.67, and θ_{13} and θ_{14} in Equation 2.68.

A graph in Figure 2.24 shows the functional connectivity between the loops. Such a graph is called a *tree* because it does not have any connections between the branches. In this graph a *node* represents a mechanism loop, and a *vertex* represents a common link between the loops. The node corresponding to the bucket is called the *root* node. It is seen that the motion of the bucket is the result of two independent inputs (prove that the mobility of this mechanism is 2).

The following data were used to simulate this compound mechanism:

Loop 1

$$\frac{r_1}{r_2} = 0.6,\quad \theta_1 = \frac{\pi}{2},\quad \frac{v_3}{r_2} = 1,\quad \text{and}\quad \frac{r_3}{r_2} = 0.76 + \frac{v_3}{r_2}t$$

where v_3 is the piston velocity and t is time.

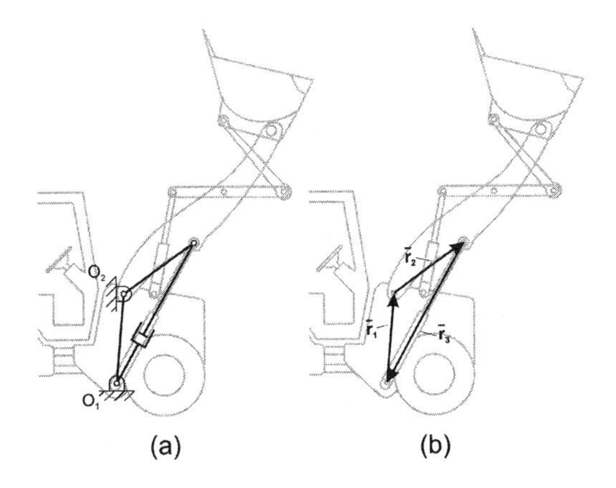

FIGURE 2.21 First mechanism (a) and corresponding vector diagram (b).

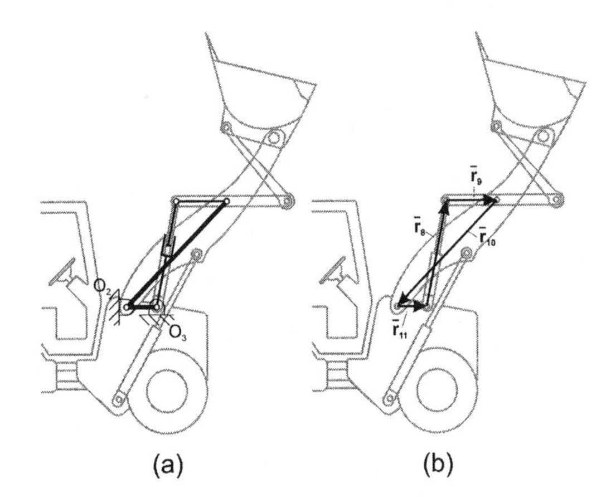

FIGURE 2.22 Second mechanism (a) and corresponding vector diagram (b).

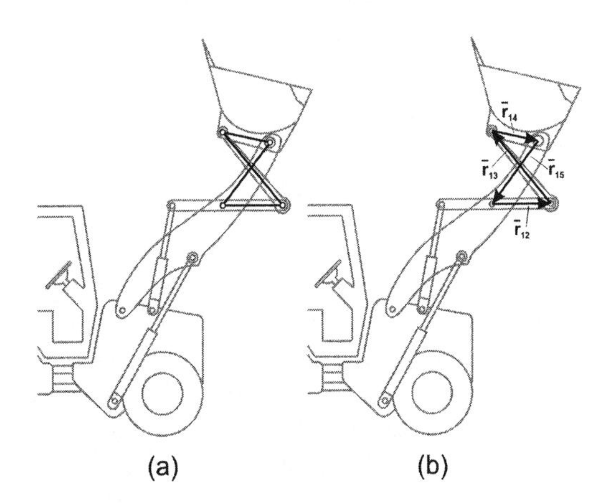

FIGURE 2.23 Third mechanism (a) and corresponding vector diagram (b).

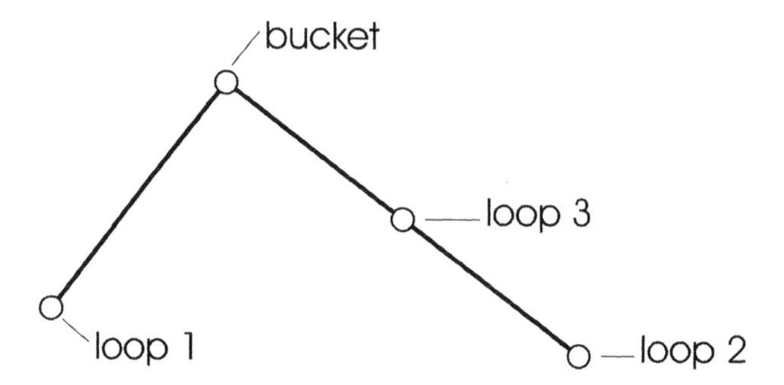

FIGURE 2.24 A graph representing functional connectivity in a loader as a compound mechanism.

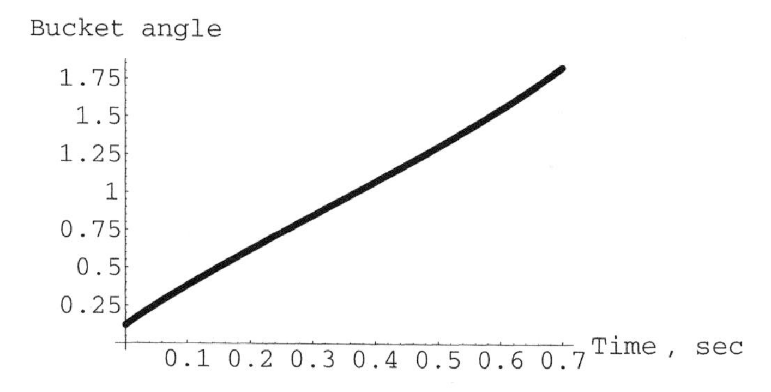

FIGURE 2.25 Change of bucket orientation in time.

Loop 2

$$\frac{r_9}{r_2} = 0.8, \ \frac{r_{10}}{r_2} = 1.6, \ \frac{r_{11}}{r_2} = 0.14, \ \theta_{11} = 0, \ \frac{v_8}{r_2} = 1, \text{ and } \frac{r_8}{r_2} = 0.76 + \frac{v_8}{r_2} t$$

Loop 3

$$\frac{r_{12}}{r_2} = 0.4, \ \frac{r_{13}}{r_2} = 1.1, \ \frac{r_{14}}{r_2} = 0.8, \ \frac{r_{15}}{r_2} = 1.2$$

In Figure 2.25 the change of the bucket orientation in time is shown. The smallest angle corresponds to the bucket in the low position.

In general, any mechanism the functionality of which is represented by a tree graph can be analyzed starting from loops containing input links, then solving equations associated with each branch, and eventually finding the motion of the root note, i.e., the output link. Note that the number of inputs in mechanisms represented functionally as trees equals the number of branches in a tree.

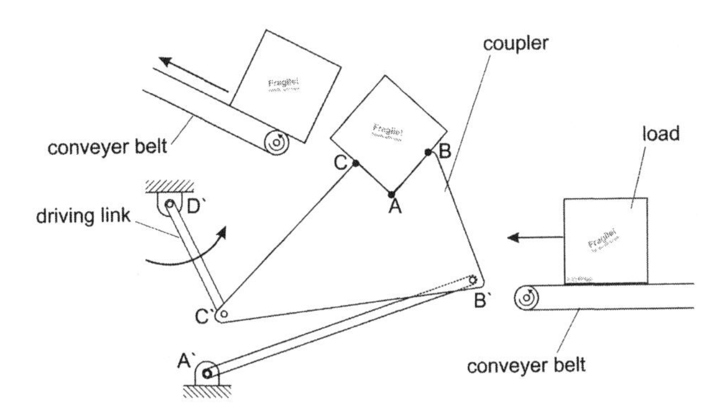

FIGURE 2.26 Material-handling mechanism.

2.3.7 Trajectory of a Point on a Mechanism

In many applications the function of the mechanism is associated with the motion of a specific point on this mechanism. In Figure 2.26 a material-handling mechanism is shown. The function of the mechanism is to transfer the load from one conveyor belt to another. There are *position, velocity*, and *acceleration requirements* to be met by the designers of this mechanism. The position requirements are to have the line *AB* on a coupler horizontal to accept the load from the right belt in one position and the line *AC* parallel to the left belt for proper unloading in another position. The velocity requirements to be met by the mechanism are that at the two extreme positions (loading/unloading) the velocities be minimal (or zero) to allow sufficient time for load transfer. At the same time, the average velocity between the two extreme positions must be synchronized with the belt velocities to maintain constant load flow rate. The acceleration requirements are to prevent jerks (sudden changes in inertial forces) in the system during the load transfer process to ensure the reliability of this operation. To meet these objectives a proper mechanism skeleton should be found first, and then a proper shape of the coupler.

A proper mechanism skeleton requirement may be to have a rocker-crank mechanism subject to size constraints. This can be achieved by satisfying Grashof's criteria (see Equation 1.2) for a rocker-crank. The needed shape of the coupler is the subject of this section. For the example shown in Figure 2.28 this problem is reduced to finding the trajectories of points identified on a coupler.

Consider the trajectory of any point *P* on a coupler (Figure 2.27). The position of the point P in the global coordinate system is given by the vector $\mathbf{R}_P$ which can be found either as

$$\mathbf{R_P} = r_2 + r_P = r_2(\cos\theta_2, \sin\theta_2)^T + r_P[\cos(\theta_3+\alpha), \sin(\theta_3+\alpha)]^T \tag{2.69}$$

or

$$\mathbf{R_P} = \mathbf{r}_2 + \mathbf{r}_D + \mathbf{r}_{DP} = r_2(\cos\theta_2, \sin\theta_2)^T + r_D(\cos\theta_3, \sin\theta_3)^T + r_{DP}\left[\cos\left(\theta_3+\frac{\pi}{2}\right), \sin\left(\theta_3+\frac{\pi}{2}\right)\right]^T \tag{2.70}$$

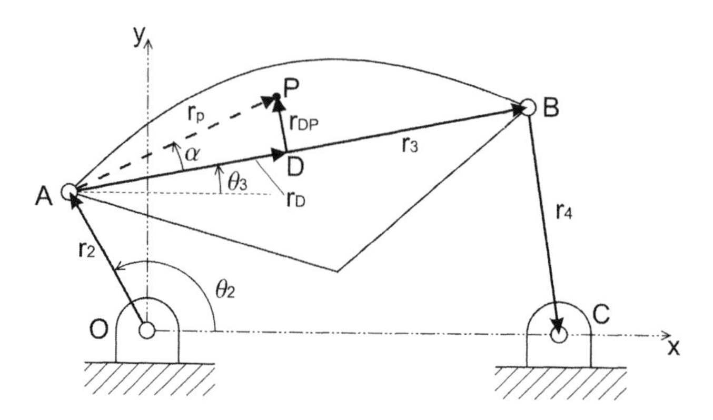

FIGURE 2.27 Point *P* on a coupler.

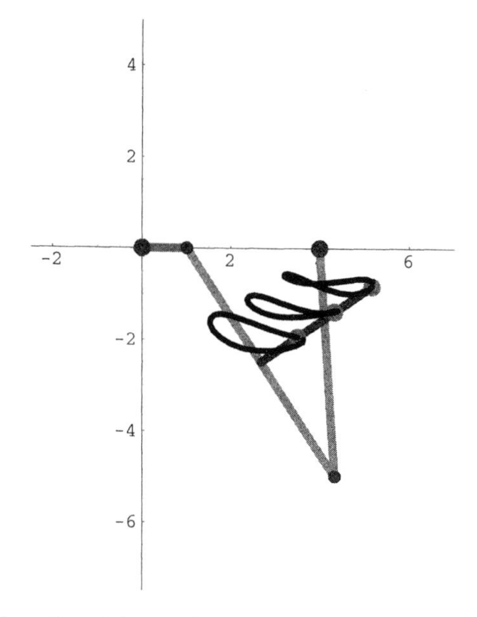

FIGURE 2.28 Trajectories of three points on a coupler.

When the position analysis of the skeleton of the mechanism is done, then for any pair of angles θ_2 and θ_3 the vector $\mathbf{R}_P$ can be found for either of the above equations. The important point is that there is an infinite number of possible trajectories and identification of the needed one is done by trial and error.

In Figure 2.28 an example of a four-bar linkage with a coupler having another bar attached to it in the middle is shown. Three points are identified on this bar and their trajectories are shown. The mechanism is of the rocker-crank type with the following nondimensional units:

$$\frac{r_1}{r_2} = 4, \ \frac{r_3}{r_2} = 6, \ \frac{r_4}{r_2} = 5$$

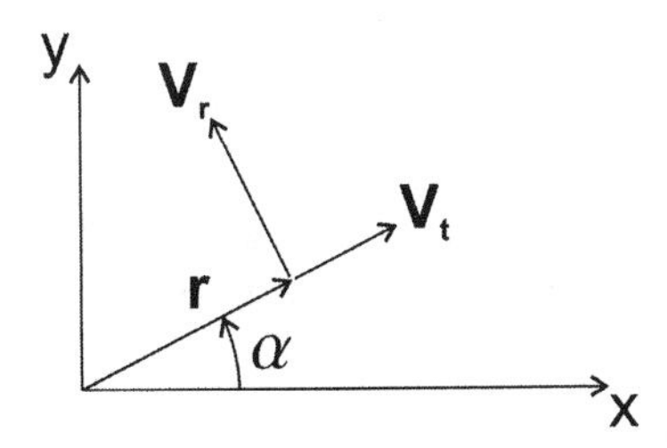

FIGURE 2.29 Two components of the velocity vector.

By choosing different points, different trajectories can be obtained: from the circle for the crank–coupler joint to the chord for the follower–coupler joint.

2.4 VELOCITY ANALYSIS

2.4.1 Velocity Vector

For any vector the magnitude and the direction of which are functions of time, *t*,

$$\mathbf{r}(t) = r(t)[\cos\theta(t), \sin\theta(t)]^T \tag{2.71}$$

the time derivative represents the velocity vector:

$$\frac{d\mathbf{r}}{dt} = \dot{\mathbf{r}} = \dot{r}(t)[\cos\theta(t), \sin\theta(t)]^T + r(t)[-\sin\theta(t), \cos\theta(t)]^T\dot{\theta}(t) \tag{2.72}$$

One can see that the velocity vector comprises two components. The first has the same direction as the original vector Equation 2.71, and it is called the *translational* (or *linear*) velocity vector, $\mathbf{V}_t$

$$\mathbf{V}_t = \dot{r}(t)[\cos\theta(t), \sin\theta(t)]^T \tag{2.73}$$

The magnitude of the translational velocity vector represents the rate of change of the vector length *r*. The second component is perpendicular to the direction of the original vector ($[-\sin\theta(t), \cos\theta(t)]^T = [\cos(\theta(t) + \pi/2), \sin(\theta(t) + \pi/2)]^T$) and it is rotated by $\pi/2$ in a counterclockwise direction. This component represents the *rotational* (or *angular*) velocity vector:

$$\mathbf{V}_r = r(t)\dot{\theta}(t)[(-\sin\theta)(t), \cos\theta(t)]^T \tag{2.74}$$

The magnitude of the rotational velocity is proportional to the angular velocity $\omega(t) = \dot{\theta}(t)$.

In Figure 2.29 the two components of the velocity vector are shown. These components can also be viewed as characterizing the velocity of the point at the tip of vector **r**. One should also note that the translational and rotational velocity components are independent components.

2.4.2 Equations for Velocities

The equations for velocities follow from the loop-closure equation for positions, Equation 2.15. Indeed, if in Equation 2.15 $r_i = r_i\,(t)$ and $\theta_i = \theta_i(t)$, then differentiating it with respect to t and using Equation 2.72 the following vector equation is obtained:

$$\sum_{i=1}^{N} \dot{r}_i(t)[\cos\theta_i(t),\ \sin\theta_i(t)]^T + r_i(t)[-\sin\theta_i(t),\ \cos\theta_i(t)]^T\omega_i(t) = 0 \tag{2.75}$$

where $\omega_i(t) = \dot{\theta}_i(t)$ is the angular velocity.

In Equation 2.75 the unknowns can be any two velocity components: two translational, two rotational, or a combination of rotational and translational. The important point is that Equation 2.75 is a linear function of velocities, and thus for any unknown velocities they are found from a system of two linear algebraic equations. How to find symbolic solutions for a linear system of algebraic equations is shown in the Appendix. These solutions, as is shown below, are also easy to derive manually. A symbolic solution allows a better understanding of position–velocity dependency in some mechanisms.

First Case

In this case (see Equation 2.16), all vectors except vector $\mathbf{r_j}$ are moved to the right-hand side. Correspondingly, Equation 2.75 will take the form (note that all variables are assumed to be time dependent):

$$\dot{r}_j[\cos(\theta_j,\ \sin\theta_j)]^T + r_j[-\sin\theta_j, \cos\theta_j]^T\omega_j = [\dot{b}_x, \dot{b}_y]^T \tag{2.76}$$

Note that in the velocity analysis it is more convenient to take vector **b** in the form shown in Equation 2.76, since this equation is linear with respect to the two unknowns, $\dot{r}_j\,(t)$ and $\omega_j\,(t)$.

One can find $\omega_j\,(t)$ by multiplying Equation 2.76 from the left by a unit vector perpendicular to the first vector on the left-hand side, i.e., by the vector $\mathbf{u_1} = [-\sin\theta_j, \cos\theta_j]$. The result is

$$r_j\omega_j = -\dot{b}_x \sin\theta_j + \dot{b}_y \cos\theta_j \tag{2.77}$$

The latter explicitly defines $\omega_j\,(t)$ as a function of variables found in position analysis. Similarly, if one multiplies Equation 2.76 from the left by a unit vector perpendicular to the second vector on the left-hand side of Equation 2.76, i.e. by the vector $\mathbf{u}_2 = [\cos\theta_j, \sin\theta_j]$, the other unknown will be found:

$$\dot{r}_j(t) = \dot{b}_x \cos\theta_j + \dot{b}_y \sin\theta_j \tag{2.78}$$

Second Case

In this case the magnitude $r_i(t)$ and the angle $\theta_j(t)$ were the two unknowns in Equation 2.28. Then Equation 2.75 takes the form

$$\dot{r}_i[\cos\theta_i, \sin\theta_i]^T + r_i[-\sin\theta_i, \cos\theta_i]^T\omega_i + \dot{r}_j[\cos\theta_j, \sin\theta_j]^T + r_j[-\sin\theta_j, \cos\theta_j]^T\omega_j = [\dot{b}_x, \dot{b}_y]^T \tag{2.79}$$

In Equation 2.79 the unknowns are $\dot{r}_i(t)$ and $\omega_j(t)$. As before, by premultiplying Equation 2.79 by unit vectors perpendicular to the first and fourth vectors on the left-hand side of this equation, the explicit expressions for the two unknowns are obtained in the form:

$$r_j\omega_j\cos(\theta_j - \theta_i) = -r_i\omega_i - \dot{r}_j\sin(\theta_j - \theta_i) - \dot{b}_x\sin\theta_i + \dot{b}_y\cos\theta_i \tag{2.80}$$

and

$$\dot{r}_i\cos(\theta_j - \theta_i) = -r_i\omega_i\sin(\theta_j - \theta_i) - \dot{r}_j + \dot{b}_x\cos\theta_j + \dot{b}_y\sin\theta_j \tag{2.81}$$

Third Case

Since in this case the magnitudes of two vectors, r_i and r_j were the unknowns in Equation 2.28, the corresponding velocity unknowns in Equation 2.79 are $\dot{r}_i(t)$ and $\dot{r}_j(t)$. These two unknowns can be found using the same procedure as above, and the corresponding expressions are

$$\dot{r}_i\sin(\theta_i - \theta_j) = -r_i\omega_i\cos(\theta_i - \theta_j) - r_j\omega_j - \dot{b}_x\sin\theta_j + \dot{b}_y\cos\theta_j \tag{2.82}$$

and

$$\dot{r}_j\sin(\theta_j - \theta_i) = -r_j\omega_j\cos(\theta_j - \theta_i) - r_i\omega_i - \dot{b}_x\sin\theta_i + \dot{b}_y\cos\theta_i \tag{2.83}$$

Note that the last equation is obtained from the previous one by interchanging indices i and j. An alternative way to derive the expressions for velocities in this case is to differentiate Equations 2.34 and 2.35.

Fourth Case

In this case the two unknown angular velocities $\dot{\theta}_i(t)$ and $\dot{\theta}_j(t)$ are found from Equation 2.79. The corresponding expressions are as follows:

$$\omega_i r_i\sin(\theta_j - \theta_i) = -\dot{r}_i\cos(\theta_j - \theta_i) - \dot{r}_j + \dot{b}_x\cos\theta_j + \dot{b}_y\sin\theta_j \tag{2.84}$$

and

$$\omega_j r_j \sin(\theta_i - \theta_j) = -\dot{r}_j \cos(\theta_i - \theta_j) - \dot{r}_i + \dot{b}_x \cos\theta_i + \dot{b}_y \sin\theta_i \tag{2.85}$$

Fifth Case

The loop-closure Equation 2.45 is differentiated under the assumption that all parameters, except γ and β, are time dependent.

$$\begin{aligned} &\dot{r}_i [\cos\theta_i, \sin\theta_i]^T + r_i[-\sin\theta_i, \cos\theta_i]^T \omega_i \\ &+ \dot{r}_j[\cos(\theta_i - \gamma), \sin(\theta_i - \gamma)]^T + r_j[-\sin(\theta_i - \gamma), \cos(\theta_i - \gamma)]^T \omega_\iota \\ &+ \dot{r}_k[\cos(\theta_i - \beta), \sin(\theta_i - \beta)]^T + r_k[-\sin(\theta_i - \beta), \cos(\theta_i - \beta)]^T \omega_\iota = [\dot{b}_x, \dot{b}_y]^T \end{aligned} \tag{2.86}$$

Recall that the unknowns in this case are $\omega_i(t)$ and $\dot{r}_j(t)$. Collect similar terms in the latter equation.

$$\begin{aligned} &\dot{r}_i[\cos\theta_i, \sin\theta_i]^T + [d_x, d_y]^T \omega_i + \dot{r}_j[\cos(\theta_i - \gamma), \sin(\theta_i - \gamma)]^T \\ &+\dot{r}_k[\cos(\theta_i - \beta), \sin(\theta_i - \beta)]^T = [\dot{b}_x, \dot{b}_y]^T \end{aligned} \tag{2.87}$$

where it is denoted

$$d_x = -r_i \sin\theta_i - r_j \sin(\theta_i - \gamma) - r_k \sin(\theta_i - \beta) \tag{2.88}$$

and

$$d_y = r_i \cos\theta_i + r_j \cos(\theta_i - \gamma) + r_k \cos(\theta_i - \beta) \tag{2.89}$$

Now the two unknowns can be found in the usual way by identifying the unit vectors perpendicular to the vectors $[\cos(\theta_i - \gamma), \sin(\theta_i - \gamma)]^T$ and $[d_x, d_y]^T$, and premultiplying Equation 2.87 by these vectors (note that the vector perpendicular to the latter vector is $[-d_{4444}, d_x]^T$). As a result, the following expressions are obtained:

$$\begin{aligned} (-d_x \sin(\theta_i - \gamma) + d_y \cos(\theta_i - \gamma))\omega_i &= -\dot{r}_i \sin\gamma - \dot{r}_k \sin(\gamma - \beta) \\ &-\dot{b}_x \sin(\theta_i - \gamma) + \dot{b}_y \cos(\theta_i - \gamma) \end{aligned} \tag{2.90}$$

and

$$\begin{aligned} (-d_y \cos(\theta_i - \gamma) + d_x \sin(\theta_i - \gamma))\dot{r}_j &= -\dot{r}_i + (-d_y \cos\theta_i + d_x \sin\theta_i) \\ &-\dot{r}_k(-d_y \cos(\theta_i - \beta) + d_x \sin(\theta_i - \beta)) + (-d_y \dot{b}_x + d_x \dot{b}_y) \end{aligned} \tag{2.91}$$

2.4.3 Applications to Simple Mechanisms

Slider-Crank Inversions (Figure 1.14)

- Figure 1.14a with the driving crank

The inversion in Figure 1.14a falls into the second case if the crank is the driver (see Equation 2.54 for the case when a crank is the driver). The corresponding velocity equations are found from Equation 2.79 by taking that $i = 1$, $j = 3$, $b_x = -r_2\cos\theta_2$, and $b_y = -r_2\sin\theta_2$ (see Equation 2.54). Now one can use the general solutions given by Equations 2.80 and 2.81 for this specific case, taking into account that $\dot{b}_x = r_2\omega_2 \sin\theta_2$, $\dot{b}_y = -r_2\omega_2 \cos\theta_2$, $r_2 =$ const., $r_3 =$ const. and $\theta_1 = \pi$.

$$-r_3\omega_3\cos\theta_3 = r_2\omega_2\cos\theta_2 \tag{2.92}$$

Thus, the angular velocity of the connecting rod is

$$\omega_3 = -\omega_2\frac{r_2\cos\theta_2}{r_3\cos\theta_3} \tag{2.93}$$

Similarly, for the slider velocity $\dot{r}_1(t)$,

$$\dot{r}_1(t) = -r_2\omega_2\frac{\sin(\theta_2 - \theta_3)}{\cos\theta_3} \tag{2.94}$$

As one can see, even if the crank angular velocity is constant, the velocities of the piston and the connecting rod are periodic functions which, in turn, depend on the crank and connecting rod lengths. Note also that when $\theta_2 - \theta_3 = 0, \pi, 2\pi$ the piston velocity becomes zero. These extreme piston positions are called *dead points*.

In Figures 2.30 and 2.31 the variations of the angular velocity of the connecting rod and the translational velocity of the piston during one cycle of crank rotation are shown. As one can see from Figure 2.31 the extreme positions of the piston are at $\theta_2 = 0, \pi, 2\pi$, while the corresponding angles θ_3 are 0, 0, 0 (see Figure 2.7). Note that, since ω_3 is a derivative of θ_3, the maximums of the former are at the crank positions corresponding to the minimums of the latter.

- Figure 1.14a with the driving piston

If in the inversion shown in Figure 1.14a the piston is the driver, then, as is known, this situation belongs to the fourth case category. In this case $i = 2$, $j = 3$, $\dot{b}_x = \dot{r}_1$, $\dot{b}_y = 0$, and $\theta_1 = \pi$. The equations for the unknown angular velocities $\omega_2 = \dot{\theta}_2$ and $\omega_3 = \dot{\theta}_3$ follow from Equations 2.84 and 2.85.

$$\omega_2 = \frac{\dot{r}_1\cos\theta_3}{r_2\sin(\theta_3 - \theta_2)} \tag{2.95}$$

and

$$\omega_3 = \frac{\dot{r}_1\cos\theta_2}{r_3\sin(\theta_2 - \theta_3)} \tag{2.96}$$

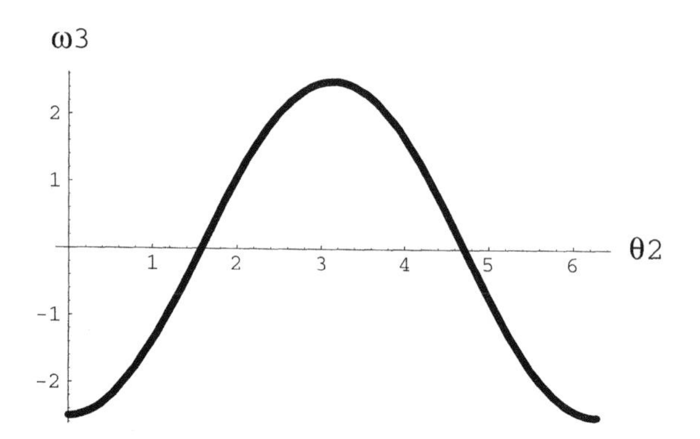

FIGURE 2.30 Angular velocity of the connecting rod during one cycle of crank rotation.

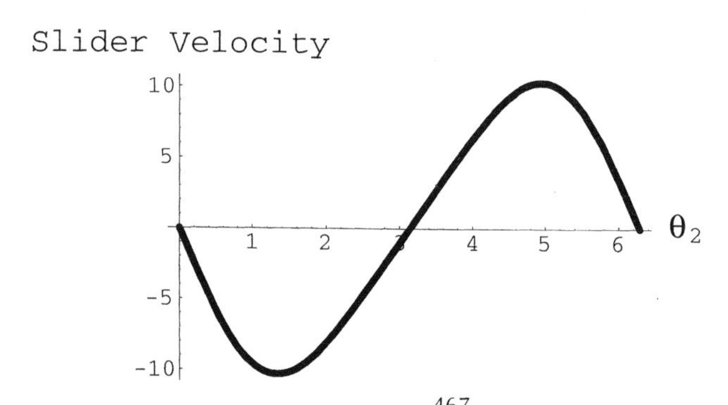

467

FIGURE 2.31 Slider velocity during one cycle of crank rotation.

It is seen from the two equations that when $\theta_2 - \theta_3 = 0$ or $\theta_2 - \theta_3 = \pi$ the denominators in both equations become zero. However, at these dead points the piston velocity is also zero. Thus, the velocities become undetermined. The uncertainty can be resolved if the function $r_1(t)$ is at least twice differentiable so that L'Hopital's rule for resolving the 0/0-uncertainty can be used.

In Figures 2.32 and 2.33 the angular velocities of the crank and connecting rod are shown as functions of the piston stroke. As one can see, both angular velocities are finite at the dead points. The velocity of the piston is taken as being constant during one stroke. Note that the velocity of the crank is not constant when the piston is the driver.

- Figure 1.14b with link 3 as a driver

This inversion falls into the first case category and the equations for velocities are given by Equations 2.77 and 2.78. The specific equations are obtained by taking $j = 1$, $\theta_2 = \pi/2$, and $\mathbf{b} = (b_x, b_y)^T$, where $b_x = -r_3 \cos\theta_3(t)$ and $b_y = -r_2 - r_3 \sin\theta_3(t)$. Thus, Equations 2.77 and 2.78 are reduced to

$$\omega_1 = -\frac{r_3}{r_1}\omega_3 \cos(\theta_3 - \theta_1) \tag{2.97}$$

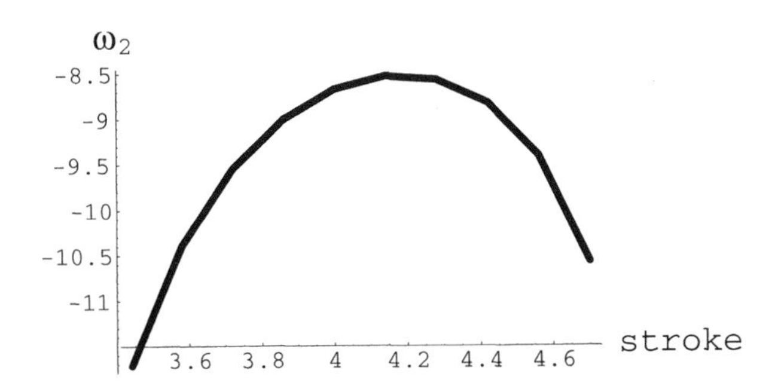

FIGURE 2.32 Angular velocity of link 2 vs. piston stroke.

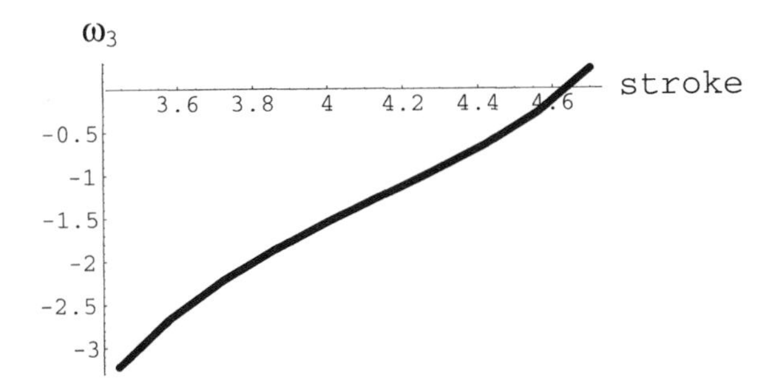

FIGURE 2.33 Angular velocity of the hydraulic rod vs. piston stroke.

and

$$\dot{r}_1(t) = r_3\omega_3\sin(\theta_3 - \theta_1) \tag{2.98}$$

Four-Bar Mechanism (Figure 1.4)

The solutions are given by Equations 2.84 and 2.85. If crank 2 is the driver, then $i = 3$, $j = 4$, $b_x = r_1 - r_2\cos\theta_2$, $b_y = -r_2\sin\theta_2$ (assuming that $\theta_1 = \pi$), and the equations for angular velocities are

$$\omega_3 = \omega_2\frac{r_2\sin(\theta_2 - \theta_4)}{r_3\sin(\theta_4 - \theta_3)} \tag{2.99}$$

$$\omega_4 = \omega_2\frac{r_2\sin(\theta_2 - \theta_3)}{r_4\sin(\theta_3 - \theta_4)} \tag{2.100}$$

The angular velocities ω_3 and ω_4 as functions of the crank angle are shown in Figures 2.34 and 2.35 for the same dimensions as in the position analysis.

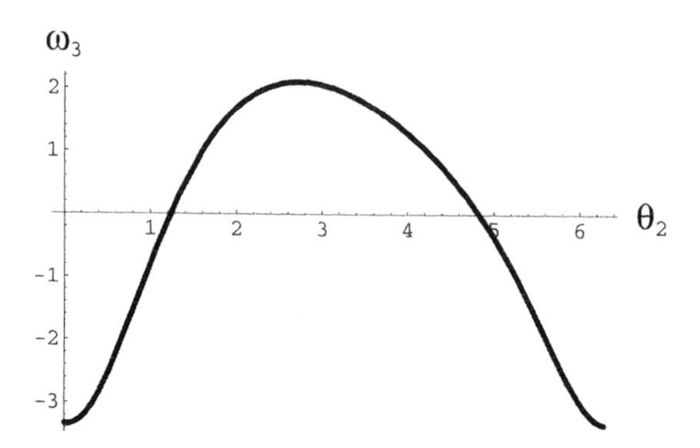

FIGURE 2.34 Angular velocity of the coupler vs. crank angle.

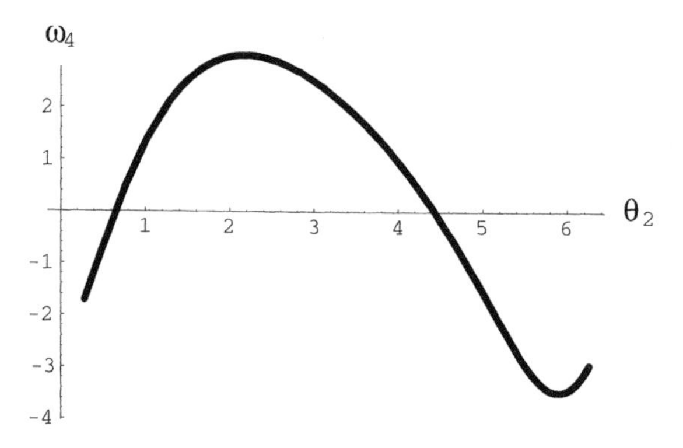

FIGURE 2.35 Angular velocity of the follower vs. crank angle.

Five-Bar Mechanism (Figure 2.14a)

The solutions are given by Equations 2.90 and 2.91. If crank 2 is the driver, then in these solutions $i = 3$, $j = 4$, $k = 5$, $\gamma = \pi/2$, and $\beta = \pi$. In this case (see Equation 2.62) $b_x = r_1 - r_2 \cos\theta_2$, $b_y = -r_2 \sin\theta_2$, and (see Equations 2.88 and 2.89) $d_x = (r_5 - r_3) \sin\theta_3 + r_4 \cos\theta_3$, $d_y = -(r_5 - r_3) \cos\theta_3 + r_4 \sin\theta_3$, where it was taken that $\theta_1 = \pi$. Recall also that r_2, r_3, and r_5 are all constant. Thus, for the case of Figure 2.14a the angular velocity of link 3 and the sliding velocity of link 4 are, respectively,

$$\dot{\theta}_3 = \omega_3 = \omega_2 \frac{r_2 \sin(\theta_2 - \theta_3)}{r_4} \tag{2.101}$$

$$\dot{r}_4 = r_2 \omega_2 \left[\cos(\theta_3 - \theta_2) + \frac{r_5 - r_3}{r_4} \sin(\theta_3 - \theta_2) \right] \tag{2.102}$$

In Figures 2.36 and 2.37 the angular velocity of link 3 and the translational velocity of the slider are shown during one cycle of crank rotation. The numerical

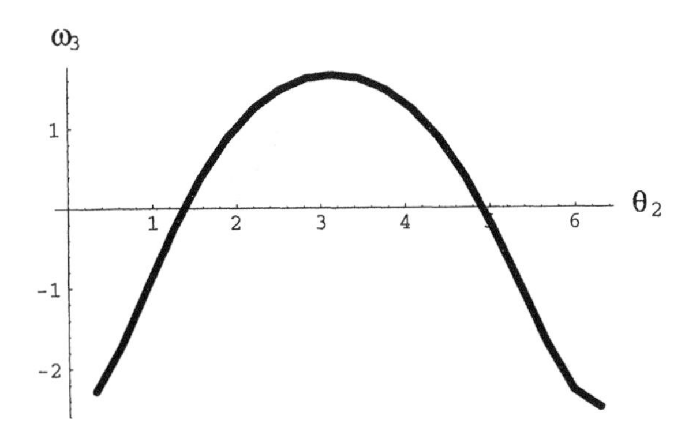

FIGURE 2.36 Angular velocity of link 3 in a five-bar mechanism.

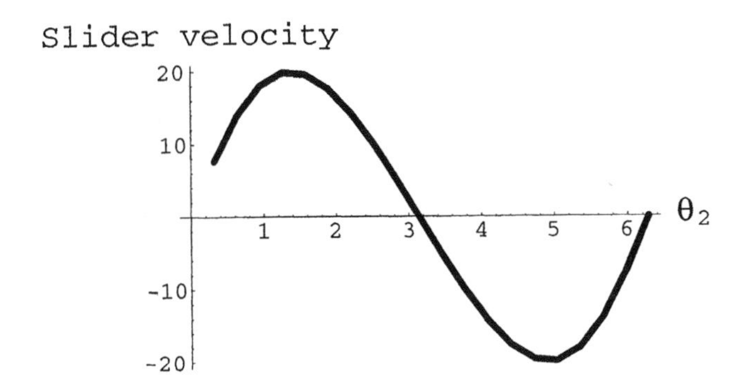

FIGURE 2.37 Translational velocity of the slider in a five-bar mechanism.

data were the same as in the position analysis, while the angular velocity was taken to be $\omega_2 = 5r_2$ rad/s.

Scotch Yoke Mechanism (Figure 2.18a)

The solution is given by Equations 2.82 and 2.83 in which $i = 3$, $j = 4$, $\theta_1 = \pi$, $\theta_3 = 0$, $\theta_4 = -\pi/2$, $b_x = r_1 - r_2 \cos\theta_2$, and $b_y = -r_2 \sin\theta_2$. After substitution,

$$\dot{r}_3 = -r_2\omega_2 \sin\theta_2 \tag{2.103}$$

$$\dot{r}_4 = -r_2\omega_2 \cos\theta_2 \tag{2.104}$$

In Figure 2.38 the velocities of sliders are shown during one cycle of crank rotation for the crank angular velocity $\omega_2 = 5r_2$ rad/s.

2.4.4 Applications to Compound Mechanisms

For a compound mechanism in which each loop is either in series or in parallel to other loops, the velocity equations are reduced to those for simple mechanisms. Below an example of a loader is considered.

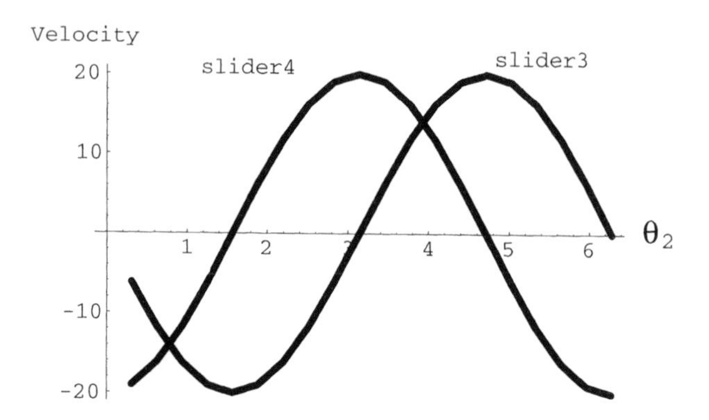

FIGURE 2.38 Velocities of two sliders during one cycle.

Loader

The loader comprises three mechanisms the loop-closure equations for which were given in Equations 2.66 to 2.68. They all fall into the fourth case category, so that the angular velocities are given by Equations 2.84 and 2.85.

Loop 1

In this case $i = 2, j = 3$, and $\dot{b}_x = \dot{b}_y = 0$. The corresponding angular velocities are

$$\omega_2 = -\frac{\dot{r}_3}{r_2 \sin(\theta_3 - \theta_2)} \tag{2.105}$$

and

$$\omega_3 = \frac{\dot{r}_3}{r_3}\cot(\theta_3 - \theta_2) \tag{2.106}$$

Loop 2

In this case $i = 8, j = 9$, $\dot{b}_x = -r_{10}\omega_2 \sin\theta_2$, and $\dot{b}_y = r_{10}\omega_2 \cos\theta_2$. The corresponding angular velocities are

$$\omega_8 = -\frac{\dot{r}_8 \cos(\theta_9 - \theta_8) + r_{10}\omega_2 \sin(\theta_2 - \theta_9)}{r_8 \sin(\theta_8 - \theta_9)} \tag{2.107}$$

and

$$\omega_9 = -\frac{\dot{r}_8 + r_{10}\omega_2 \sin(\theta_2 - \theta_8)}{r_9 \sin(\theta_8 - \theta_9)} \tag{2.108}$$

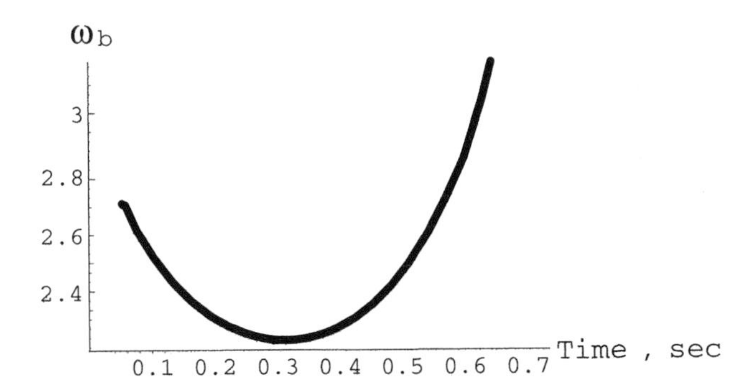

FIGURE 2.39 Angular velocity of the bucket vs. time.

Loop 3

In this case $i = 13, j = 14$, $\dot{b}_x = r_{12}\omega_9 \sin\theta_9 - r_{15}\omega_2 \sin\theta_2$, and $\dot{b}_y = -r_{12}\omega_9\cos\theta_9 + r_{15}\omega_2\cos\theta_2$. The corresponding angular velocities are

$$\omega_{13} = \frac{r_{12}\omega_9 \sin(\theta_9 - \theta_{14}) - r_{15}\omega_2 \sin(\theta_2 - \theta_{14})}{r_{13}\sin(\theta_{14} - \theta_{13})} \tag{2.109}$$

and

$$\omega_{14} = \frac{r_{12}\omega_9 \sin(\theta_9 - \theta_{13}) - r_{15}\omega_2 \sin(\theta_2 - \theta_{13})}{r_{14}\sin(\theta_{13} - \theta_{14})} \tag{2.110}$$

In Figure 2.39 the angular velocity of the bucket ($\omega_{14} = \omega_b$) vs. time is shown.

2.5 ACCELERATION ANALYSIS

2.5.1 Acceleration Vector

The acceleration vector is the time derivative of the velocity vector given by Equation 2.72

$$\frac{d^2\mathbf{r}}{dt^2} = \ddot{\mathbf{r}} = \ddot{r}[\cos\theta, \sin\theta]^T + 2\dot{r}[-\sin\theta, \cos\theta]^T\frac{d\theta}{dt} - r[\cos\theta, \sin\theta]^T\left(\frac{d\theta}{dt}\right)^2 + r[-\sin\theta, \cos\theta]^T\frac{d^2\theta}{dt^2} \tag{2.111}$$

The following notations are used: for the angular velocity, $\omega = d\theta/dt$ and for the angular acceleration, $\alpha = d^2\theta/dt^2$. With these notations Equation 2.111 has the form

$$\ddot{\mathbf{r}} = \ddot{r}[\cos\theta, \sin\theta]^T + 2\dot{r}[-\sin\theta, \cos\theta]^T\omega - r[\cos\theta, \sin\theta]^T\omega^2 + r[-\sin\theta, \cos\theta]^T\alpha \tag{2.112}$$

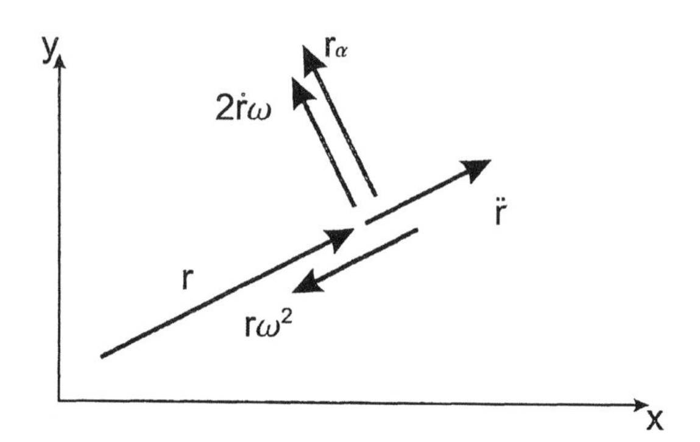

FIGURE 2.40 Components of acceleration vector.

One can rewrite the above equation in such a way that the directions of all vectors are referenced with respect to the direction of vector **r**. Then Equation 2.112 takes the form

$$\begin{aligned}\ddot{\mathbf{r}} &= \ddot{r}[\cos\theta,\ \sin\theta]^T + 2\dot{r}[-\sin\theta,\ \cos\theta]^T\omega \\ &\quad r[\cos\theta, \sin\theta]^T\omega^2 + \mathrm{r}[-\sin\theta,\ \cos\theta]^T\alpha\end{aligned} \tag{2.113}$$

The directions of four vectors and their magnitudes are shown in Figure 2.40, where one can see that the direction of first vector coincides with the direction of vector **r**; the direction of third vector is opposite to that of vector **r**, and the two other vectors are perpendicular to vector **r** whereas their directions are found by rotating vector **r** counterclockwise by $\pi/2$. The first vector in Equation 2.113 is called the *translational* component of acceleration, the third is called the *centripetal* component of acceleration, the fourth is called the *angular* component of acceleration, and the second is called the *coriolis* component of acceleration.

Note that centripetal acceleration is caused by the rotation of the vector (irrespective of whether this rotation is time dependent or time independent), whereas coriolis acceleration is caused by the rotation of a translationary moving vector. Both of these components are functions of velocities only, and thus can be found based on the velocity analysis. The other two components of acceleration, translational and angular, are found as a result of acceleration analysis.

2.5.2 Equations for Accelerations

The equations for accelerations follow from the loop-closure equation for positions, Equation 2.15, if the equation is differentiated twice with respect to time. As a result, the loop-closure equation for accelerations is obtained. Note that it is assumed that all parameters are time–dependent variables.

$$\begin{aligned}&\sum_{i=1}^{N} \ddot{r}_i[\cos\theta_i,\ \sin\theta_i]^T + 2\dot{r}_i[-\sin\theta_i,\ \cos\theta_i]^T\omega_i \\ &-r_i[\cos\theta_i,\ \sin\theta_i]^T\omega_i^2 + r_i[-\sin\theta_i,\ \cos\theta_i]^T\alpha_i = 0\end{aligned} \tag{2.114}$$

In Equation 2.114 the unknowns are $\ddot{r}_i$ and α_i, and the system defining them is linear. As before, the loop-closure equation for accelerations can have only two unknowns. This again entails five possible combinations of these unknowns. In this case the solutions for each can be found from the solutions for velocities in a straightforward manner.

First Case

The unknowns, $\ddot{\theta}_j$ and $\ddot{r}_j$, are found by differentiating Equations 2.77 and 2.78 with respect to time assuming that all the variables are time dependent. The results, taking into account Equations 2.77 and 2.78, are

$$\alpha_j = \frac{1}{r_j}[-\ddot{b}_x \sin\theta_j + \ddot{b}_y \cos\theta_j - 2\dot{r}_j \omega_j] \tag{2.115}$$

and

$$\ddot{r}_j = \ddot{b}_x \cos\theta_j + \ddot{b}_y \sin\theta_j + r_j \omega_j^2 \tag{2.116}$$

Second Case

In this case the two unknowns, $\ddot{\theta}_j$ and $\ddot{r}_i$, are found by differentiating Equations 2.80 and 2.81, respectively.

$$\alpha_j = \frac{1}{r_j \cos(\theta_j - \theta_i)} \begin{matrix} [((\omega_j - \omega_i) r_j \omega_j - \ddot{r}_j)\sin(\theta_j - \theta_i) - (2\omega_j - \omega_i)\dot{r}_j \cos(\theta_j - \theta_i) \\ -(\omega_i \dot{b}_y + \ddot{b}_x)\sin\theta_i + (-\omega_i \dot{b}_x + \ddot{b}_y)\cos\theta_i - \omega_i \dot{r}_i - r_i \alpha_i] \end{matrix} \tag{2.117}$$

and

$$\ddot{r}_i = \frac{1}{\cos(\theta_j - \theta_i)} \begin{matrix} [(\dot{r}_i(\omega_j - 2\omega_i) - r_i \alpha_i)\sin(\theta_j - \theta_i) - r_i \omega_i(\omega_j - \omega_i)\cos(\theta_j - \theta_i) - \ddot{r}_j \\ +(\ddot{b}_x + \dot{b}_y \omega_j)\cos\theta_j + (\ddot{b}_y - \dot{b}_x \omega_j)\sin\theta_j] \end{matrix} \tag{2.118}$$

Third Case

The unknowns in this case are the translational accelerations, $\ddot{r}_i$ and $\ddot{r}_j$. The expressions for them are found by differentiating Equations 2.82 and 2.83.

$$\ddot{r}_i = \frac{1}{\sin(\theta_i - \theta_j)} \begin{matrix} [-(\dot{r}_i(2\omega_i - \omega_j) + r_i \alpha_i)\cos(\theta_i - \theta_j) + r_i \omega_i(\omega_i - \omega_j)\sin(\theta_i - \theta_j) \\ -\dot{r}_j \omega_j - r_j \alpha_j - (\ddot{b}_x + \dot{b}_y \omega_j)\sin\theta_j + (\ddot{b}_y - \dot{b}_x \omega_j)\cos\theta_j] \end{matrix} \tag{2.119}$$

and the equation for $\ddot{r}_j$ is obtained by interchanging indices i and j in the above.

$$\ddot{r}_j = \frac{1}{\sin(\theta_j - \theta_i)} \begin{matrix} [-(\dot{r}_j(2\omega_j - \omega_i) + r_j \alpha_j)\cos(\theta_j - \theta_i) + r_j \omega_j(\omega_j - \omega_i)\sin(\theta_j - \theta_i) \\ -\dot{r}_i \omega_i - r_i \alpha_i - (\ddot{b}_x + \dot{b}_y \omega_i)\sin\theta_i + (\ddot{b}_y - \dot{b}_x \omega_i)\cos\theta_i] \end{matrix} \tag{2.120}$$

Fourth Case

The angular accelerations are found by differentiating Equations 2.84 and 2.85.

$$\alpha_i = \frac{1}{r_i \sin(\theta_j - \theta_i)} \begin{array}{c} [(-\omega_i r_i(\omega_j - \omega_i) - \ddot{r}_i)\cos(\theta_j - \theta_i) + \dot{r}_i(\omega_j - 2\omega_i)\sin(\theta_j - \theta_i) \\ -\ddot{r}_j + (\ddot{b}_x + \dot{b}_y\omega_j)\cos\theta_j + (\ddot{b}_y - \dot{b}_x\omega_j)\sin\theta_j] \end{array} \tag{2.121}$$

and the equation for α_j is obtained by interchanging indices i and j in the above.

$$\alpha_j = \frac{1}{r_j \sin(\theta_i - \theta_j)} \begin{array}{c} [(-\omega_j r_j(\omega_i - \omega_j) - \ddot{r}_j)\cos(\theta_i - \theta_j) + \dot{r}_j(\omega_i - 2\omega_j)\sin(\theta_i - \theta_j) \\ -\ddot{r}_i + (\ddot{b}_x + \dot{b}_y\omega_j)\cos\theta_i + (\ddot{b}_y - \dot{b}_x\omega_i)\sin\theta_i] \end{array} \tag{2.122}$$

Fifth Case

In this case the accelerations, $\ddot{\theta}_i$ and $\ddot{r}_j$, are found by differentiating Equations 2.90 and 2.91 taking into account that γ and β are constants.

$$\alpha_i = \frac{A_i\cos(\theta_i - \gamma) - B_i\sin(\theta_i - \gamma) - \ddot{r}_i\sin\gamma - \ddot{r}_k\sin(\gamma - \beta)}{-d_x\sin(\theta_i - \gamma) + d_y\cos(\theta_i - \gamma)} \tag{2.123}$$

and

$$\ddot{r}_j = \frac{1}{T_j} \begin{array}{c} [-A_j\cos(\theta_i - \gamma) - B_j\sin(\theta_i - \gamma) + C_j\cos\theta_i + D_j\sin\theta_i \\ + K_j\cos(\theta_i - \beta) + L_j\sin(\theta_i - \beta) + Q_j] \end{array} \tag{2.124}$$

where it is denoted

$$A_i = -\dot{d}_y\omega_i + d_x\omega_i^2 - \dot{b}_x\omega_i + \ddot{b}_y \tag{2.125}$$

$$B_i = -\dot{d}_x\omega_i + d_y\omega_i^2 - \dot{b}_y\omega_i + \ddot{b}_x \tag{2.126}$$

$$A_j = (-\dot{d}_y + d_x\omega_i)\dot{r}_j \tag{2.127}$$

$$B_j = (\dot{d}_x + d_y\omega_i)\dot{r}_j \tag{2.128}$$

$$C_j = \ddot{r}_i d_y + \dot{r}_i\dot{d}_i - \dot{r}_i d_x\omega_i \tag{2.129}$$

$$D_j = -\ddot{r}_i d_x - \dot{r}_i\dot{d}_x - \dot{r}_i d_y\omega_i \tag{2.130}$$

$$K_j = \ddot{r}_k d_y + \dot{r}_k\dot{d}_y - \dot{r}_k d_x\omega_i \tag{2.131}$$

$$L_j = -\ddot{r}_k d_x - \dot{r}_k\dot{d}_x - \dot{r}_k d_y\omega_i \tag{2.132}$$

$$Q_j = -\dot{d}_y\dot{b}_x - d_y\ddot{b}_x + \dot{d}_x\dot{b}_y + d_x\ddot{b}_y \tag{2.133}$$

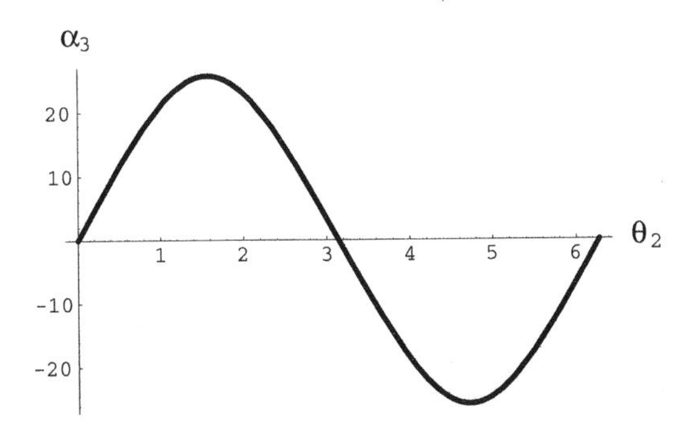

FIGURE 2.41 Angular acceleration of the connecting rod vs. crank angle.

$$T_j = -d_y \cos(\theta_i - \gamma) + d_y \sin(\theta_i - \gamma) \tag{2.134}$$

where b_x, b_y are given by Equation 2.45, and d_x, d_y are defined by Equations 2.88 and 2.89, respectively.

2.5.3 Applications to Simple Mechanisms

Slider-Crank Inversions (Figure 1.14)

- Figure 1.14a with the driving crank

This mechanism falls into the second case category. The solutions are given by Equations 2.117 and 2.118. In this case $i = 1$, $j = 3$, $b_x = -r_2 \cos\theta_2$, $b_y = -r_2 \sin\theta_2$, $\dot{b}_x = r_2\dot{\theta}_2 \sin\theta_2$, $\dot{b}_y = -r_2\dot{\theta}_2 \cos\theta_2$, $r_2 =$ const., $r_3 =$ const., and $\theta_1 = \pi$. Assume also that the crank rotates with constant angular velocity, i.e., $\omega_2 =$ const. Then, $\ddot{b}_x = r_2\omega_2^2 \cos\theta_2$, and $\ddot{b}_y = r_2\omega_2^2 \sin\theta_2$. Taking all this into account, the general formulas, Equations 2.117 and 2.118, are reduced to

$$\alpha_3 = \frac{r_2\omega_2^2 \sin\theta_2 + r_3\omega_3^2 \sin\theta_3}{r_3 \cos\theta_3} \tag{2.135}$$

and

$$\ddot{r}_1 = \frac{1}{\cos\theta_3}[\dot{r}_1\omega_3 \sin\theta_3 - r_2\omega_2(\omega_2 - \omega_3)\cos(\theta_2 - \theta_3)] \tag{2.136}$$

The above equations can also be obtained by differentiating the expressions for the corresponding velocities (Equations 2.93 and 2.94).

A plot of the change of the angular acceleration of the connecting rod with the crank angle is shown in Figure 2.41. A change of the slider acceleration with the crank angle is shown in Figure 2.42.

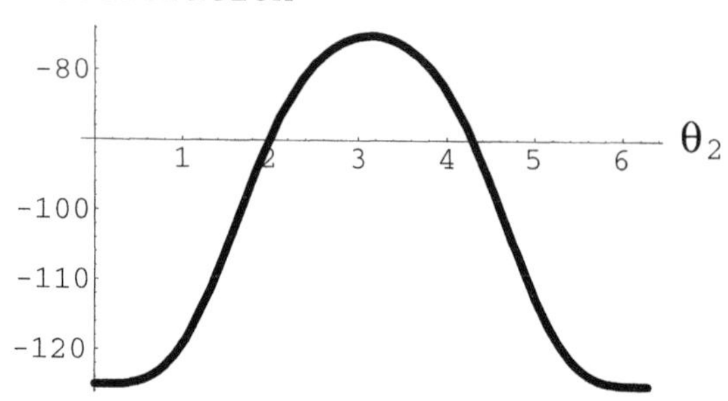

FIGURE 2.42 Slider acceleration vs. crank angle.

- Figure 1.14a with the driving piston

The angular acceleration of link 2 is obtained by differentiating Equations 2.95 and 2.96, taking into account that θ_3 is time dependent,

$$\alpha_2 = \frac{-r_2\omega_2(\omega_3-\omega_2)\cos(\theta_3-\theta_2)+\ddot{r}_1\cos\theta_3-\dot{r}_1\omega_3\sin\theta_3}{r_2\sin(\theta_3-\theta_2)} \tag{2.137}$$

and

$$\alpha_3 = \frac{-r_3\omega_3(\omega_2-\omega_3)\cos(\theta_2-\theta_3)+\ddot{r}_1\cos\theta_2-\dot{r}_1\omega_2\sin\theta_2}{r_3\sin(\theta_2-\theta_3)} \tag{2.138}$$

The angular accelerations of links 2 and 3 are shown in Figures 2.43 and 2.44.

- Figure 1.14b with link 3 as a driver

The angular acceleration of the cylinder and the translational acceleration of the piston are obtained by differentiating Equations 2.97 and 2.98, respectively, taking into account that θ_3 and ω_3 are time dependent,

$$\alpha_1 = \frac{-\dot{r}_1\omega_1+r_3\omega_3(\omega_3-\omega_1)\sin(\theta_3-\theta_1)-r_3\alpha_3\cos(\theta_3-\theta_1)}{r_1} \tag{2.139}$$

and

$$\ddot{r}_1 = r_3\alpha_3\sin(\theta_3-\theta_1)+r_3\omega_3(\omega_3-\omega_1)\cos(\theta_3-\theta_1) \tag{2.140}$$

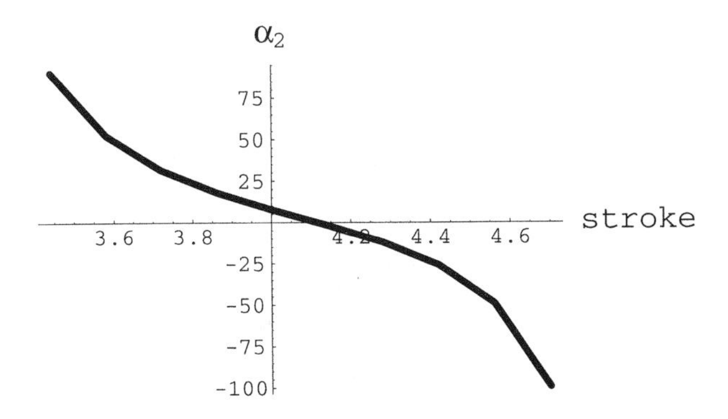

FIGURE 2.43 Angular acceleration of link 2 vs. piston stroke.

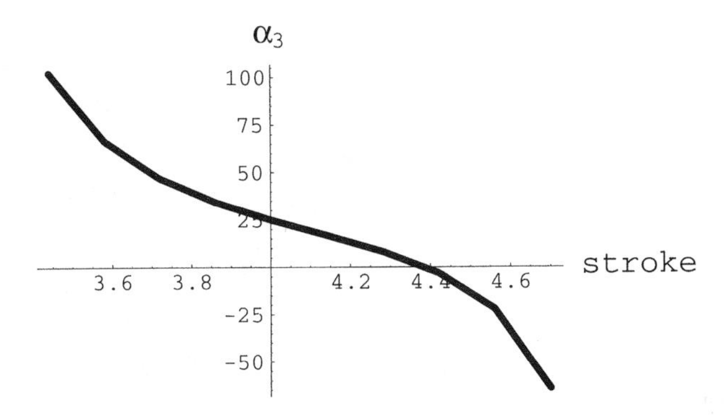

FIGURE 2.44 Angular acceleration of the hydraulic rod vs. piston stroke.

Four-Bar Mechanism (Figure 1.4)

The angular accelerations of the coupler and the follower are obtained by differentiating Equations 2.99 and 2.100, respectively, taking into account that ω_2 is time dependent,

$$\alpha_3 = \frac{-r_3\omega_3(\omega_4-\omega_3)\cos(\theta_4-\theta_3)+r_2\alpha_2\sin(\theta_2-\theta_4)+r_2\omega_2(\omega_2-\omega_4)\cos(\theta_2-\theta_4)}{r_3\sin(\theta_4-\theta_3)} \tag{2.141}$$

and

$$\alpha_4 = \frac{-r_4\omega_4(\omega_3-\omega_4)\cos(\theta_3-\theta_4)+r_2\alpha_2\sin(\theta_2-\theta_3)+r_2\omega_2(\omega_2-\omega_3)\cos(\theta_2-\theta_3)}{r_4\sin(\theta_3-\theta_4)} \tag{2.142}$$

The angular accelerations of the coupler and the follower are shown in Figures 2.45 and 2.46.

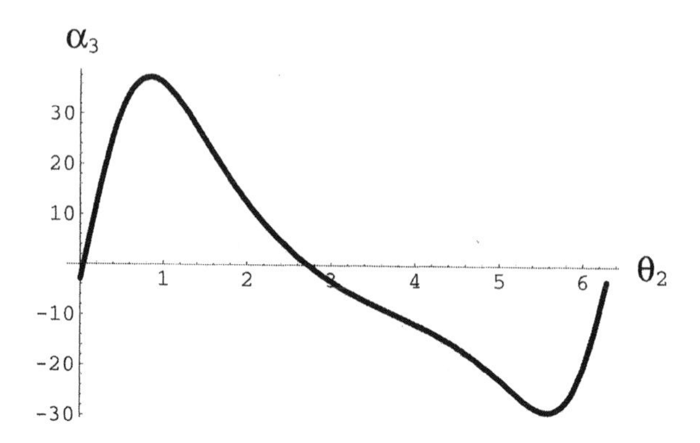

FIGURE 2.45 Angular acceleration of the coupler in a four-bar linkage.

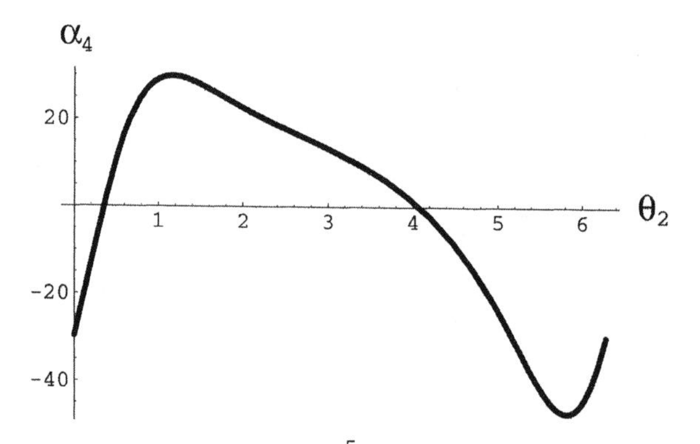

5

FIGURE 2.46 Angular acceleration of the follower in a four-bar linkage.

Five-Bar Mechanism (Figure 2.14a)

The angular acceleration of link 3 and the translational acceleration of the slider are obtained by differentiating Equations 2.101 and 2.102, taking into account that ω_2 is time dependent,

$$\alpha_3 = \frac{1}{r_4}[-\omega_3\dot{r}_4 + r_2\alpha_2\sin(\theta_2-\theta_3) + r_2\omega_2(\omega_2-\omega_3)\cos(\theta_2-\theta_3)] \tag{2.143}$$

and

$$\begin{aligned}\ddot{r}_4 &= r_2\alpha_2\left[\cos(\theta_3-\theta_2) + \frac{r_5-r_3}{r_4}\sin(\theta_3-\theta_2)\right] \\ &+ r_2\omega_2(\omega_3-\omega_2)\left[-\sin(\theta_3-\theta_2) + \frac{r_5-r_3}{r_4}\cos(\theta_3-\theta_2)\right]\end{aligned} \tag{2.144}$$

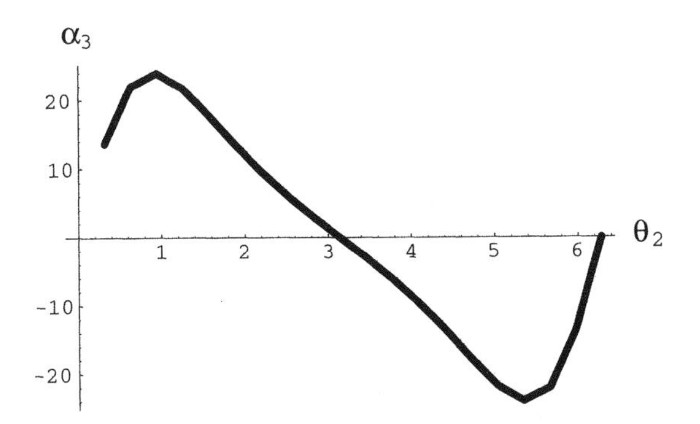

FIGURE 2.47 Angular acceleration of link 3 in a five-bar linkage.

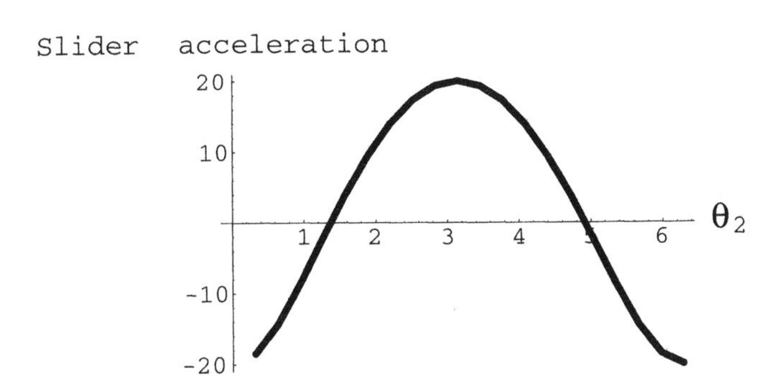

FIGURE 2.48 Slider acceleration in a five-bar linkage.

The angular acceleration of the link 3 and the translational acceleration of the slider are shown in Figures 2.47 and 2.48.

Scotch Yoke Mechanism (Figure 2.18a)

One can use Equations 2.119 and 2.120, in which $i = 3$, $j = 4$, $\theta_3 = 0$, $\theta_4 = 3\pi/2$, and b_x, b_y are defined by Equation 2.65, to find the two unknown accelerations.

$$\ddot{r}_3 = -r_2\alpha_2 \sin\theta_2 - r_2\omega_2^2 \cos\theta_2 \tag{2.145}$$

and

$$\ddot{r}_4 = -r_2\alpha_2 \cos\theta_2 + r_2\omega_2^2 \sin\theta_2 \tag{2.146}$$

Note that the above equations can be obtained by differentiating expressions for velocities, Equations 2.103 and 2.104. The translational accelerations of sliders 3 and 4 are shown in Figure 2.49.

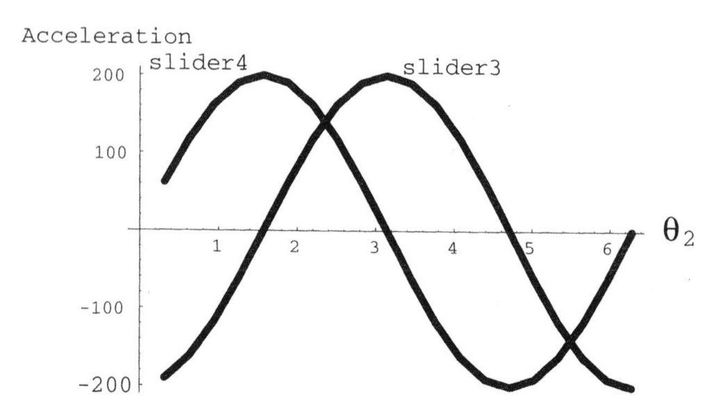

FIGURE 2.49 Accelerations of two sliders during one cycle.

2.6 INTERMITTENT-MOTION MECHANISMS: GENEVA WHEEL

Very often there is a need to transform a continuous rotation of the driver into an intermittent motion of the follower, for example, in such applications as film advances, indexing, motion along the production line, etc. One of the mechanisms able to accomplish such a transformation is called the *Geneva wheel.* In Figures 2.50 and 2.51 sketches of the mechanism are shown in two positions. The driver is wheel 2 with a pin *P*, and the driven element is slotted wheel 3. The rotation of the latter takes place only when the pin is engaged with the slot. In Figure 2.50 the mechanism is shown in a locked position; i.e., wheel 3 is not rotated while the driver is. To prevent wheel 3 from any rotation (to lock it into position), the convex surface of plate 2 matches the concave surface of wheel 3 until pin *P* becomes engaged. At this moment wheel 3 starts rotating (Figure 2.51).

From the point of view of motion transfer during the engagement, the Geneva mechanism can be reduced to a slider-crank mechanism in which the rotation of the crank is limited to some specified angle. The skeleton of the equivalent slider-crank mechanism is shown in Figure 2.52 in two extreme positions of engagement and disengagement.

As opposed to the conventional slider-crank mechanisms discussed earlier, the mechanism shown in Figure 2.52 must meet some constraints on the dimension of links, and also must relate these dimensions to the number of slots in wheel 3. To make the engagement and disengagement as smooth as possible, the angle between the crank r_2 and the slot must be 90° at these positions. This is the first requirement to be met by the mechanism design, which leads to a relationship between the crank and slotted wheel radii, and the center distance r_1.

The second requirement concerns the kinematics, relating the crank and wheel rotations. The problem of designing a Geneva wheel is as follows. Given the crank speed of rotation ω_2, how many slots are needed to accomplish a single intermittent motion in time τ? If the number of slots is N, then the angle γ, corresponding to the wheel rotation, equals

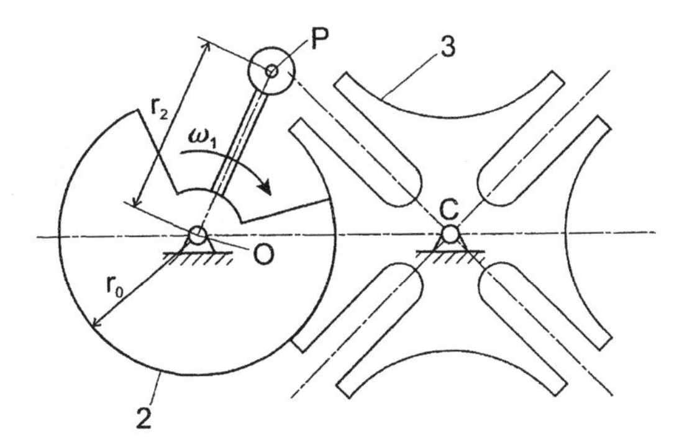

FIGURE 2.50 Geneva mechanism in a locked position.

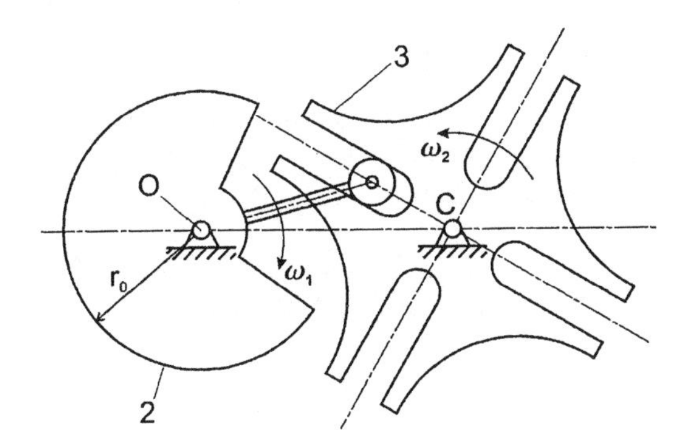

FIGURE 2.51 Geneva mechanism in an unlocked position.

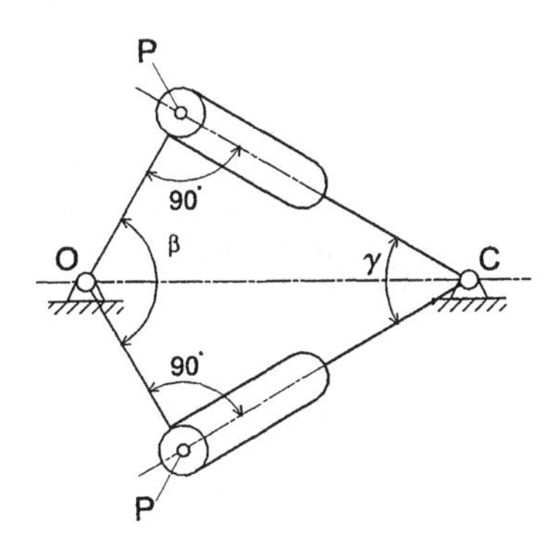

FIGURE 2.52 Crank and slot at two extreme positions during engagement.

$$\gamma^{\max} = \frac{2\pi}{N} \tag{2.147}$$

The minimum number of slots is $N_{\min} = 3$. From the triangle in Figure 2.52 the relationship between β and γ follows:

$$\beta^{max} = \pi - \gamma^{max} \tag{2.148}$$

Since the rotations through the angles β^{max} and γ^{max} are accomplished during the same time τ, from the relationship $\beta^{max} = \omega_2\tau$ it follows, using Equations 2.147 and 2.148,

$$\tau = \pi\frac{N-2}{N\omega_2} \tag{2.149}$$

Note that the above equation makes clear why the number of slots must be not less than three.

Analysis of Geneva Wheel Motion

Since the Geneva wheel is an inversion of the slider-crank mechanism, Figure 1.14b, in which the crank is the driver, the loop-closure equation is given by Equation 2.16 and the solution by Equation 2.27, in which $j = 3$,

$$r_3 = b = \sqrt{r_1^2 + r_2^2 - 2r_1r_2\cos\theta_2} \tag{2.150}$$

and

$$\cos\alpha = \cos\theta_3 = -\frac{r_2\cos\theta_2 - r_1}{b}, \quad \sin\alpha = \sin\theta_3 = -\frac{r_2\sin\theta_2}{b} \tag{2.151}$$

In Equations 2.150 and 2.151, one should take into account that

$$-\frac{\beta^{max}}{2} \leq \theta_2 \leq \frac{\beta^{max}}{2} \quad \text{and} \quad \theta_3 = 2\pi - \gamma$$

In Figure 2.53 the angle of rotation of a four-slotted Geneva wheel as a function of angle β for the case of

$$\frac{r_2}{r_1} = \cos\frac{\beta^{max}}{2} = \frac{1}{\sqrt{2}}$$

is shown. The latter relationship follows from the right triangle in Figure 2.52. Note also that for the four-slotted wheel $\beta^{max} = \gamma^{max} = \pi/2$.

In Figures 2.54 and 2.55 the angular velocity and acceleration of the Geneva wheel are shown during the driver rotation through angle β.

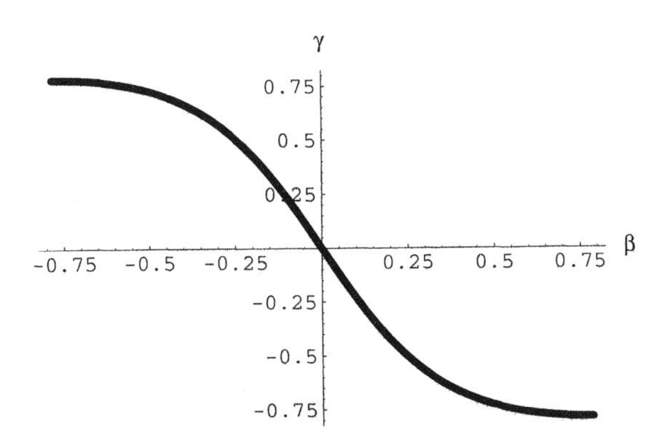

FIGURE 2.53 Angle of rotation of four-slotted Geneva wheel.

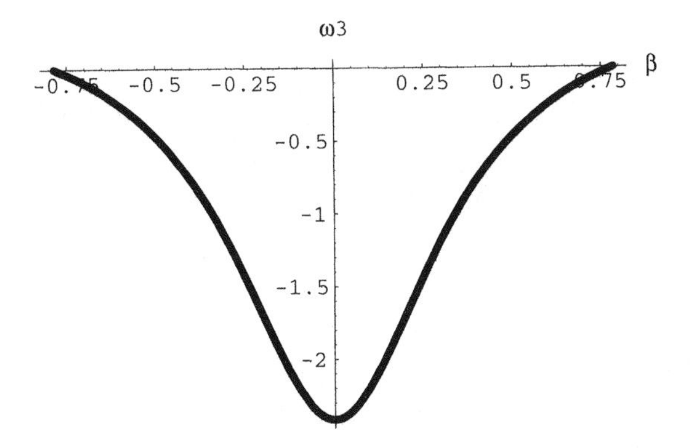

FIGURE 2.54 Angular velocity of four-slotted Geneva wheel.

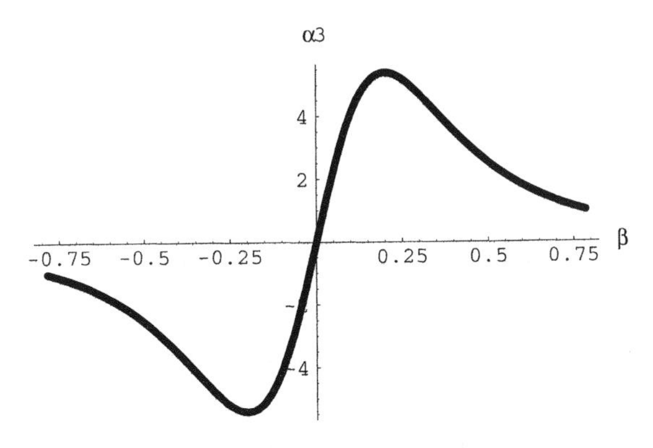

FIGURE 2.55 Angular acceleration of four-slotted Geneva wheel.

PROBLEMS AND EXERCISES

PROBLEMS

1. In Figure 1.14c an inverted slider-crank mechanism is shown.
   a. Write a loop-closure equation for this mechanism.
   b. If the input is the displacement of the cylinder, what are the unknowns?
   c. Solve the equation for the unknowns.
2. In Figure 1.14d an inverted slider-crank mechanism is shown.
   a. Write a loop-closure equation for this mechanism.
   b. If the input is the displacement of the cylinder, what are the unknowns?
   c. Solve the equation for the unknowns.
3. In Figure P2.1 an inverted slider-crank mechanism is shown.
   a. Write a loop-closure equation for this mechanism.
   b. If the input is the crank angle, what are the unknowns?
   c. Solve the equation for the unknowns.
   d. Express the position of point P in terms of the input angle.

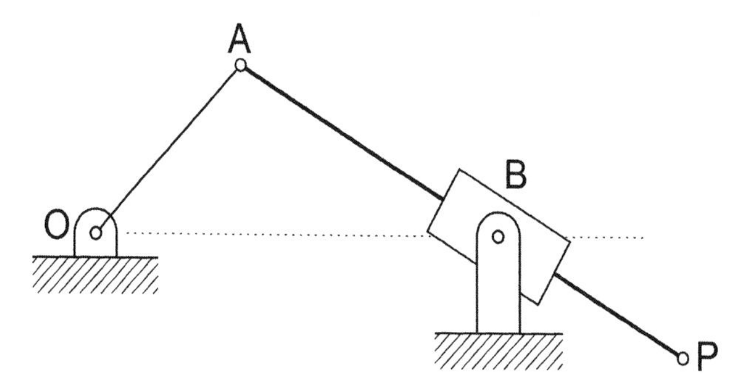

FIGURE P2.1

4. In Figure P.2.2 a five-bar linkage is shown, in which AB is parallel and equal to O_2C.
   a. Write a loop-closure equation for this mechanism.
   b. If the input is the crank angle, what are the unknowns?
   c. Solve the equation for the unknowns.
   d. Express the position of point P in terms of the input angle.
5. In Figure P2.3 a slider-crank mechanism is shown.
   a. Write a loop-closure equation for this mechanism.
   b. If the input is the crank angle, what are the unknowns?
   c. Solve the equation for the unknowns.
   d. Express the position of point P in terms of the input angle.

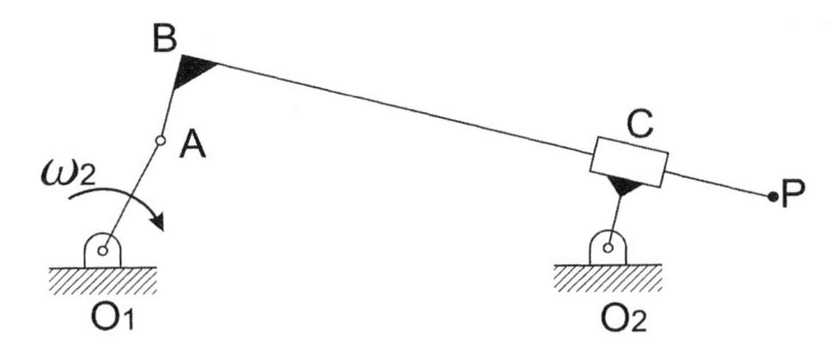

FIGURE P2.2

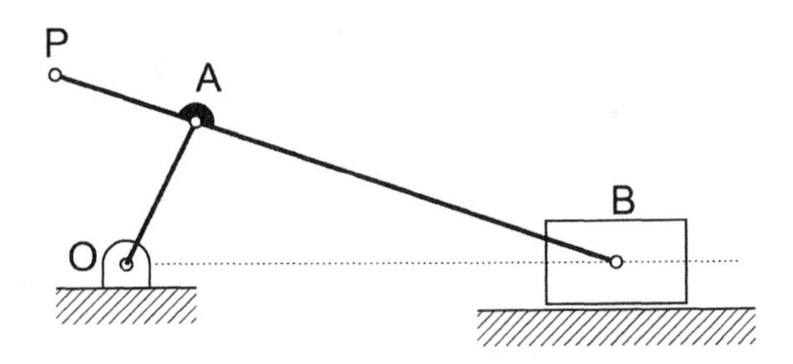

FIGURE P2.3

6. In Figure P2.4 a four-bar linkage mechanism is shown.
   a. Write a loop-closure equation for this mechanism.
   b. If the input is the crank angle, what are the unknowns?
   c. Solve the equation for the unknowns.
   d. Express the positions of points P_1 and P_2 in terms of the input angle.

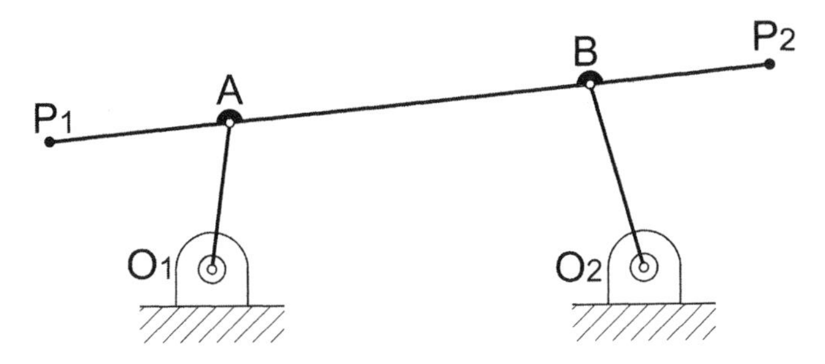

FIGURE P2.4

7. For the mechanism in Figure P2.4,
   a. Formulate the requirement for dimensions in order for the mechanism to be a crank-rocker.
   b. If the follower is to rock within angle γ, is there a unique set of dimensions to meet this requirement? Outline the solution procedure.

8. In Figure 2.4 find the maximum angle of rocking of the connecting rod, if r_2 and r_3 are given.

9. For a variable-stroke drive mechanism in Figure P1.7,

   a. Identify two loops.
   b. For a coordinate system such that axis x is directed from the center of the driving shaft to the center of the output shaft, outline the solution procedure using solutions in the book for a four-bar linkage.

10. For the double-toggle mechanism in Figure P1.6,

    a. Identify two loops.
    b. For a coordinate system such that the x-axis passes through the center of the crankshaft and is directed to the right, outline the solution procedure using solutions in the book for a four-bar linkage and a slider-crank mechanism.

11. For the constant velocity mechanism shown in Figure P1.5,

    a. Identify two loops.
    b. For a coordinate system such that the x-axis is directed along the cylinder to the right, outline the solution procedure using solutions in the book for a four-bar linkage and a slider-crank mechanism.

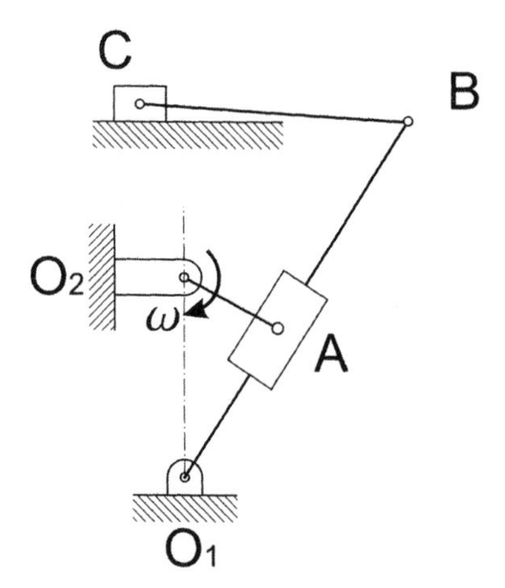

FIGURE P2.5

12. A quick-return mechanism, shown in Figure P2.5, is used in machine tools. For a constant angular velocity of the driving crank, it produces slow velocity during the cutting phase and then fast return.

    a. Identify two loops.
    b. Define a coordinate system.

c. For a coordinate system such that the y-axis is going through the points O_1 and O_2 upwards, outline the solution procedure using solutions in the book for a slider-crank mechanism.
d. Define the time ratio (time of cutting stroke to the time of return stroke) in terms of the distances O_1O_2 and O_2A.

13. For the mechanism in Figure P2.1, find

    a. the velocities.
    b. the accelerations of point P.

    Assume that the position, velocity, and acceleration analyses of the skeleton have been done.

14. For the mechanism in Figure P2.2, find

    a. the velocities.
    b. the accelerations of point P.

    Assume that the position, velocity, and acceleration analyses of the skeleton have been done.

15. For the mechanism in Figure P2.3, find

    a. the velocities.
    b. the accelerations of point P.

    Assume that the position, velocity, and acceleration analyses of the skeleton have been done.

16. Assume that dimensions of all links in Figure P1.4 are known.

    a. What would be the configuration of the links when the pliers are closest?
    b. Assume that the coupler link is parallel to the frame link, and that the frame link is 10 cm, the coupler link is 7 cm, the link connecting the two jaws is 2 cm, and the fourth link is 3 cm. What is the angle by which the coupler link rotates from the initial to the extreme position?

17. Consider the double-rocker mechanism in Figure P2.6.

    a. What should the relationship between the links dimensions be in order for the connecting link to make a complete rotation while the arms are rocking?
    b. Assume $2a = 1$ cm, $h = 0.5$ cm, $2d = 10$ cm, and that the arms have equal length 7 cm. What is the angle of arm 1 rocking?

18. For an eight-slot Geneva mechanism

    a. Find the distance between the centers of rotation, r_1, given the radius of the crank r_2 (distance from the center of the driving plate to the center of the pin) and assuming that the pin enters and leaves the slot smoothly.
    b. Find the length of the slot.

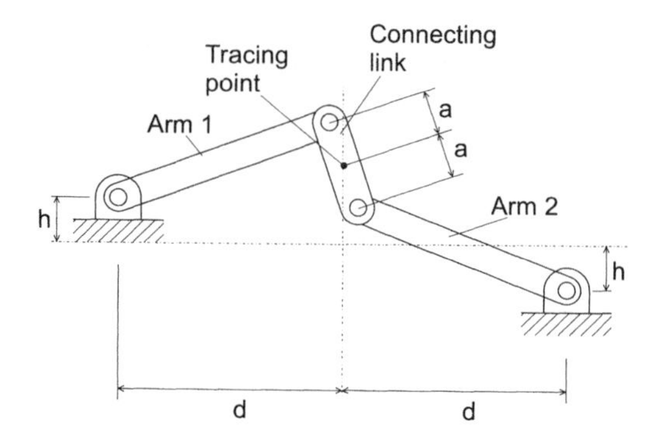

FIGURE P2.6

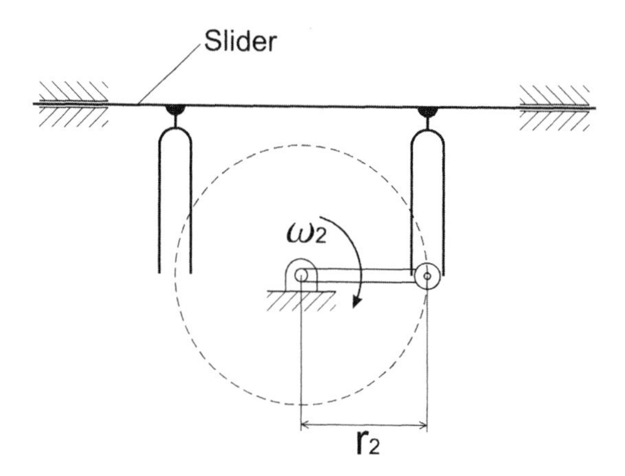

FIGURE P2.7

19. In Figure P2.7 a linear intermittent motion mechanism is shown. Given the angular velocity of the crank and its radius, r_2, how would velocity and acceleration of the slider change during the cycle of the crank?

Exercises (Projects) with Mathematica

1. A motor drives a film-advancing mechanism with constant velocity ω (Figure P2.8) Link 2 is the driver. The path of point C should be as indicated, so that during the engagement with the film point C moves along a straight line.
   a. Find (by trial and error) such dimensions of the mechanism that the needed trajectory of point C is achieved.
   b. Animate the motion.
   c. Plot the velocity and acceleration of point C over the cycle. Find the velocities and accelerations during engagement and disengagement with the film. Does the velocity remain constant during the engagement?

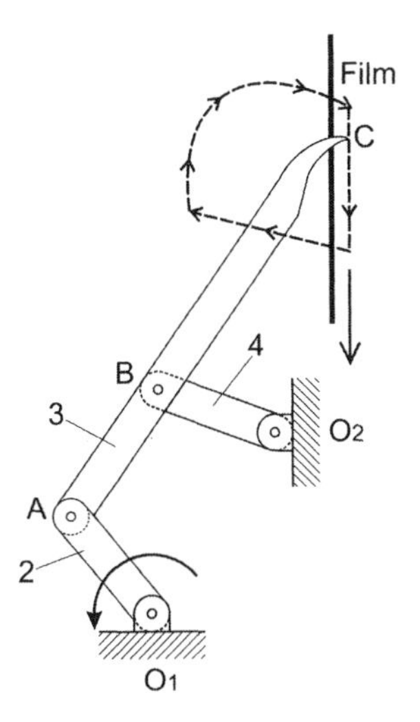

FIGURE P2.8

2. A motor drives a carrier mechanism with constant velocity ω (see Figure 2.26). The path of point A should be such that at the lowest position the line AB is parallel to the horizontal conveyor, and at the highest position the line AC should be parallel to the inclined conveyor (so that the load can be transferred from one conveyor to another).

   a. Find (by trial and error) such dimensions of the mechanism (for the assumed positions of the conveyor belts) that the needed trajectory of point A is achieved. (*Hint*: At extreme positions the motion of point A is reversed, which means that at these positions the links DC and CB are collinear.)
   b. Animate the motion.
   c. Plot the velocity and acceleration of point A over the cycle. What are the velocities and accelerations at the points of load transfer?

3. For the dimensions of the double-rocker given in problem 17:

   a. Animate the motion.
   b. Plot the trajectory, velocity, and acceleration of the tracing point over the cycle.

4. The oscillating drive-arm in Figure P2.9 has the maximum operating angle $\pi/6$. For a relatively short guideway, the reciprocating output stroke is large and it follows a straight line.

   a. Find (by trial and error) such dimensions of the mechanism that the needed trajectory of the tracing point is achieved.
   b. Animate the motion.
   c. Plot the velocity and acceleration of point P over the cycle. What are the velocities and accelerations at the end points of the stroke?

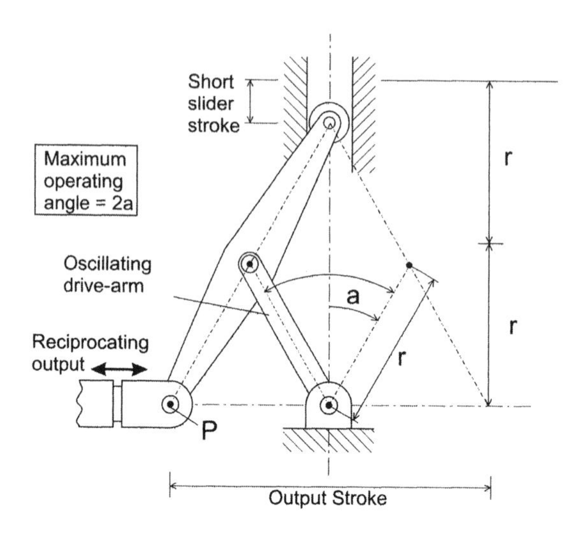

FIGURE P2.9

5. For a complex mechanism operating a dump truck in Figure 1.6 assume that the bed moves from a horizontal to a $\pi/3$ position.

   a. Find (by trial and error) such dimensions of the mechanism that the needed trajectory of the bed is achieved.
   b. Animate the motion.

6. In Figure P1.6 a double-toggle puncher is shown. When the drive crank rotates clockwise, the second toggle begins to straighten to create a strong punching force.

   a. Find link dimensions such that the desired motion is achieved.
   b. Animate the motion.
   c. Plot velocities and accelerations of the point *P* over the cycle.

7. In Figure P1.7 a variable-stroke drive, which is a combination of two four-bar linkages, is shown. The driving member rotates the eccentric, which, through the linkage, causes the output link to rotate a fixed amount. The ratchet on the output shaft transfers motion in one direction only. Thus, on the return stroke, the output link overrides the output shaft. As a result, a pulsating motion is transmitted to the output shaft, which is needed in many applications, such as feeders and mixers. A smoother drive can be produced by mounting on the same shaft the same device but with some shift in phase with respect to the first one. A continuously variable drive can be designed by mounting a few of such devices on the same shaft and using the control link to change the position of the adjustable pivot.

   a. Perform a complete kinematic analysis for one position of the adjustable pivot. Plot velocities and accelerations of the output link over the cycle.
   b. Animate the motion.

8. In Figure P1.3 an adjustable stroke mechanism is shown. The output link slides along the horizontal line, and the stroke is controlled by the position of the pivot point.

   a. Perform complete kinematic analysis for one position of the adjustable pivot. Plot velocities and accelerations of the output link over the cycle.
   b. Animate the motion.

9. For the loader shown in Figure 2.3, assume that rods 5 and 6 move during the same time interval (synchronized motion), but their velocities may be different.

   a. Find by trial and error such dimensions that the bucket rotates over the $\pi/2$ range from the lowest to the highest position.
   b. Perform a complete kinematic analysis. Plot angular velocity and acceleration of the bucket when it moves from the lowest to the highest position.
   c. Animate the motion.

3 Force Analysis of Mechanisms

3.1 INTRODUCTION

The design of mechanisms and their components requires information about forces acting on these components. Some mechanisms are designed to perform a specific kinematic function (like the windshield wiper mechanism, Figure 1.4), others to transfer energy (like the internal combustion engine). However, in any mechanism, identification of forces is needed to determine the proper dimensions of components.

The power supplied to the input link flows through the mechanism to the output link. Associated with this power flow is a force flow. The objective of the force analysis of mechanisms is to find the *transformation of forces* from the input to the output links. This transformation of forces depends on the position of the mechanism; in other words, it is a function of time. Thus, it is important to find out how these forces change during one cycle in order to find their maxima.

One should differentiate between two types of forces: *external* and *internal*. The former are forces that are applied to the links from external (with respect to the mechanism) sources — driving forces, resistance forces — whereas the latter are forces acting between the joints (they are called *constraint* or *reaction* forces).

The motion of a mechanism is caused by the known external forces, and can be found by formulating and solving the differential equation describing the dynamic equilibrium of the mechanism at any moment in time. This approach to motion analysis is called *direct dynamics*. An alternative approach is to assume that the motion is known (in other words, the motion of the input link is given as a function of time). Then, as a result of kinematic analysis, the accelerations of all links are known, and thus the inertial forces associated with these links. These inertial forces can be treated as known external forces, and the force analysis is then reduced to solving equilibrium equations for the mechanism at any given position. This approach to force analysis is called *inverse dynamics*. It is important to keep in mind that inverse dynamics is based on the assumption of known motion, whereas in fact such motion can be found only from direct dynamics analysis. However, in many situations the much simpler inverse dynamics approach is sufficient as a first approximation. This approach is considered in this book.

To summarize, it is assumed here that the forces acting on the input link are given as a function of time (or link position) and the inertial (dynamic) forces are also known as a result of kinematic analysis of motion. The objective of force analysis then is to find the internal and resistance forces. The method of solution is to perform static analysis of a mechanism in a number of fixed positions over the region of input link motion.

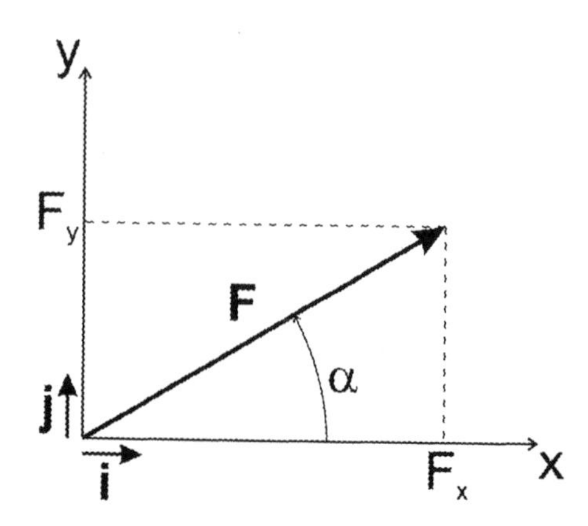

FIGURE 3.1 Force vector.

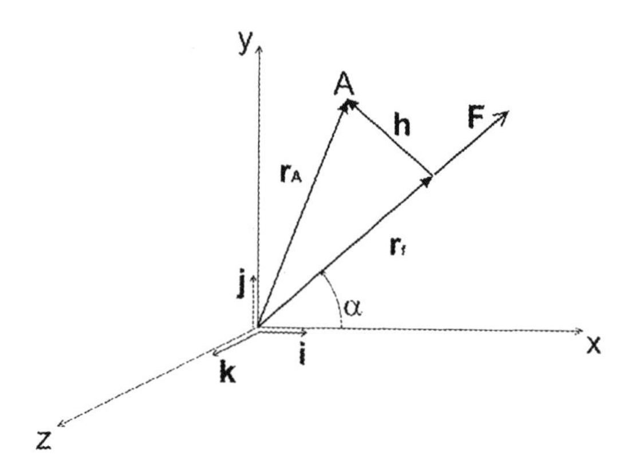

FIGURE 3.2 Illustration of the force moment.

3.2 FORCE AND MOMENT VECTORS

A force is characterized by its magnitude and direction, and thus is a vector. In an (x, y)-plane the force vector, $\mathbf{F}$, can be represented in different forms

$$\mathbf{F} = [F_x, F_y] = F\,[\cos\alpha, \sin\alpha]^{\mathrm{T}} = F\,(\mathbf{i}\cos\alpha + \mathbf{j}\sin\alpha) \tag{3.1}$$

where F_x, F_y are the x- and y-components of the vector (Figure 3.1), α indicates force direction (positive α is measured counterclockwise), and $\mathbf{i}$ and $\mathbf{j}$ are the unit vectors directed along the x- and y-axis, correspondingly.

A moment of the force $\mathbf{F}$ with respect to a point A (Figure 3.2) is a vector found as a cross-product of two vectors:

$$\mathbf{M} = \mathbf{r}_A \times \mathbf{F} \tag{3.2}$$

This vector is directed along the line perpendicular to the plane made by vectors $\mathbf{r}_A$ and $\mathbf{F}$, which in this case is the (x,y)-plane. In Equation 3.2 $\mathbf{r}_A$ is a vector associated with point A and can be represented as a sum of two vectors: one collinear with $\mathbf{F}$ ($\mathbf{r}_F$) and another perpendicular to it ($\mathbf{h}$) (Figure 3.2). Then Equation 3.2 is reduced to

$$\mathbf{M} = \mathbf{h} \times \mathbf{F} \tag{3.3}$$

since the moment of the collinear component is zero (see Equation 2.9). The magnitude of vector $\mathbf{h}$ is the distance from point A to the line of force $\mathbf{F}$, whereas its direction is toward point A. Thus, vector $\mathbf{h}$ has the following forms:

$$\mathbf{h} = h\left[\cos\left(\alpha + \frac{\pi}{2}\right), \sin\left(\alpha + \frac{\pi}{2}\right)\right]^T = h[-\sin\alpha, \cos\alpha]^T = h(-\mathbf{i}\sin\alpha + \mathbf{j}\cos\alpha) \tag{3.4}$$

Substituting Equations 3.1 and 3.4 into Equation 3.3, one obtains (see Equation 2.7)

$$\mathbf{M} = hF(\cos(\alpha + \pi/2)\sin\alpha - \sin(\alpha + \pi/2)\cos\alpha)\mathbf{k} = -hF\mathbf{k} \tag{3.5}$$

Thus, the moment vector is directed along the z-axis in such a way that for an observer on the tip of vector $\mathbf{M}$ the rotation from $\mathbf{h}$ to $\mathbf{F}$ is counterclockwise. It is important to note that its sign is determined by the convention for measuring the direction of vectors (or by choosing the right-hand coordinate system).

3.3 FREE-BODY DIAGRAM FOR A LINK

A diagram of a link with all forces (external and internal) applied to it is called a *free-body diagram*. Under the action of all forces (static and inertial), the link must be in equilibrium. This requirement results in relationships between the known and unknown forces for a single link.

The internal forces originate in joints since joints constrain the relative motion between the connected links. In the case of a revolute joint, in general, both the magnitude and the direction of the constraint force are unknown, whereas in the case of a prismatic joint only the magnitude of the constraint force is unknown. This is because the latter force is always directed along the normal to the axis of the slider (note that if the friction forces are taken into account, then their magnitudes and directions are assumed to be known functions of normal forces).

For each link, the vector equilibrium equations (for a planar problem) can be written:

$$\sum_{i=1}^{n} \mathbf{F}_i = 0 \text{ and } \sum_{j=1}^{m} \mathbf{M}_j = 0 \tag{3.6}$$

where n is the number of forces and m is the number of moments.

In global coordinates the system of equations (Equation 3.6) written for each link is coupled, which means that the constraint forces for each link are interdependent

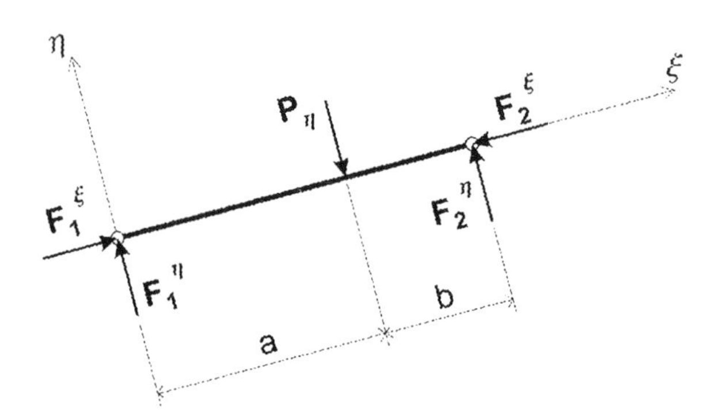

FIGURE 3.3 Free-body diagram for case 1.

and are functions of external forces and external moments applied to a mechanism. However, in local coordinates when the x-axis is directed along the link, some of the components of the constraint forces can be uncoupled, thus reducing the number of unknowns in a coupled system. This allows one to solve the coupled system of equations analytically for most mechanisms. Consider various loading situations for a single link.

Case 1. *In the case when the external force is perpendicular to a link with two revolute joints, the components of the constraint forces perpendicular to the link are uniquely defined, whereas the components of the constraint forces acting along the link are equal and opposite to each other.*

Indeed, from the equilibrium equations written in a local coordinate system (ξ,η) embedded into the link (Figure 3.3),

$$F_1^\eta + F_2^\eta - P_\eta = 0 \tag{3.7}$$

$$F_1^\xi - F_2^\xi = 0 \tag{3.8}$$

and

$$-(a+b)F_1^\eta + bP_\eta = 0 \tag{3.9}$$

It follows that F_1^η and F_2^η are found from Equations 3.7 and 3.9, whereas F_1^ξ and F_2^ξ are equal and opposite but remain unknown.

Case 2. *In the case of the external force parallel to a link, the components of the constraint forces are directed along the link.*

The system of equilibrium equations in this case is reduced to one (Figure 3.4):

$$F_1^\xi - F_2^\xi + P_\xi = 0 \tag{3.10}$$

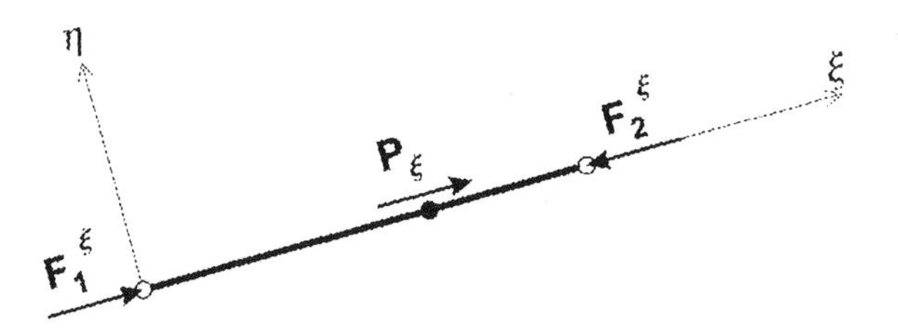

FIGURE 3.4 Free-body diagram for case 2.

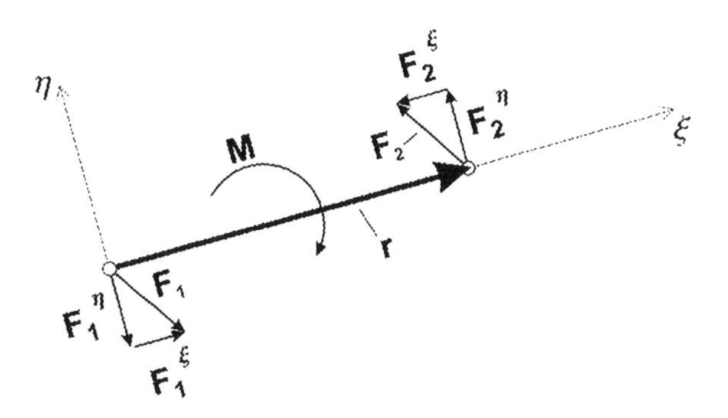

FIGURE 3.5 Free-body diagram for case 3.

Case 3. *In the case when only an external moment is applied to a link with two revolute joints, the constraint forces form a couple.*

The equilibrium equations in this case are (Figure 3.5)

$$\mathbf{F}_1 + \mathbf{F}_2 = 0 \tag{3.11}$$

and

$$\mathbf{r} \times \mathbf{F}_1 - \mathbf{M} = 0 \tag{3.12}$$

If each of the constraint forces is resolved into two components, parallel and perpendicular to the link (see Figure 3.5), then the magnitudes of the perpendicular to the link components are completely defined by the acting moment:

$$-F_1^\eta = F_2^\eta = \frac{M}{r} \tag{3.13}$$

while the components parallel to the link are equal but remain unknown:

$$F_1^\xi = F_2^\xi \tag{3.14}$$

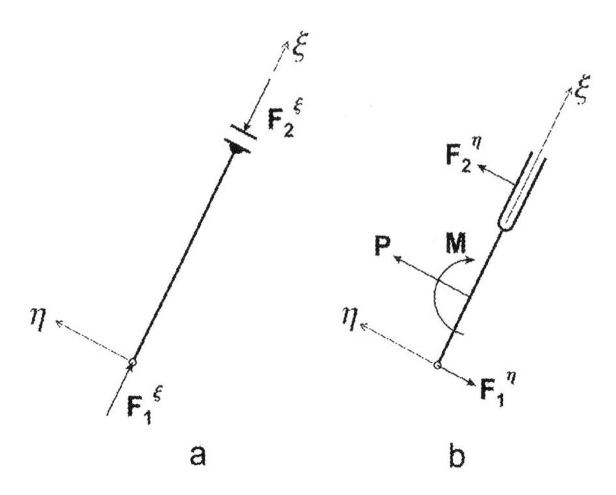

FIGURE 3.6 Free-body diagrams for cases 5 and 6.

Case 4. *In the case when there are no external forces acting on a link with two revolute joints, the constraint forces act along the link and are equal and opposite to each other.*

This is a particular situation of case 1. Indeed, if $P_\eta = 0$, then $F_1^\eta = 0$ and $F_2^\eta = 0$ (see Equations 3.7 and 3.9).

Case 5. *In the case of a link with one revolute joint and another sliding joint whose axis is perpendicular to the link, the constraint forces act along the link and are equal and opposite to each other.*

This is clear from Figure 3.6a in the case when the friction forces in the sliding joint are neglected.

Case 6. *In the case of a link with one revolute joint and another sliding joint whose axis coincides with the link axis the constraint forces act normally to the link.*

This is clear from Figure 3.6b in the case when the friction forces in the sliding joint are neglected. The constraint forces are determined by Equations 3.11 and 3.12.

What follows from all of the above cases is that for a link with two revolute joints:

- For any external load acting on a link there is only one unknown associated with this link.
- The unknown is the component of the constraint force directed along the link.
- The two components of the constraint force directed along the link are always equal and opposite to each other.

Now one can split the unknown constraint forces for each link into two components: one parallel and the other perpendicular to the link. As a result, the system of equilibrium equations for each link, Equation 3.6, will be split into two systems: one, independent from the equations for other links, and the other, coupled with the equations for other links. *The order of the coupled system is equal to the number of links in the mechanism, and thus constitutes the minimum system of equations.*

3.4 INERTIAL FORCES

The inertial forces are generated by nonzero translational and angular accelerations. If the mass center for each link j is identified, then the motion equations for this link as a free body are

$$m_j\ddot{\mathbf{r}}_{cj} = \sum \mathbf{F}_{\text{ext}} \tag{3.15}$$

and

$$I_j\ddot{\boldsymbol{\theta}}_j = \sum \mathbf{M}_{ext} \tag{3.16}$$

where $\mathbf{r}_{cj}$ is the position vector of the mass center of link j in the global coordinate system, θ_j is the angular coordinate of the link, m_j is the mass of the link, I_j is the moment of inertia of the link, and $\mathbf{F}_{\text{ext}}$ and $\mathbf{M}_{\text{ext}}$ are the external forces and moments, respectively. Note that $\mathbf{F}_{\text{ext}}$ and $\mathbf{M}_{\text{ext}}$ include constraint forces applied to the joints.

If one assumes that the accelerations are known from the kinematic analysis, the inertial forces can be considered as known external forces so that Equations 3.15 and 3.16 can be written in a standard form:

$$\sum \mathbf{F}_{\text{ext}} + \mathbf{F}_{\text{inert}} = 0 \tag{3.17}$$

and

$$\sum \mathbf{M}_{\text{ext}} + \mathbf{M}_{\text{inert}} = 0 \tag{3.18}$$

where

$$\mathbf{F}_{\text{inert}} = -m_j\ddot{\mathbf{r}}_{cj} \text{ and } \mathbf{M}_{inert} = -I_j\ddot{\boldsymbol{\theta}}_j \tag{3.19}$$

The problem of finding the unknown constraint forces in a moving mechanism is thus reduced to a problem of static equilibrium with the additional inertial forces given by Equation 3.19 treated as known external forces. This is known as *D'Alembert's principle*.

In the following, if the direction of the internal force is known, then it is convenient to represent this force in a directional form $\mathbf{F}_{\text{ij}} = F_{ij}[\cos\alpha_{ij}, \sin\alpha_{ij}]^T$; otherwise a component form $\mathbf{F}_{\text{ij}} = [F_{ij}^x, F_{ij}^y]^T$ will be used.

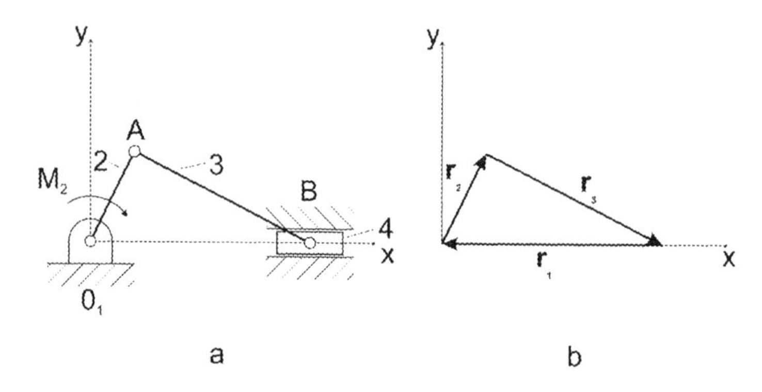

FIGURE 3.7 Slider-crank mechanism driven by moment $\mathbf{M}_2$.

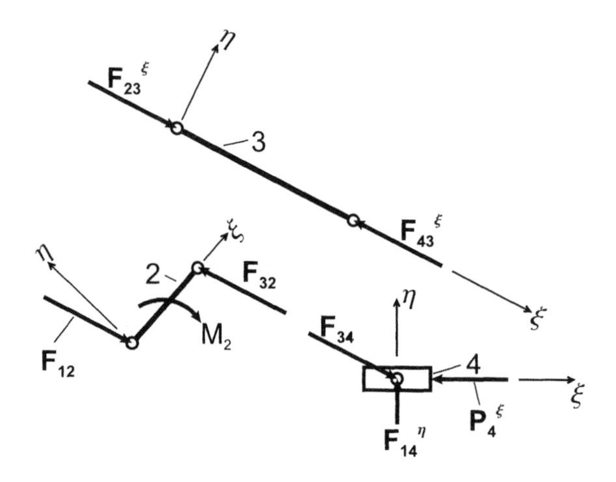

FIGURE 3.8 Free-body diagrams of links for the slider-crank mechanism with negligible inertial forces.

3.5 APPLICATION TO SIMPLE MECHANISMS

Here a few examples of simple mechanisms will be considered: the slider-crank mechanism, the four-bar linkage, the five-bar mechanism, and the Scotch yoke mechanism.

3.5.1 Slider-Crank Mechanism: The Case of Negligibly Small Inertial Forces

The skeleton of the mechanism and its vector representation are shown in Figure 3.7a and b, and the free-body diagrams of links are shown in Figure 3.8.

Assume that $\mathbf{M}_2$ is the known external moment, $\mathbf{P}_4^{\xi}$ is the unknown resistance force, and the inertial forces can be neglected. All other forces shown in Figure 3.8 are internal (constraint) forces. Each internal force is identified by two indices: the first one indicates the adjoining link, and the second one the link to which the force is applied. It follows from the equilibrium of forces applied to a joint that

$$\mathbf{F}_{ij} + \mathbf{F}_{ji} = 0 \tag{3.20}$$

Link 3 is the only one to which external forces are not applied. It follows then that for this link the internal forces are directed along the link and they are equal in magnitude ($F_{23}^{\xi} = F_{43}^{\xi}$) and opposite to each other. This reduces the system of equations and simplifies the solution. Thus, the forces in the joints of link 3 contain only the unknown magnitude, since the directions of these forces are known. Assume that the direction of vector $\mathbf{F}_{23}$ coincides with that of vector $\mathbf{r}_3$ (if the assumption is wrong, the solution will have the opposite sign for this force).

$$\mathbf{F}_{23} = F_{23}^{\xi}[\cos\theta_3, \sin\theta_3]^T \tag{3.21}$$

Then all other unknown vectors are as follows:

$$\mathbf{F}_{43} = \mathbf{F}_{32} = -\mathbf{F}_{34} = -\mathbf{F}_{12} = -\mathbf{F}_{23} \tag{3.22}$$

where it was taken into account that forces $\mathbf{F}_{32}$ and $\mathbf{F}_{12}$ form a couple.

Thus, there are three unknowns in this problem, which are the magnitudes of three vectors: $\mathbf{F}_{23}$, $\mathbf{F}_{14}$, and $\mathbf{P}_4$. The first one is found from the moment equation for link 2, and the other two from the equilibrium equation for link 4.

Link 2 Use the vector form of the moment equation, taking into account that the external moment is negative

$$\mathbf{r}_2 \times \mathbf{F}_{32} - \mathbf{M}_2 = 0 \tag{3.23}$$

If one substitutes in the above $\mathbf{r}_2 = r_2[\cos\theta_2, \sin\theta_2]^T$, $\mathbf{M}_2 = -M_2\,\mathbf{k}$, and takes into account Equations 2.9 and 3.21, one obtains a scalar equation for the unknown F_{32}:

$$r_2 F_{32}\sin(\theta_3 - \theta_2) - M_2 = 0 \tag{3.24}$$

Thus, the unknown force magnitude is equal to

$$F_{32} = \frac{M_2}{r_2\sin(\theta_3 - \theta_2)} \tag{3.25}$$

Note that the force magnitude is supposed to be positive by definition. If it becomes negative, then it means that all forces in Equation 3.24 change directions.

Link 4

$$\mathbf{F}_{34} + \mathbf{F}_{14} + \mathbf{P}_4 = 0 \tag{3.26}$$

In Equation 3.26 the forces are as follows:

$$\mathbf{F}_{14} = F_{14}\left[\cos\frac{\pi}{2}, \sin\frac{\pi}{2}\right]^T = F_{14}[0, 1]^T \tag{3.27}$$

$$\mathbf{P}_4 = P_4[\cos\pi, \sin\pi]^T = P_4[-1, 0]^T \tag{3.28}$$

and

$$\mathbf{F}_{34} = F_{32}(\cos\theta_3, \sin\theta_3)^T \tag{3.29}$$

After substituting the above forces into Equation 3.26 and equating the x- and y-components of the vector equation to zero, the expressions for the two unknowns are obtained:

$$P_4 = F_{32}\cos\theta_3 \tag{3.30}$$

and

$$F_{14} = -F_{32}\sin\theta_3 \tag{3.31}$$

3.5.2 Slider-Crank Mechanism: The Case of Significant Inertial Forces

The skeleton of the mechanism is shown in Figure 3.9a, where the inertial forces are identified with the superscript i. In Figure 3.9b the free-body diagrams of links are shown, where C_2 and C_3 are the centers of gravity of links 2 and 3.

Assume, as in the previous case, that $\mathbf{M}_2$ is an applied moment, and $\mathbf{P}_4$ is an unknown resistance force. One can see from Figure 3.9b that there are eight unknowns in this problem: $\mathbf{F}_{12}$, $\mathbf{F}_{32}$, $\mathbf{F}_{43}$, $\mathbf{F}_{14}$, and $\mathbf{P}_4$ (remember that the direction of the last two forces is known), and thus eight scalar equations are needed to find the unknowns. However, as is known, only the components of constraint forces collinear with the links are coupled. One can then resolve each inertial force into two components: one parallel and the other perpendicular to the corresponding link. Then the free-body diagrams for links 2 and 3 in Figure 3.9b can be seen as the superposition of free-body diagrams in Figures 3.3 through 3.5, where the superscripts ξ and η identify force components that are collinear and normal to the corresponding links. It should be pointed out that the direction of forces and moments can be considered with respect to the direction of the vector identifying the link (in other words, in a local coordinate system). Then, since all forces normal to the link and

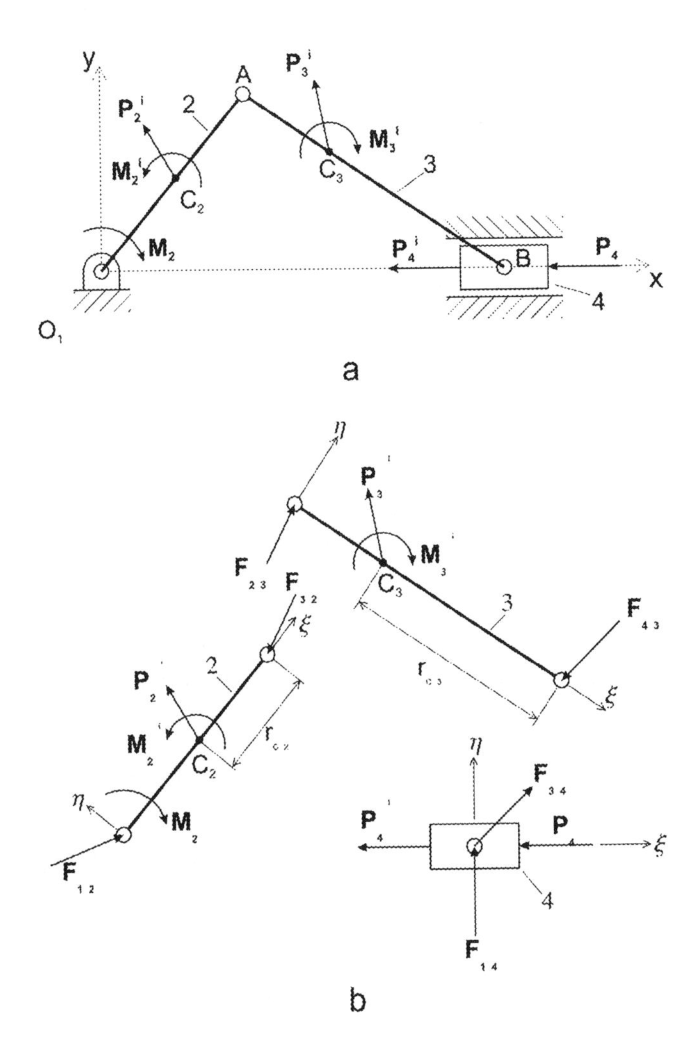

FIGURE 3.9 (a) Slider-crank mechanism with inertial forces. (b) Free-body diagrams for links in a slider-crank mechanism with significant inertial forces.

all forces parallel to the link are collinear, the positive directions of forces can be chosen arbitrarily. In addition, in the local coordinate system one can treat these forces as scalars since their directions are known. Later on, when all the unknowns are found, they can be transformed into a global coordinate system.

First consider the case of normal forces in links shown in Figure 3.9b.

Link 2 The two equilibrium equations are

$$-F_{12}^{\eta} + P_2^{i\eta} - F_{32}^{\eta} = 0 \tag{3.32}$$

and

$$r_2 F_{12}^{\eta} - r_{C2} P_2^{i\eta} - M_2 + M_2^i = 0 \tag{3.33}$$

where r_2 and r_{C2} are the length of link 2 and the distance C_2A (see Figure 3.9a). It follows from Equation 3.33 that

$$F_{12}^{\eta} = \frac{M_2 - M_2^i + r_{C2} P_2^{i\eta}}{r_2} \tag{3.34}$$

and from Equation 3.32 that

$$F_{32}^{\eta} = -F_{12}^{\eta} + P_2^{i\eta} \tag{3.35}$$

Equations 3.34 and 3.35 define the normal components of joint forces in link 2 explicitly.

Link 3 The equilibrium equations are

$$F_{23}^{\eta} + P_3^{i\eta} - F_{43}^{\eta} = 0 \tag{3.36}$$

and

$$-r_3 F_{23}^{\eta} - r_{C3} P_3^{i\eta} - M_3^i = 0 \tag{3.37}$$

where r_3 and r_{C3} are the length of link 3 and the distance C_3B (see Figure 3.9a). It follows from Equation 3.37 that

$$F_{23}^{\eta} = \frac{-M_3^i - r_{C3} P_3^{i\eta}}{r_3} \tag{3.38}$$

and from Equation 3.36 that

$$F_{43}^{\eta} = F_{23}^{\eta} + P_3^{i\eta} \tag{3.39}$$

Equations 3.38 and 3.39 define the normal components of joint forces in link 3.

Now consider forces acting along the link (see Figure 3.9b). For these forces there is only one equilibrium equation per link.

Link 2

$$F_{12}^{\xi} + P_2^{i\xi} - F_{32}^{\xi} = 0 \tag{3.40}$$

Link 3

$$-F_{23}^{\xi} - P_3^{i\xi} - F_{43}^{\xi} = 0 \tag{3.41}$$

As far as link 4 is concerned, the two equilibrium equations for both the x- and y-components of the forces contain three unknown constraint forces and one unknown resistance force (see Figure 3.9b). Since none of the unknowns can be found explicitly from these equations, they must be solved together with Equations 3.40 and 3.41.

Link 4

$$-P_4^i - P_4 + F_{34}^{\xi} = 0 \tag{3.42}$$

and

$$F_{34}^{\eta} + F_{14}^{\eta} = 0 \tag{3.43}$$

Recall that Equations 3.40 through 3.43 are written in local coordinate systems so that, for example, $F_{32}^{\xi} \neq F_{23}^{\xi}$. Because of this, the number of unknowns in the four equations (Equations 3.40 through 3.43) is eight. Thus, it is necessary to supplement Equations 3.40 through 3.43 with four more equations. These come from the requirement that the sum of constraint forces in a joint must be zero. Thus, there are two more vector equations for the joints connecting links 2 and 3, and links 3 and 4.

$$\mathbf{F}_{32} + \mathbf{F}_{23} = 0 \tag{3.44}$$

and

$$\mathbf{F}_{43} + \mathbf{F}_{34} = 0 \tag{3.45}$$

All forces in Equations 3.44 and 3.45 are expressed in the global coordinate system. Now it is necessary to express these vector forces through the components of these forces in local coordinate systems. Such a transformation is easily performed. Consider, for example, vector $\mathbf{F}_{12}$ comprising two components $\mathbf{F}_{12}^{x}$ and $\mathbf{F}_{12}^{y}$ in the global coordinate system. In the local coordinate system (ξ,η), $\mathbf{F}_{12}^{\xi}$ is directed along link 2 and thus has the same direction as vector $\mathbf{r}_2$, whereas vector $\mathbf{F}_{12}^{\eta}$ is normal to vector $\mathbf{r}_2$ and is being rotated by the angle $-\pi/2$ in the negative (counterclockwise) direction. Thus, vector $\mathbf{F}_{12}$ is equal to

$$\mathbf{F}_{12} = F_{12}^{\xi}[\cos\theta_2, \sin\theta_2]^T + F_{12}^{\eta}\left[\cos\left(\theta_2 - \frac{\pi}{2}\right), \sin\left(\theta_2 - \frac{\pi}{2}\right)\right]^T \tag{3.46}$$

or

$$\mathbf{F}_{12} = F_{12}^{\xi}[\cos\theta_2, \sin\theta_2]^T + F_{12}^{\eta}[\sin\theta_2, -\cos\theta_2]^T \tag{3.47}$$

Similarly,

$$\mathbf{F}_{32} = -F_{32}^{\xi}[\cos\theta_2, \sin\theta_2]^T + F_{32}^{\eta}[\sin\theta_2, -\cos\theta_2]^T \tag{3.48}$$

$$\mathbf{F}_{23} = -F_{23}^{\xi}[\cos\theta_3, \sin\theta_3]^T + F_{23}^{\eta}[-\sin\theta_3, \cos\theta_3]^T \tag{3.49}$$

$$\mathbf{F}_{43} = -F_{43}^{\xi}[\cos\theta_3, \sin\theta_3]^T + F_{43}^{\eta}[\sin\theta_3, -\cos\theta_3]^T \tag{3.50}$$

$$\mathbf{F}_{34} = F_{34}^{\xi}[\cos 0, \sin 0]^T + F_{34}^{\eta}\left[\cos\frac{\pi}{2}, \sin\frac{\pi}{2}\right]^T = F_{34}^{\xi}[1, 0]^T + F_{34}^{\eta}[0, 1]^T \tag{3.51}$$

Now, substitute vectors given by Equations 3.48 through 3.51 into Equations 3.44 and 3.45:

$$\begin{aligned} &-F_{32}^{\xi}[\cos\theta_2, \sin\theta_2]^T + F_{32}^{\eta}[\sin\theta_2, -\cos\theta_2]^T \\ &-F_{23}^{\xi}[\cos\theta_3, \sin\theta_3]^T + F_{23}^{\eta}[-\sin\theta_3, \cos\theta_3]^T = 0 \end{aligned} \tag{3.52}$$

and

$$-F_{43}^{\xi}[\cos\theta_3, \sin\theta_3]^T + F_{43}^{\eta}[\sin\theta_3, -\cos\theta_3]^T + F_{34}^{\xi}[1, 0]^T + F_{34}^{\eta}[0, 1]^T = 0 \tag{3.53}$$

The system of six equations (Equations 3.40 through 3.43 and Equations 3.52 and 3.53) defines eight unknowns: $F_{12}^{\xi}, F_{32}^{\xi}, F_{23}^{\xi}, F_{43}^{\xi}, F_{34}^{\xi}, F_{34}^{\eta}, F_{14}^{\eta}$, and P_4. From the first four equations (Equations 3.40 through 3.43) some of the unknowns are easily expressed through others:

$$F_{32}^{\xi} = F_{12}^{\xi} + P_2^{i\xi} \tag{3.54}$$

$$F_{23}^{\xi} = -P_3^{i\xi} - F_{43}^{\xi} \tag{3.55}$$

$$F_{34}^{\xi} = P_4^{i\xi} + P_4 \tag{3.56}$$

$$F_{14}^{\eta} = -F_{34}^{\eta} \tag{3.57}$$

After the above expressions are substituted into Equations 3.52 and 3.53, the latter system of four equations will have only four unknowns: $F_{12}^{\xi}, F_{43}^{\xi}, F_{34}^{\eta}$, and P_4:

$$(F_{12}^{\xi} + P_2^{i\xi})[\cos\theta_2, \sin\theta_2]^T + F_{32}^{\eta}[\sin\theta_2, -\cos\theta_2]^T$$
$$+ (F_{43}^{\xi} + P_3^{i\xi})[\cos\theta_3, \sin\theta_3]^T + F_{23}^{\eta}[-\sin\theta_3, \cos\theta_3]^T = 0 \quad (3.58)$$

and

$$-F_{43}^{\xi}[\cos\theta_3, \sin\theta_3]^T + F_{43}^{\eta}[\sin\theta_3, -\cos\theta_3]^T$$
$$+ (P_4^i + P_4)[1, 0]^T + F_{34}^{\eta}[0, 1]^T = 0 \quad (3.59)$$

In Equation 3.58 the two unknowns, F_{12}^{ξ} and F_{43}^{ξ}, are uncoupled from other unknowns, and thus they can be found by solving this equation. By multiplying Equation 3.58 from the left by the unit vector $\mathbf{u}_1 = [-\sin\theta_2, \cos\theta_2]$ this equation is reduced to

$$-F_{32}^{\eta} + (F_{43}^{\xi} + P_3^{i\xi})\sin(\theta_3 - \theta_2) - F_{23}^{\eta}\cos(\theta_3 - \theta_2) = 0 \quad (3.60)$$

From the above equation the unknown F_{43}^{ξ} is found to be

$$F_{43}^{\xi} = -P_3^{i\xi} - \frac{F_{23}^{\eta}\cos(\theta_3 - \theta_2) - F_{32}^{\eta}}{\sin(\theta_3 - \theta_2)} \quad (3.61)$$

Now, premultiply Equation 3.58 by the unit vector $\mathbf{u}_2 = [-\sin\theta_3, \cos\theta_3]$. The result is

$$(F_{12}^{\xi} + P_2^{i\xi})\sin(\theta_2 - \theta_3) - F_{32}^{\eta}\cos(\theta_2 - \theta_3) + F_{23}^{\eta} = 0 \quad (3.62)$$

From the above equation the unknown F_{12}^{ξ} is found to be

$$F_{12}^{\xi} = -P_2^{i\xi} + \frac{F_{32}^{\eta}\cos(\theta_2 - \theta_3) - F_{23}^{\eta}}{\sin(\theta_2 - \theta_3)} \quad (3.63)$$

Now, since F_{43}^{ξ} is given by Equation 3.61, the remaining unknowns, F_{34}^{η} and P_4, can be found from Equation 3.59. First consider the ξ-components in Equation 3.59. As a result, the unknown P_4 is found:

$$P_4 = -P_4^{i\xi} + F_{43}^{\xi}\cos\theta_3 - F_{43}^{\eta}\sin\theta_3 \quad (3.64)$$

Similarly, by considering the η-components in Equation 3.59, the unknown F_{34}^{η} is found

$$F_{34}^{\eta} = F_{43}^{\xi}\sin\theta_3 + F_{43}^{\eta}\cos\theta_3 \quad (3.65)$$

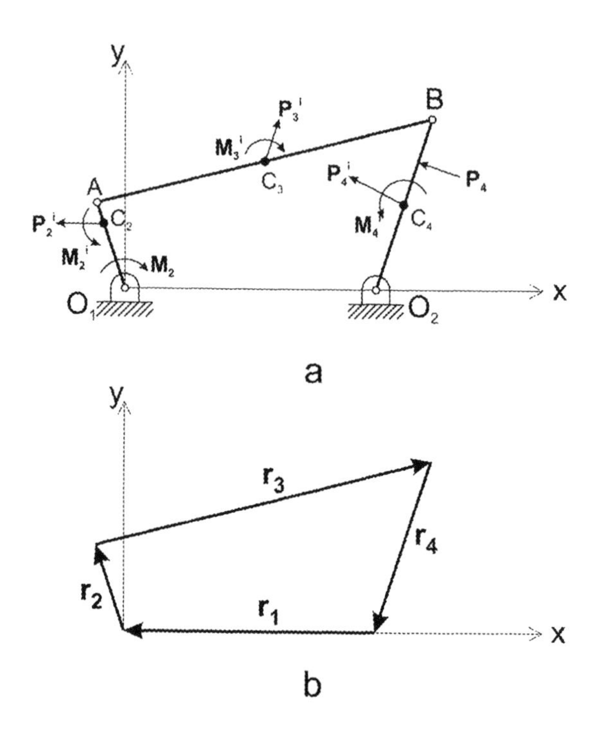

FIGURE 3.10 Four-bar linkage driven by the moment $\mathbf{M}_2$.

Thus, the unknown joint forces and the external resistance force are found for any crank position and any moment applied to it. The explicit formulas for all 12 unknowns are given in Equations 3.34, 3.35, 3.38, 3.39, 3.54 through 3.57, 3.61, and 3.63 through 3.65.

In conclusion, note that by representing the constraint forces in joints in local coordinate systems the number of unknowns in these joints doubled, so that instead of a total of eight unknowns in a global coordinate system one solved for 12 unknowns. However, the above representation of joint forces allowed one to decouple the equations and thus to simplify the solution.

3.5.3 Four-Bar Mechanism: The Case of Significant Inertial Forces

The mechanism is shown in Figure 3.10a, its vector representation in Figure 3.10b, and the free-body diagrams in Figure 3.11, where the inertial forces and moments are identified by the superscript i.

One can see that the free-body diagrams for links 2 and 3 are identical, in terms of loading, to those for the slider-crank mechanism, Figure 3.9b. The only link that is different is link 4, which means that Equations 3.44 and 3.45 for the joints are valid for this mechanism. However, vector $\mathbf{F}_{34}$ is different in this case. Namely, it is equal to

$$\mathbf{F}_{34} = -F_{34}^{\xi}[\cos\theta_4, \sin\theta_4]^T + F_{34}^{\eta}[-\sin\theta_4, \cos\theta_4]^T \tag{3.66}$$

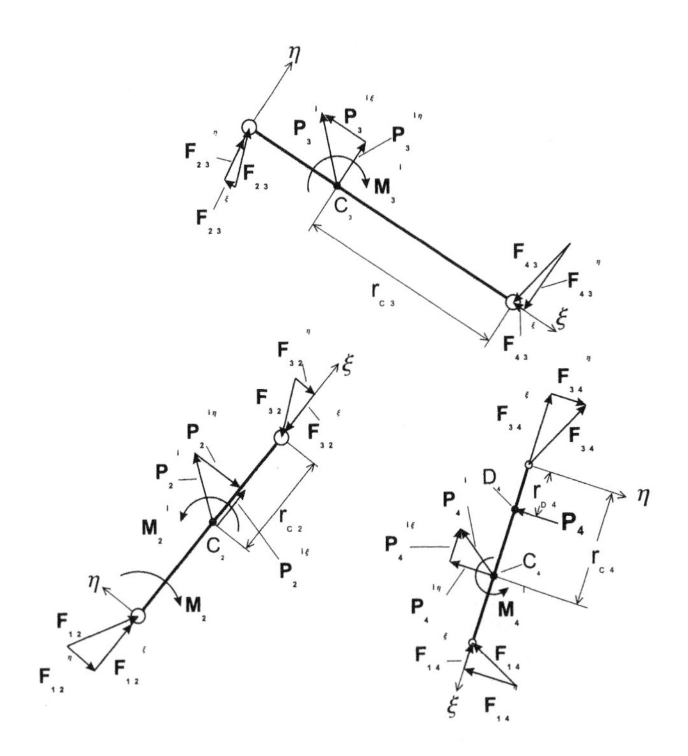

FIGURE 3.11 Free-body diagrams of links for a four-bar linkage.

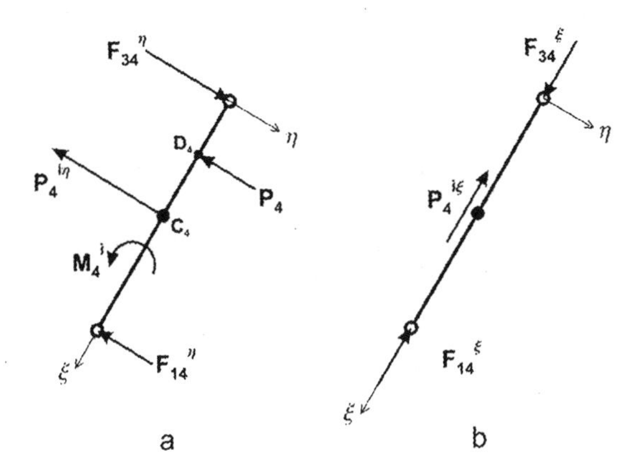

FIGURE 3.12 Free-body diagram for the link 4 in a four-bar linkage.

The vector $\mathbf{F}_{14}$ is also different

$$\mathbf{F}_{14} = -F_{14}^{\xi}[\cos\theta_4, \sin\theta_4]^T + F_{14}^{\eta}[-\sin\theta_4, \cos\theta_4]^T \tag{3.67}$$

The free-body diagrams for the normal and along the link forces in link 4 are shown in Figure 3.12. The corresponding scalar equilibrium equations are

$$-F_{34}^{\eta} + F_{14}^{\eta} - P_4^{i\eta} - P_4 = 0 \quad (3.68)$$

$$-r_4 F_{14}^{\eta} + M_4^{i} - r_{C4} P_4^{i\eta} - r_{D4} P_4 = 0 \quad (3.69)$$

and

$$-F_{14}^{\xi} - F_{34}^{\xi} - P_4^{i\xi} = 0 \quad (3.70)$$

There are three unknowns in Equations 3.68 and 3.69: F_{34}^{η}, F_{14}^{η}, and P_4. Thus, one can express the internal forces through the unknown external force:

$$F_{14}^{\eta} = \frac{M_4^{i} - r_{C4} P_4^{i\eta} - r_{D4} P_4}{r_4} \quad (3.71)$$

and

$$F_{34}^{\eta} = F_{14}^{\eta} - P_4^{i\eta} - P_4 \quad (3.72)$$

There are a total of seven remaining unknowns: F_{12}^{ξ}, F_{32}^{ξ}, F_{23}^{ξ}, F_{43}^{ξ}, F_{34}^{ξ}, F_{14}^{ξ}, and P_4. These unknowns are defined by the following system of equations: Equations 3.40 and 3.41 for links 2 and 3, Equation 3.70 for link 4, and Equations 3.44 and 3.45 in which vector $\mathbf{F}_{34}$ is as defined by Equation 3.66. The two expressions (Equations 3.54 and 3.55) are valid in this case as well. Another relationship between the unknowns comes from Equation 3.70:

$$F_{34}^{\xi} = -F_{14}^{\xi} - P_4^{i\xi} \quad (3.73)$$

If Equations 3.54 and 3.55 and Equation 3.73 are now substituted into Equations 3.44 and 3.45, one will obtain two vector equations. The first one, corresponding to Equation 3.44, is Equation 3.58 and thus will have the same solutions: Equations 3.61 and 3.63. The second vector equation is different, and it is as follows:

$$\begin{aligned} &-F_{43}^{\xi}[\cos\theta_3, \sin\theta_3]^T + F_{43}^{\eta}[\sin\theta_3, -\cos\theta_3]^T \\ &-F_{34}^{\xi}[\cos\theta_4, \sin\theta_4]^T + F_{34}^{\eta}[-\sin\theta_4, \cos\theta_4]^T = 0 \end{aligned} \quad (3.74)$$

There are two unknowns in Equation 3.74, F_{34}^{ξ} and F_{34}^{η}, which can be easily found. Premultiply Equation 3.74 by a unit vector $\mathbf{u}_1 = [-\sin\theta_4, \cos\theta_4]$. As a result, the unknown F_{34}^{η} is found

$$F_{34}^{\eta} = F_{43}^{\xi}\sin(\theta_3 - \theta_4) + F_{43}^{\eta}\cos(\theta_3 - \theta_4) \quad (3.75)$$

Now premultiply Equation 3.74 by a unit vector $\mathbf{u}_2 = [\cos\theta_4, \sin\theta_4]$. As a result, the unknown F_{34}^{ξ} is found:

$$F_{34}^{\xi} = F_{43}^{\xi}\cos(\theta_4 - \theta_3) + F_{43}^{\eta}\sin(\theta_4 - \theta_3) \tag{3.76}$$

Now, substitute Equations 3.71 and 3.75 into Equation 3.72 and solve for P_4. The result is

$$P_4 = \frac{-r_4 F_{43}^{\xi}\sin(\theta_3 - \theta_4) - r_4 F_{43}^{\eta}\cos(\theta_3 - \theta_4) + M_4^{i} - (r_4 + r_{C4})P_4^{i\eta}}{r_4 + r_{D4}} \tag{3.77}$$

Thus, the explicit formulas for all 13 unknowns — F_{12}^{ξ}, F_{12}^{η}, F_{32}^{ξ}, F_{32}^{η}, F_{23}^{ξ}, F_{23}^{η}, F_{43}^{ξ}, F_{43}^{η}, F_{34}^{ξ}, F_{34}^{η}, F_{14}^{ξ}, F_{14}^{η}, and P_4 — expressed in local coordinate systems are given by Equations 3.34, 3.35, 3.38, 3.39, 3.54, 3.55, 3.61, 3.63, 3.71, 3.73, and 3.75 through 3.77.

Note that, if the inertial forces can be neglected, then the corresponding expressions for the forces are obtained as a particular case of the above solution.

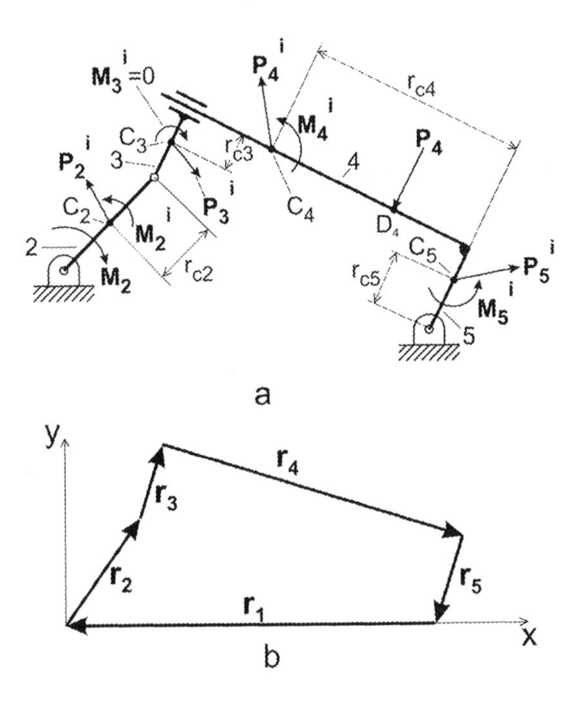

FIGURE 3.13 Five-bar mechanism driven by moment $\mathbf{M}_2$.

3.5.4 Five-Bar Mechanism: The Case of Significant Inertial Forces

The linkage with applied external moment $\mathbf{M}_2$ and inertial forces is shown in Figure 3.13. The resistance force $\mathbf{P}_4$ is assumed to be normal to link 4 during the motion. Note that links 4 and 5 are treated as different bodies for the sake of convenience, but in fact they constitute one mechanism element. The external and constraint forces directed along the links are shown in Figure 3.14a, and those normal to the links in Figure 3.14b. Note that because links 4 and 5 are rigidly connected, there is a constraint moment

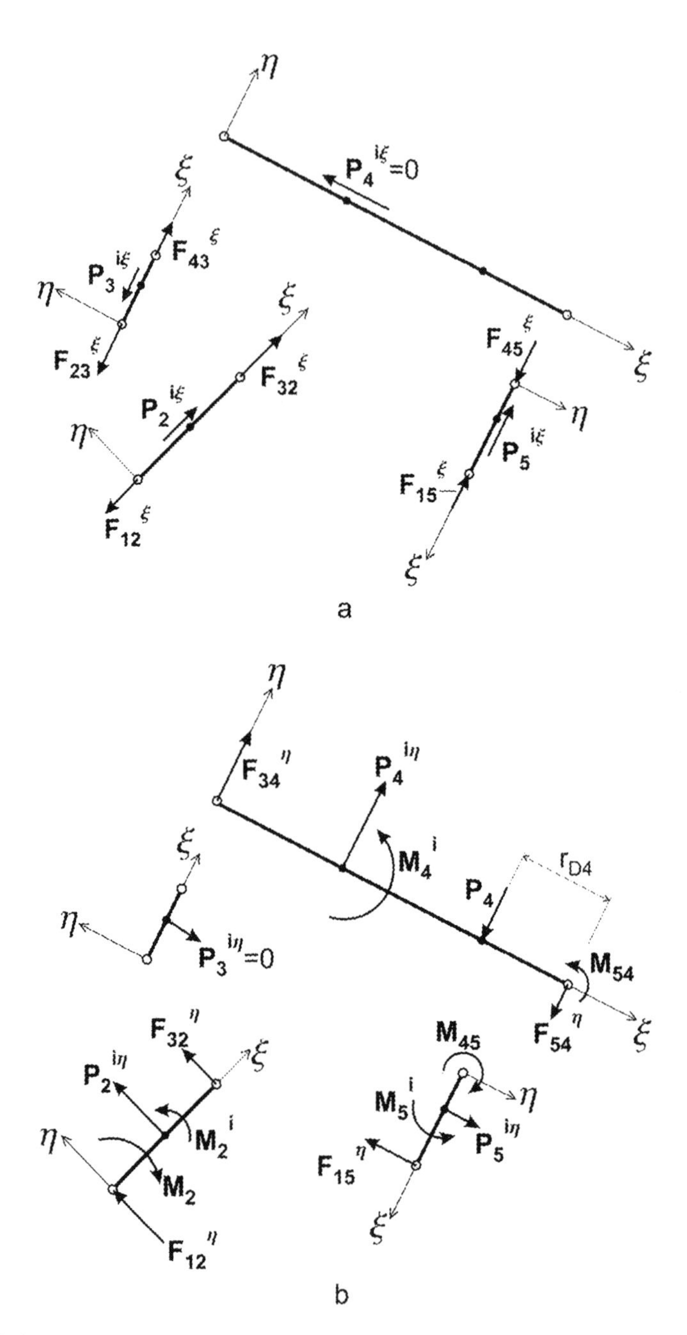

FIGURE 3.14 (a) Free-body diagrams for links subjected to along-the-link forces in a five-bar mechanism. (b) Free-body diagrams for links subjected to normal to the link forces in a five-bar mechanism.

at this interface, $M_{54} = M_{45}$. Also assume an ideal sliding joint, i.e., friction forces are zero.

First, consider the equilibrium of links subjected to normal forces.

Link 2

$$F_{12}^{\eta} + F_{32}^{\eta} + P_2^{i\eta} = 0 \tag{3.78}$$

and

$$-r_2 F_{12}^{\eta} - M_2 + M_2^{i} - r_{C2} P_2^{i\eta} = 0 \tag{3.79}$$

From the above equation,

$$F_{12}^{\eta} = \frac{-M_2 + M_2^{i} - r_{C2} P_2^{i\eta}}{r_2} \tag{3.80}$$

and from Equation 3.78,

$$F_{32}^{\eta} = -P_2^{i\eta} - F_{12}^{\eta} \tag{3.81}$$

Link 3

If the friction forces are neglected, then, as was found above, this link cannot have any normal reaction forces.

Link 4

$$F_{34}^{\eta} - F_{54}^{\eta} + P_4^{i\eta} - P_4 = 0 \tag{3.82}$$

and

$$-r_4 F_{34}^{\eta} + M_4^{i} - r_{C4} P_4^{i\eta} + M_{54} + r_{D4} P_4 = 0 \tag{3.83}$$

From the above equation,

$$F_{34}^{\eta} = \frac{M_{54} + M_4^{i} - r_{C4} P_4^{i\eta} + r_{D4} P_4}{r_4} \tag{3.84}$$

and from Equation 3.82,

$$F_{54}^{\eta} = P_4^{i\eta} + F_{34}^{\eta} - P_4 \tag{3.85}$$

Note that the unknown internal moment M_{54} and the external resistance force P_4 cannot be found from the equilibrium requirements for this link.

Link 5

$$-F_{15}^{\eta}+P_{5}^{i\eta}=0 \tag{3.86}$$

and

$$M_{5}^{i}-r_{C5}P_{5}^{i\eta}-M_{45}=0 \tag{3.87}$$

From the latter equation,

$$M_{45}=M_{5}^{i}-r_{C5}P_{5}^{i\eta} \tag{3.88}$$

and from Equation 3.86

$$F_{15}^{\eta}=P_{5}^{i\eta} \tag{3.89}$$

Thus, the only unknown left from the analysis of normal forces is the external resistance force P_4.

Now, consider the equilibrium of links subjected to along-the-link forces.

Link 2

$$-F_{12}^{\xi}+F_{32}^{\xi}=0 \tag{3.90}$$

Link 3

$$-F_{23}^{\xi}+F_{43}^{\xi}-P_{3}^{i\xi}=0 \tag{3.91}$$

Link 4

This link is not subjected to any longitudinal forces, if the friction forces are neglected.

Link 5

$$F_{45}^{\xi}-F_{15}^{\xi}-P_{5}^{i\xi}=0 \tag{3.92}$$

As can be seen, Equations 3.90 through 3.92 contain six unknowns: F_{12}^{ξ}, F_{32}^{ξ}, F_{23}^{ξ}, F_{43}^{ξ}, F_{45}^{ξ}, and F_{15}^{ξ}. Recall that the external resistance force P_4 is another unknown. Thus, in addition to three equations (Equations 3.90 through 3.92) four more equations are needed to find the above unknowns. These come from the requirements of joints equilibrium.

Joint 2–3

$$F_{32}^{\xi}[\cos\theta_2,\ \sin\theta_2]^T+F_{32}^{\eta}[-\sin\theta_2,\ \cos\theta_2]^T-F_{23}^{\xi}[\cos\theta_3,\ \sin\theta_3]^T=0 \tag{3.93}$$

Joint 3–4

Since link 3 is normal to link 4, it follows that the corresponding magnitudes of internal forces are equal:

$$F_{43}^{\xi} = F_{34}^{\eta} \tag{3.94}$$

Joint 4–5

Since links 4 and 5 are perpendicular to each other, the following magnitudes of forces are equal:

$$F_{45}^{\xi} = F_{54}^{\eta} \tag{3.95}$$

From the last two equations and Equations 3.84 and 3.85 it follows that F_{43}^{ξ} and F_{45}^{ξ} are functions of the external force P_4.

$$F_{43}^{\xi} = \frac{M_{54} + M_4^i - r_{C4}P_4^{i\eta} + r_{D4}P_4}{r_4} \tag{3.96}$$

and

$$F_{45}^{\xi} = \frac{M_{54} + M_4^i + (r_4 - r_{C4})P_4^{i\eta} - (r_4 - r_{D4})P_4}{r_4} \tag{3.97}$$

From Equation 3.93 the two unknowns F_{32}^{ξ} and F_{23}^{ξ} are found by first premultiplying it by the unit vector $\mathbf{u}_1 = [-\sin\theta_3, \cos\theta_3]$ to obtain

$$F_{32}^{\xi} = -F_{32}^{\eta}\cot(\theta_2 - \theta_3) \tag{3.98}$$

and then by premultiplying it by the unit vector $\mathbf{u}_2 = [-\sin\theta_2, \cos\theta_2]$ to obtain

$$F_{23}^{\xi} = \frac{F_{32}^{\eta}}{\sin(\theta_3 - \theta_2)} \tag{3.99}$$

With F_{32}^{ξ} and F_{23}^{ξ} given by the last two equations, F_{12}^{ξ} and F_{43}^{ξ} are found from Equations 3.90 and 3.91. Recall that F_{32}^{η} is given by Equations 3.80 and 3.81. Now by substituting F_{23}^{ξ} and F_{43}^{ξ} from Equations 3.99 and 3.96 into Equation 3.91, the unknown external force P_4 is found:

$$P_4 = \frac{r_4}{r_{D4}}\left(\frac{-M_{54} - M_4^i + r_{C4}P_4^{i\eta}}{r_4} + P_3^{i\xi} + \frac{F_{32}^{\eta}}{\sin(\theta_3 - \theta_2)}\right) \tag{3.100}$$

The internal force F_{45}^{ξ} can now be found from Equation 3.97, and then F_{15}^{ξ} from Equation 3.92.

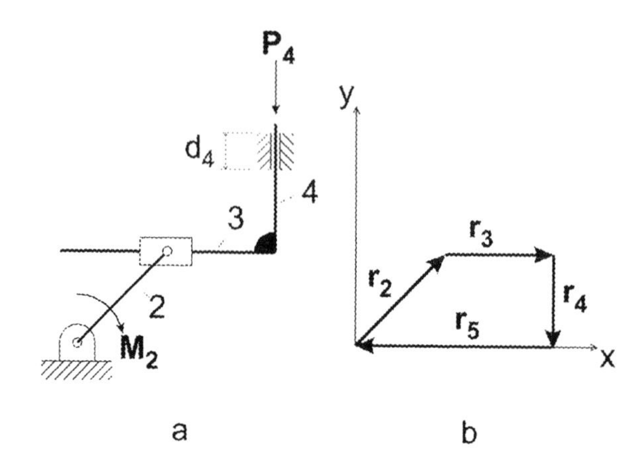

FIGURE 3.15 Scotch yoke mechanism driven by moment M_2.

Now, transform the internal forces from the local to the global coordinate system taking into account the directions of local forces in the global coordinate system. For example, for the force $\mathbf{F}_{12}$,

$$\mathbf{F}_{12} = -F_{12}^{\xi}[\cos\theta_2, \sin\theta_2]^T + F_{12}^{\eta}[-\sin\theta_2, \cos\theta_2]^T \tag{3.101}$$

3.5.5 Scotch Yoke Mechanism: The Case of Significant Inertial Forces

The mechanism with the driving moment $\mathbf{M}_2$ and resistance force $\mathbf{P}_4$ is shown in Figure 3.15. Note that links 3 and 4 constitute one mechanism element. However, it is convenient, as was done earlier in the five-bar linkage analysis, to represent it as comprising two perpendicular links.

In Figures 3.16a and b the free-body diagrams of links under the action of forces parallel and normal to links are shown. First, one notices that link 3 does not experience any along-the-link forces (Figure 3.16a), if friction forces in the slider are neglected. Second, since a moment is transferred through a rigid joint connecting links 3 and 4, this moment is counterbalanced by a couple in the sliding support of link 4. The distance d_4 is assumed to be the length of the support. And, finally, if the friction forces are neglected, then no such couple is generated between link 3 and the slider because the slider is connected to link 2 by a revolute joint.

The equilibrium equations for the along-the-link forces are as follows:

Link 2

$$F_{12}^{\xi} - F_{32}^{\xi} + P_2^{i\xi} = 0 \tag{3.102}$$

Link 4

$$-F_{34}^{\xi} - P_4^{i\xi} + P_4 = 0 \tag{3.103}$$

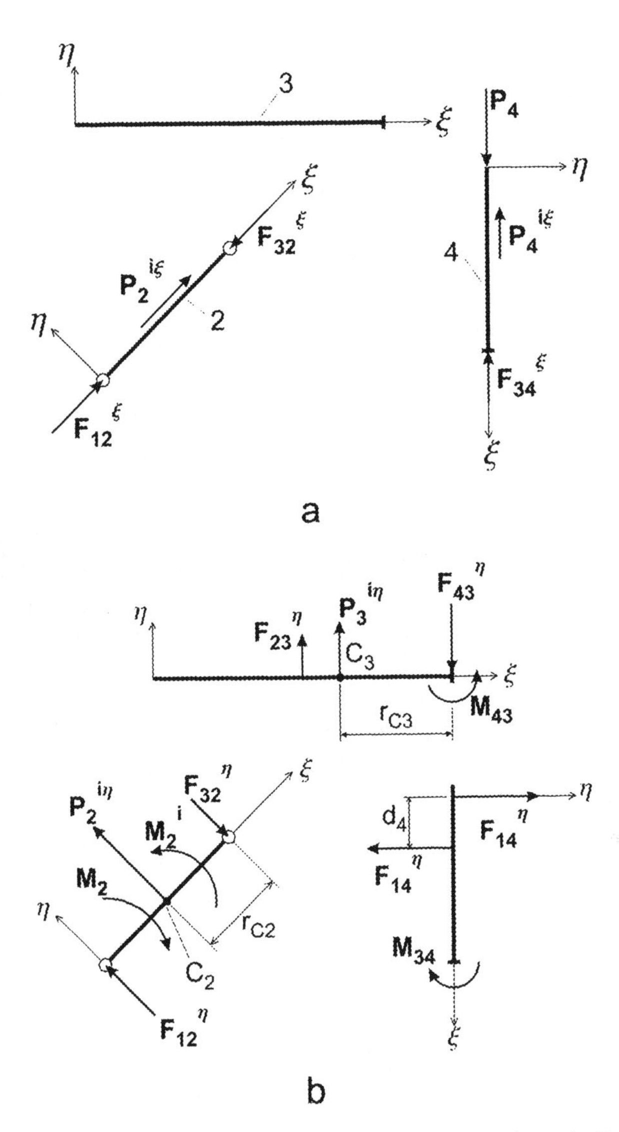

FIGURE 3.16 (a) Free-body diagrams for links subjected to along-the-link forces in a Scotch yoke mechanism. (b) Free-body diagrams for links subjected to forces normal to the link in a Scotch yoke mechanism.

The equilibrium equations for the forces normal to the link are as follows:

Link 2

$$F_{12}^{\eta} + P_2^{i\eta} - F_{32}^{\eta} = 0 \tag{3.104}$$

and

$$-r_2F_{12}^{\eta} + M_2^i - M_2 - r_{C2}P_2^{i\eta} = 0 \quad (3.105)$$

Link 3

$$F_{23}^{\eta} + P_3^{i\eta} - F_{43}^{\eta} = 0 \quad (3.106)$$

and

$$-r_3F_{23}^{\eta} - r_{C3}P_3^{i\eta} + M_{43} = 0 \quad (3.107)$$

Link 4

$$-d_4F_{14}^{\eta} - M_{34} = 0 \quad (3.108)$$

In seven equations (Equations 3.102 through 3.108) there are 10 unknowns. Thus three more conditions are needed to obtain a complete system of equations. These come from the balance of internal forces and moments in joints. The directions of forces and moments are according to those shown in Figure 3.16a and b.

Revolute joint 2–3

$$-F_{32}^{\xi}[\cos\theta_2, \sin\theta_2]^T + F_{32}^{\eta}[\sin\theta_2, -\cos\theta_2]^T + F_{23}^{\eta}[-\sin\theta_3, \cos\theta_3]^T = 0 \quad (3.109)$$

Rigid joint 3–4

$$F_{43}^{\eta}\left[\cos\left(-\frac{\pi}{2}\right), \sin\left(-\frac{\pi}{2}\right)\right]^T + F_{34}^{\xi}\left[\cos\left(\frac{\pi}{2}\right), \sin\left(\frac{\pi}{2}\right)\right]^T = 0 \quad (3.110)$$

and

$$M_{43} - M_{34} = 0 \quad (3.111)$$

From Equation 3.109 the two unknowns, F_{32}^{ξ} and F_{23}^{η}, are found by premultiplying it first by a unit vector $\mathbf{u}_1 = [\cos\theta_3, \sin\theta_3]$, and then by a unit vector $\mathbf{u}_2 = [-\sin\theta_2, \cos\theta_2]$, taking into account that $\theta_3 = 0$.

$$F_{32}^{\xi} = F_{32}^{\eta}\tan\theta_2 \quad (3.112)$$

and

$$F_{23}^{\eta} = \frac{F_{32}^{\eta}}{\cos\theta_2} \quad (3.113)$$

From Equation 3.107,

$$F_{23}^{\eta} = \frac{-r_{C3}P_3^{i\eta} + M_{43}}{r_3} \tag{3.114}$$

and then from the above and Equation 3.106,

$$F_{43}^{\eta} = \frac{(r_3 - r_{C3})P_3^{i\eta} + M_{43}}{r_3} \tag{3.115}$$

Equation 3.110 confirms that $F_{34}^{\xi} = F_{43}^{\eta}$. Then by substituting into the above the corresponding expressions for these forces, Equations 3.103 and 3.115, the following relationship between the two unknowns, P_4 and M_{43}, is obtained:

$$\mathbf{P}_4 = P_4^{i\xi} + \frac{M_{43} + (r_3 - r_{C3})P_3^{i\eta}}{r_3} \tag{3.116}$$

By solving Equations 3.104 and 3.105 for F_{12}^{η} and F_{32}^{η},

$$F_{12}^{\eta} = \frac{M_2^i - M_2 - r_{C2}P_2^{i\eta}}{r_2} \tag{3.117}$$

and

$$F_{32}^{\eta} = \frac{M_2^i - M_2 + (r_2 - r_{C2})P_2^{i\eta}}{r_2} \tag{3.118}$$

And, finally, M_{43} is found from Equation 3.113 by substituting corresponding forces from Equations 3.114 and 3.118:

$$M_{43} = r_3 F_{32}^{\eta} \operatorname{cosec}\theta_2 + r_{C3}P_3^{i\eta} \tag{3.119}$$

The unknown force P_4 can now be found from Equation 3.116.

PROBLEMS AND EXERCISES

PROBLEMS

1. Prove that if the external force is perpendicular to a link, the constraint forces perpendicular to a link are uniquely defined.
2. Prove that if the external moment is applied to a link, the constraint forces form a couple.
3. Prove that if there are no external forces applied to a link, the constraint forces act along the link and are equal and opposite to each other.

4. Prove that in the case of a link with one revolute joint and another one sliding whose axis is perpendicular to the link, the constraint forces act along the link and are equal and opposite to each other.

5. Prove that in the case of a link with one revolute joint and another one sliding whose axis is parallel to the link, the constraint forces are perpendicular to the link.

6. Explain D'Alembert's principle and how it is used in dynamic analysis of mechanisms.

7. Consider a punch mechanism in Figure 1.3. Neglect inertial forces and draw free-body diagrams of links 3 and 4, taking into account that a resistance force acts along link 4. Is there a constraint force perpendicular to link 4? Explain.

8. Consider a windshield wiper mechanism in Figure 1.4. Neglect inertial forces and draw a free-body diagram of the link *DBE*, taking into account a resistance force acting on the wiper. Identify on the diagram the directions of the rocker and crank couplers.

9. Consider a dump truck mechanism in Figure 1.6. Neglect inertial forces and draw a free-body diagram of link 4. Identify on the diagram the directions of links 3 and 5.

10. For the slider-crank mechanism of the internal combustion engine shown in Figure 2.4, assume that the external force is applied to the piston and that inertial forces are significant. Then,

    a. Draw free-body diagrams for all links.
    b. Write equilibrium equations.
    c. Identify the number of unknown forces and relate it to the number of equations.

11. For the Scotch yoke mechanism shown in Figure 2.18a assume that the external force is applied to the vertical part of link 3 (along the vector $\mathbf{r}_4$ in Figure 2.18b) and that inertial forces are significant. Then,

    a. Draw free-body diagrams for all links.
    b. Write equilibrium equations.
    c. Identify the number of unknown forces and relate it to the number of equations.

12. For the material-handling mechanism in Figure 2.26 assume that the inertial forces are insignificant. Then,

    a. Draw free-body diagrams for all links.
    b. Write equilibrium equations.
    c. Identify the number of unknown forces and relate it to the number of equations.

Exercises (Projects) with Mathematica

The following projects are a continuation of those outlined in Chapter 2.

1. A motor drives a film-advancing mechanism with constant velocity (Figure P2.8). Assume that a constant resistance force is applied to the coupler only during its engagement with the film. Neglect inertial forces and plot the external moment required to rotate the crank with constant speed vs. angle of rotation.

2. A motor drives a material-handling mechanism with constant velocity ω (Figure 2.26). Assume the mass of the load and take into account the corresponding inertial force associated with this mass. Neglect inertial forces associated with links. Plot the external moment required to rotate the crank with constant speed vs. angle of rotation.

3. Do the force analysis for the double-rocker mechanism described in exercise 3 in Chapter 2. Assume that a constant horizontal force, encountered only when the tracing point moves from left to right, is given, and also that the coupler rotates with a constant angular velocity. Neglect inertial forces associated with links and plot the external moment applied to the arm 1 needed to rotate the coupler as a function of the coupler angle of rotation.

4. For the complex mechanism operating a dump truck in Figure 1.6, where the bed moves from a horizontal to a $\pi/4$ position, assume some load in the truck bed and plot the force needed to lift the bed. Neglect inertial forces.

5. For a double-toggle puncher shown in Figure P1.6, assume that a constant resistance force acts at point P only during the downward part of the cycling motion. Neglect inertial forces associated with links and plot the external moment required to rotate the crank with constant speed vs. angle of rotation.

6. For a variable-stroke drive in Figure P1.7, assume that a constant resistance moment acts on the output link only when the latter rotates clockwise. Neglect inertial forces associated with links and plot the external moment required to rotate the crank with constant speed vs. angle of rotation.

7. For an adjustable-stroke mechanism shown in Figure P1.3, assume that a constant resistance force acts on the slider only during one half of the cycle. Neglect inertial forces associated with links and plot the external moment required to rotate the crank with constant speed vs. angle of rotation.

4 Cams

4.1 INTRODUCTION

A cam mechanism is a two-link system in which the cam is always a driving link. This mechanism transforms a rotational or translational motion of the cam into a prescribed translational or angular motion of the follower. An example of a cam mechanism is shown in Figure 4.1 where 1 is the cam, 2 is the follower, 3 is the spring, and 4 is the camshaft. In Figure 4.1 the cam is a plate and the follower is a pin. The transfer of motion is achieved through the contact between the cam and the follower. As is clear, this contact exists only if there is a compressive force between the cam and the follower. The function of the prestressed spring in this respect is to ensure that this contact is maintained during the cycle of cam rotation. The motion of the follower reflects the shape of the cam profile, and thus *the main objective of cam design is to find a cam profile needed to obtain a desired follower motion.*

The follower in Figure 4.1 is called the *knife-edge follower.* The specific feature of this mechanism is that at the cam–follower interface a relative motion (sliding) takes place. Since the transfer of motion involves transfer of forces through the cam–follower interface, the contact stresses at this interface in the presence of sliding may be unacceptable from the point of view of wear. Given that the contact stresses vary during the rotation, the wear is not uniform along the profile, thus leading to a deviation from the designed follower motion. Such potential for the deterioration of motion transformation in systems with knife-edge followers has led to cam mechanisms with *flat-faced* (Figure 4.2) or *roller* (Figure 4.3) followers. With the flat-faced follower, the contact stresses are lower while the sliding takes place. With the roller follower the sliding is eliminated at the expense of a more complex design. Note that the axis of the follower may pass through the camshaft center (Figure 4.1) or be at some distance from this center (Figure 4.3).

In all cases shown in Figures 4.1 through 4.3 the rotational motion of the cam is transformed into the reciprocating (oscillating) follower motion. In Figure 4.4 another transformation of motion, namely, from rotational to angular oscillation, is shown. Instead of the roller follower shown in Figure 4.4, a flat-faced follower can be used. Another type of cam mechanism is shown in Figures 4.5 and 4.6, where a reciprocating motion of the cam is transformed into either an angular oscillation (Figure 4.5) or translational oscillation (Figure 4.6).

The functional role of the spring in the above cam designs is to ensure a constant contact between the cam and the follower. The presence of the spring complicates the design and results in increased contact stresses. An alternative design solution is to insert the roller inside a groove, Figure 4.7. The outside and inside profiles of this groove are in fact the profiles of cams, one moving the follower up and another down. Since clearance between the roller and the groove is needed to allow for free roller rotation, the transition from up to down motion of the follower may be associated with discontinuity in motion, additional parasitic forces, and noise.

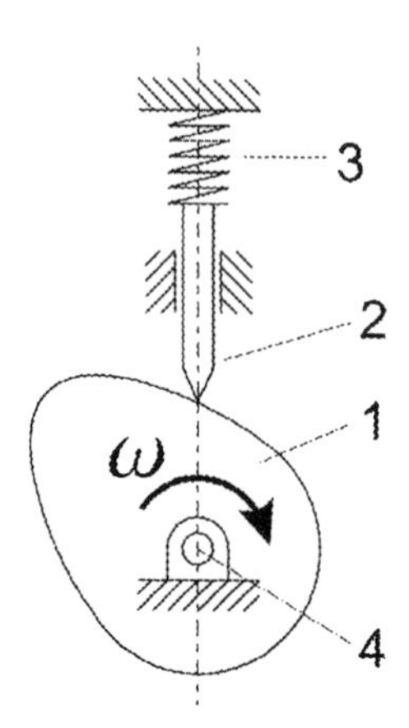

FIGURE 4.1 Cam with a knife-edge follower.

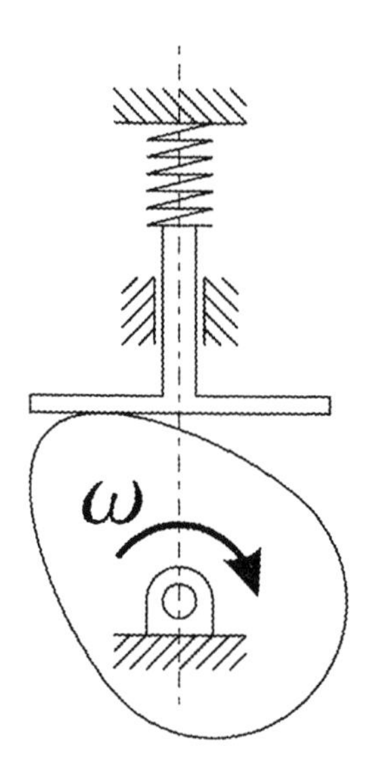

FIGURE 4.2 Cam with a flat-faced follower.

4.2 CIRCULAR CAM PROFILE

A *circular cam* is made by mounting a circular plate on a camshaft at some distance d away from the circle center (Figure 4.8). This gives the simplest cam profile. The eccentric attachment of the circular plate produces a reciprocating motion of the follower. The problem of direct analysis is to find the follower displacement given the rotation of a cam with known radius R and eccentricity d.

Place the origin of the global coordinate system in the camshaft center. Then, the distance D from this center (Figure 4.9) characterizes the follower position. The tip of the follower can also be reached following the vectors $\mathbf{d}$ and $\mathbf{R}$. The three

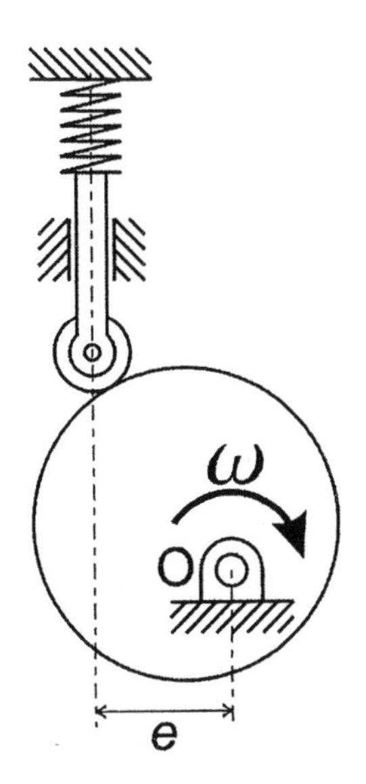

FIGURE 4.3 Cam with a roller follower.

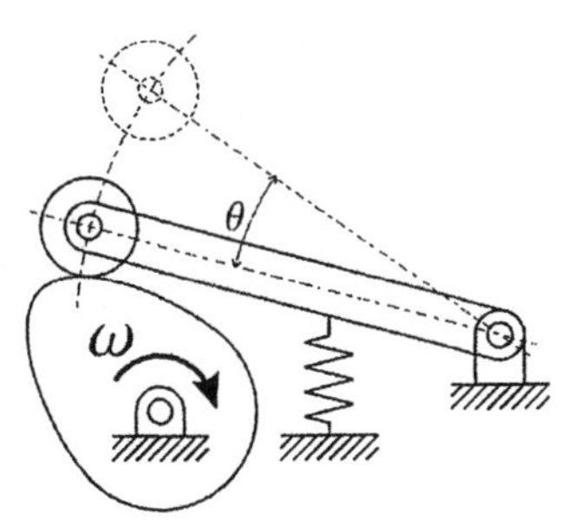

FIGURE 4.4 Cam with a rocking follower.

vectors, **d**, **R**, and **D**, form a loop at any cam position. This loop is identical to a loop for a slider-crank mechanism, in which d plays the role of a crank, R of a connecting rod, and D characterizes the slider position. This analogy means that the analysis of this cam mechanism is identical to that of the slider-crank mechanism. Indeed, the loop-closure equation in this case is

$$d[\cos\gamma, \sin\gamma]^T + R[\cos\theta, \sin\theta]^T + D\left[\cos\left(\frac{3\pi}{2}\right), \sin\left(\frac{3\pi}{2}\right)\right]^T = 0 \qquad (4.1)$$

where the angles θ and γ are shown in Figure 4.9. In the above equation the unknowns are the distance D and the angle θ. Thus, this equation falls into the second case category according to the analysis of various cases in Chapter 2. From the equivalency of Equations 2.28 and 4.1, it follows that the corresponding solutions for the former, namely, Equations 2.31 and 2.32, can be used in this case. It is necessary

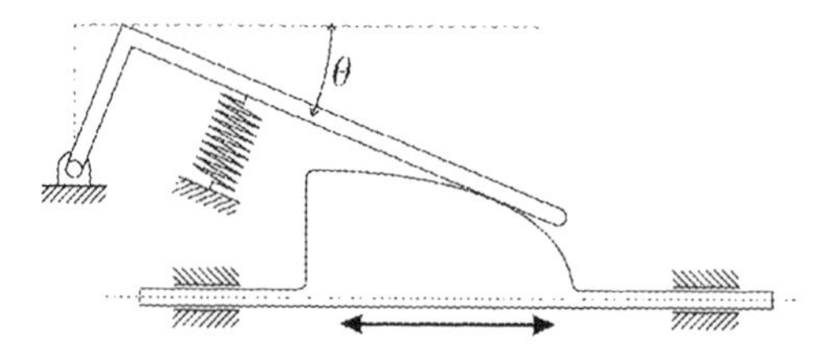

FIGURE 4.5 Reciprocating cam with a flat follower.

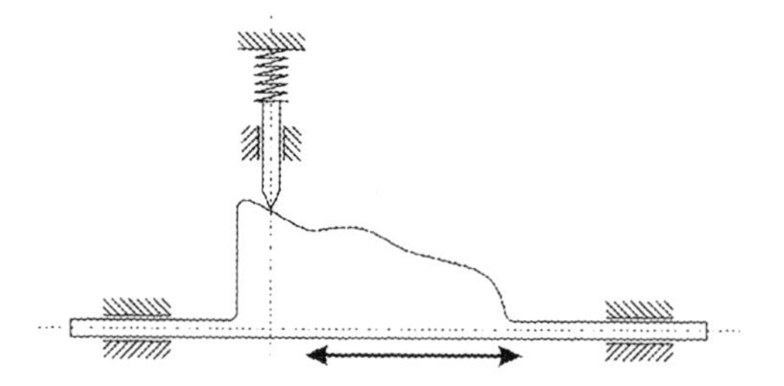

FIGURE 4.6 Reciprocating cam with a knife follower.

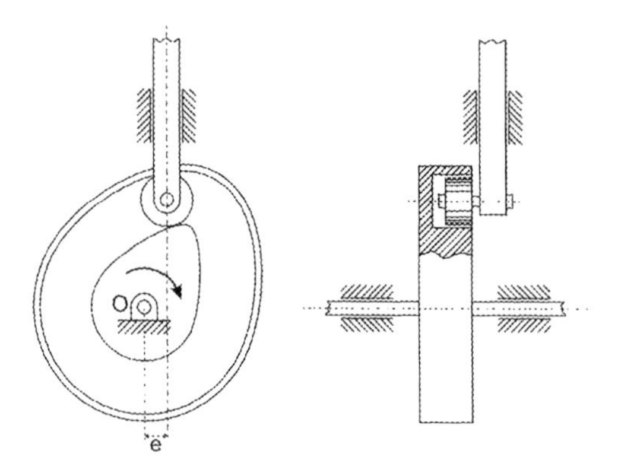

FIGURE 4.7 Cam with a roller inside a groove.

to substitute in these solutions r_i by D, θ_j by θ, θ_i by $3\pi/2$, α by γ, r_j by R, and b by $-d$. As a result, the solution for the follower displacement is

$$D = d\sin\gamma \pm \sqrt{R^2 - d^2\cos^2\gamma} \tag{4.2}$$

and the angle θ is

$$\theta = \begin{cases} \theta^* & \text{if } -\sin\theta > 0 \text{ and } \cos\theta < 0 \\ \pi - \theta^* & \text{if } -\sin\theta < 0 \text{ and } \cos\theta < 0 \\ \pi + \theta^* & \text{if } -\sin\theta < 0 \text{ and } \cos\theta > 0 \\ 2\pi - \theta^* & \text{if } -\sin\theta > 0 \text{ and } \cos\theta > 0 \end{cases} \tag{4.3}$$

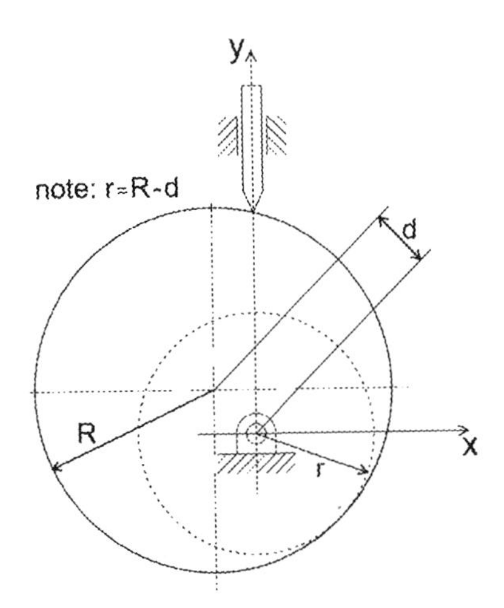

FIGURE 4.8 Circular cam.

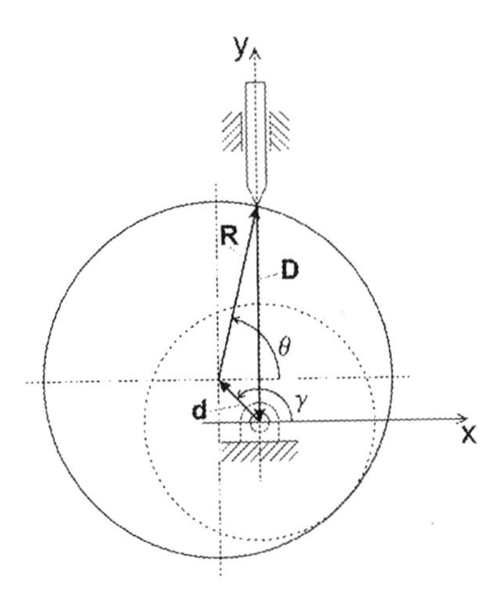

FIGURE 4.9 Loop-closure equation for a circular cam.

where

$$\theta^* = \frac{3\pi}{2} + \arcsin\left|\frac{d\cos\gamma}{R}\right| \tag{4.4}$$

$$\cos\theta = -\frac{d}{R}\cos\gamma \tag{4.5}$$

and

$$\sin\theta = -\frac{d}{R}\sin\gamma + \frac{D}{R} \tag{4.6}$$

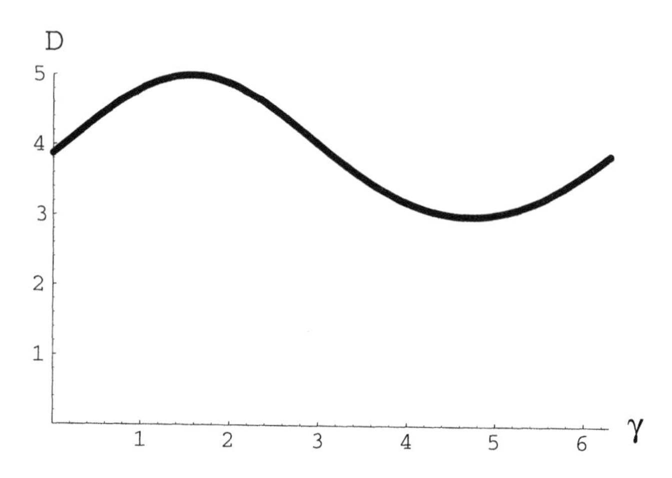

FIGURE 4.10 Position of the follower during one cycle of circular cam rotation.

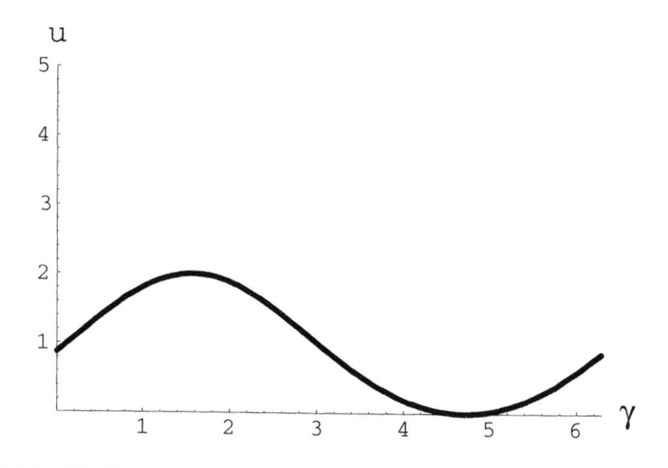

FIGURE 4.11 Displacement diagram for a circular cam.

If, at the extreme, $d = R$, then Equation 4.2 gives $D = R\sin\theta \pm R\sin\theta$. It follows from the above that the correct sign in Equation 4.2 must be plus for this case. At the other extreme, if $d << R$, then the second term under the square root can be neglected and the result is $D = d\sin\gamma \pm R$. Again, the correct sign must be plus for D to be positive. Thus, one can assume that the sign in Equation 4.2 must be plus for any value of d.

$$D = d\sin\gamma + \sqrt{R^2 - d^2\cos^2\gamma} \tag{4.7}$$

In Figure 4.10 the position of the follower, D, is shown for one cycle of cam rotation for the case of $d = 1$ cm and $R = 4$ cm. The difference between the maximum follower displacement and its minimum is called the *lift*. If the minimum follower coordinate (position) is subtracted from its current position, the resulting diagram

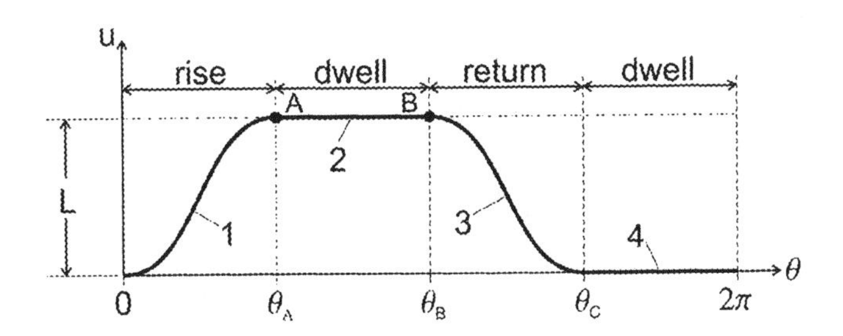

FIGURE 4.12 Typical displacement diagram.

is called the *displacement diagram*. Such a diagram for a circular cam is shown in Figure 4.11.

A minimum follower position constitutes a reference level for the follower displacement. This reference level is a circle with a constant radius $r_b = D - D_{min}$, which is called the *base radius*. Thus, a cam profile can be viewed as a displacement diagram wrapped around the base circle.

4.3 DISPLACEMENT DIAGRAM

The displacement diagram serves as an input into the cam mechanism design. Consider the cam mechanism in Figure 1.1. The function of the cam might be, as is the case in the internal combustion engine, to open the valve, to keep it open during some part of the cycle (this is called *dwell*), and then to close it and to keep it closed for some time (to dwell again). A generic displacement diagram may look as shown in Figure 4.12. The requirements of how long it should take to rise, to dwell, to return, and to dwell again, and also of what the lift should be define the size and the shape of the cam. The function depicted in Figure 4.12 is a piecewise function, which means that special attention should be paid to transition from one continuous function to another, for example, from rise to dwell. This represents another objective of cam design, *to ensure a smooth transition of the follower from one part of the displacement diagram to another.*

Consider, for example, a transition from rise to dwell in Figure 4.12. The rise curve is described by some function $u_1(\theta)$, while the dwell is described by another function $u_2(\theta) = \text{const}$. For a smooth transition from the rise to dwell, it is needed that at $\theta = \theta_A$,

$$u_1(\theta_A) = u_2(\theta_A), \frac{du_1(\theta)}{d\theta} = 0, \text{ and } \frac{d^2u_1(\theta)}{d\theta^2} = 0 \tag{4.8}$$

Since $\theta = \omega t$ (where ω is the angular velocity of the cam, and t is the time), the requirements for the equality of first and second derivatives is equivalent to the requirements that the velocity and acceleration of the follower does not experience jumps at the point of transition. The same requirement should be met at the other transitional points in Figure 4.12: $\theta = 0$, $\theta = \theta_B$, and $\theta = \theta_C$. These latter requirements put limitations on what type of functions can be used to generate the

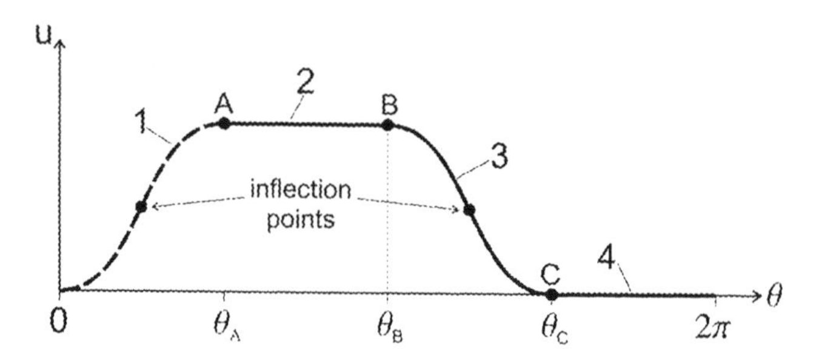

FIGURE 4.13 Inflection points on the displacement diagram.

cam profile for a given displacement diagram. The requirements (Equation 4.8) are called *smoothness requirements.*

4.4 CYCLOID, HARMONIC, AND FOUR-SPLINE CAMS

Any function that meets the type of requirements given in Equation 4.8 is a suitable cam profile function. For example, the suitable function for the rising part of the diagram must be tangential to the θ-axis at point 0 and at point A (Figure 4.13). It means that it must change the curvature from concave to convex, and thus it must have an inflection point. A few analytical functions are used to describe the rise and return parts of the displacement diagram while meeting the above smoothness conditions. These are cycloid, harmonic, and polynomial functions. In the following an application of all these functions to the displacement diagram design is considered.

A cycloid is a curve traced by a point on a circle rolling along a straight line. In Figure 4.14 an example of a cycloid function is shown, where point A is embedded into the circle, and angle α corresponds to arc αr, which is the distance traveled by the circle. The x- and y-coordinates of the cycloid are parametric functions of angle α

$$x = r(\alpha - \sin\alpha) \tag{4.9}$$

and

$$y = r(1 - \cos\alpha) \tag{4.10}$$

Both functions can be used in cam profile design. A cam in which the first function, Equation 4.9, is used is called the *cycloid cam,* while a cam in which the second function, Equation 4.10, is used is called the *harmonic cam.*

4.4.1 Cycloid Cams

The Rise Part of the Displacement Diagram

One can utilize the function given by Equation 4.9 to describe the rise of the follower from 0 to point A in Figure 4.13. Note that Equation 4.9 comprises two components: $r\alpha$ and $-r\sin\alpha$. Thus, it is a superposition of a straight line and a sinusoidal function.

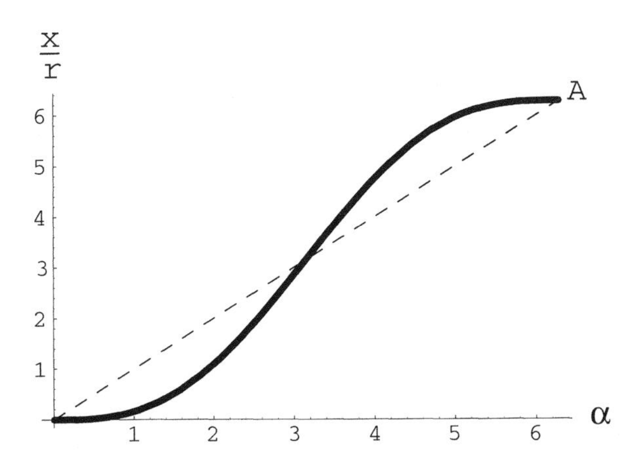

FIGURE 4.14 Normalized functions α (dashed line) and (α – sin α) (solid line).

In normalized coordinates, x/r, these functions are shown in Figure 4.14 in the interval 0 to 2π.

The straight line must go from 0 to point A, and then the sinusoidal function is drawn with respect to this line (see Figure 4.14). It can now be shown that this function meets the above smoothness requirements. But first it is necessary to transform Equation 4.9 from the (x,α)- to the (u,θ)-coordinates. The needed transformation is achieved by mapping the α-range $(0\text{–}2\pi)$ onto the θ-range $(0\text{–}\theta_A)$. Such mapping follows from the relationship

$$\frac{\alpha_{\text{rise}}}{2\pi} = \frac{\theta}{\theta_A} \tag{4.11}$$

where α_{rise} is the transformed angle α.

Now substituting α_{rise} from Equation 4.11 into Equation 4.9 gives the transformed cycloid equation:

$$u = r\left(2\pi\frac{\theta}{\theta_A} - \sin 2\pi\frac{\theta}{\theta_A}\right) \tag{4.12}$$

The radius of the circle r in Equation 4.12 remains undetermined. It should be treated as a parameter to be determined from the requirement that at $\theta = \theta_A$ $u = L$ (see Figure 4.12). Then it follows that

$$r = \frac{L}{2\pi} \tag{4.13}$$

Thus, the function describing the rise of the follower from 0 to L when the cam rotates from $\theta = 0$ to $\theta = \theta_A$ is

$$u_1 = \frac{L}{2\pi}\left(2\pi\frac{\theta}{\theta_A} - \sin 2\pi\frac{\theta}{\theta_A}\right) \tag{4.14}$$

Now, check that the requirements given by Equation 4.8 are met. The first and second derivatives of the function Equation 4.14 are

$$\frac{du_1}{d\theta} = \frac{L}{\theta_A}\left(1 - \cos 2\pi\frac{\theta}{\theta_A}\right) \tag{4.15}$$

and

$$\frac{d^2u_1}{d\theta^2} = \frac{2\pi L}{\theta_A^2}\sin 2\pi\frac{\theta}{\theta_A} \tag{4.16}$$

It is easy to see that both derivatives are equal to zero at $\theta = 0$ and $\theta = \theta_A$.

The Return Part of the Displacement Diagram

Now, one can utilize the same function given by Equation 4.9 to describe the return of the follower from θ_B to θ_C in Figure 4.13. The procedure is the same. One maps the α-range (0 to 2π) onto the θ-range (θ_B to θ_C). The coordinate transformation in this case is

$$\frac{\alpha_{\text{return}}}{2\pi} = \frac{\theta - \theta_C}{\theta_B - \theta_C} \tag{4.17}$$

where α_{return} is the transformed angle α.

Since the expression for r is known (see Equation 4.13), the return part of the displacement diagram is described by Equation 4.9, where angle α is, according to Equation 4.17,

$$u_3 = \frac{L}{2\pi}\left(2\pi\frac{\theta - \theta_C}{\theta_B - \theta_C} - \sin 2\pi\frac{\theta - \theta_C}{\theta_B - \theta_C}\right) \tag{4.18}$$

The first and second derivatives in this case are as follows:

$$\frac{du_3}{d\theta} = \frac{L}{\theta_B - \theta_C}\left(1 - \cos 2\pi\frac{\theta - \theta_C}{\theta_B - \theta_C}\right) \tag{4.19}$$

and

$$\frac{d^2u_3}{d\theta^2} = \frac{2\pi L}{(\theta_B - \theta_C)^2}\sin 2\pi\frac{\theta - \theta_C}{\theta_B - \theta_C} \tag{4.20}$$

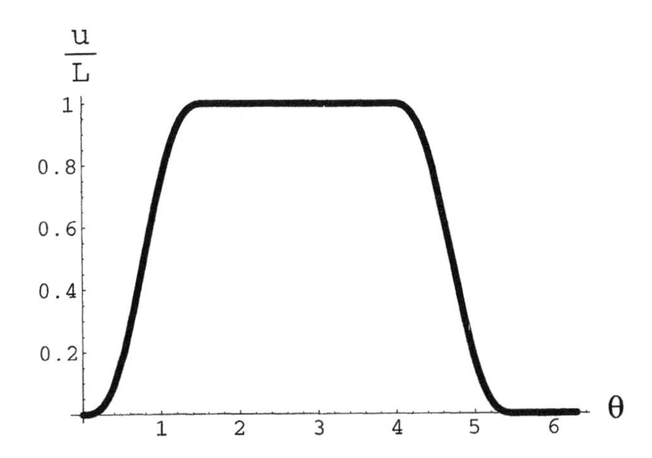

FIGURE 4.15 Normalized displacement diagram for a cycloid cam.

It is easy to check that at $\theta = \theta_C$ and $\theta = \theta_B$ both derivatives are equal to zero. Thus, the cycloid function can be used to describe the cam profile and satisfy the smoothness requirements. An example of the normalized displacement diagram for $\theta_A = \pi/2$, $\theta_B = 5\pi/4$, and $\theta_C = 7\pi/4$ is shown in Figure 4.15. In Figures 4.16 and 4.17 the corresponding normalized velocity and acceleration diagrams are shown. The normalized displacements, velocities, and accelerations are, respectively, as follows:

On the rise part:

$$\bar{u}_1 = \frac{u}{L} \tag{4.21}$$

$$\dot{\bar{u}}_1 = \frac{\dot{u}\theta_A}{L} \tag{4.22}$$

$$\ddot{\bar{u}}_1 = \frac{\ddot{u}\theta_A^2}{L} \tag{4.23}$$

On the return part:

$$\bar{u}_3 = \frac{u}{L} \tag{4.24}$$

$$\dot{\bar{u}}_3 = \frac{\dot{u}|\theta_B - \theta_C|}{L} \tag{4.25}$$

$$\ddot{\bar{u}}_3 = \frac{\ddot{u}|\theta_B - \theta_C|^2}{L} \tag{4.26}$$

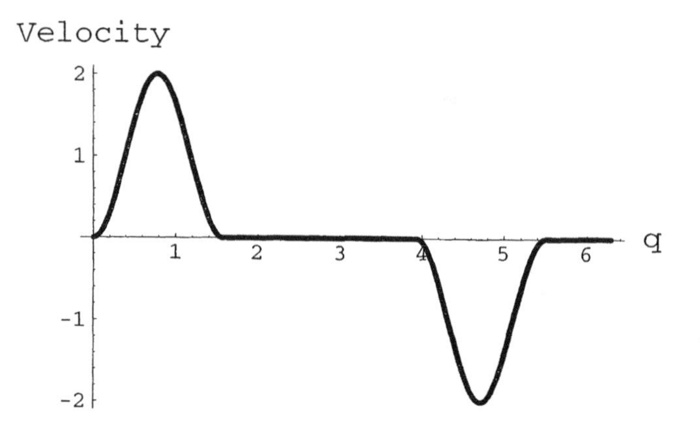

FIGURE 4.16 Normalized velocity diagram for a cycloid cam.

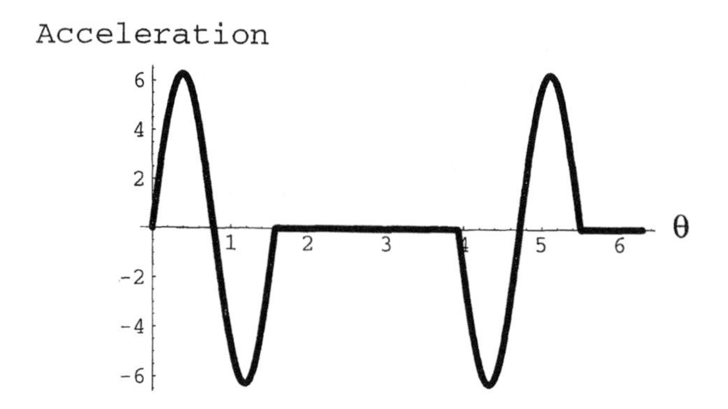

FIGURE 4.17 Normalized acceleration diagram for a cycloid cam.

It is denoted above:

$$\dot{u} = \frac{du(\theta)}{d\theta} \quad \text{and} \quad \ddot{u} = \frac{d^2u(\theta)}{d\theta^2}$$

The displacement, velocity, and acceleration of the follower, shown in Figures 4.15 through 4.17, are functions of the angle of rotation of the cam, and thus they do not depend on the angular speed of rotation. The displacement as a function of the angle of rotation allowed one to find a proper cam profile. It does not, however, answer the question of what is the real acceleration of the follower, which one has to know to choose the proper spring stiffness (see Figure 4.1). If one substitutes the angle of rotation $\theta = \omega t$ into Equations 4.14 and 4.18, one will have displacements as functions of time and angular velocity ω.

$$u_1 = \frac{L}{2\pi}\left(2\pi\frac{\omega t}{\theta_A} - \sin 2\pi\frac{\omega t}{\theta_A}\right) \tag{4.27}$$

and

$$u_3 = \frac{L}{2\pi}\left(2\pi\frac{\omega t - \theta_C}{\theta_B - \theta_C} - \sin 2\pi\frac{\omega t - \theta_C}{\theta_B - \theta_C}\right) \tag{4.28}$$

where for a single cycle the time t changes from 0 to T, and $T = 2\pi/\omega$ is the period of the cycle.

It is clear from Equations 4.17 through 4.28 that the displacement of the follower does not depend on the angular velocity of the cam. However, the velocity and acceleration of the follower do depend on it. Indeed, the velocity of the follower as a function of time is proportional to ω, since

$$\frac{du[\theta(t)]}{dt} = \frac{du[\theta(t)]}{d\theta}\frac{d\theta(t)}{dt} = \frac{du[\theta]}{d\theta}\omega \tag{4.29}$$

and $du[\theta]/(d\theta)$ is the velocity as the function of the angle of rotation. Similarly, assuming that the angular velocity is constant, the acceleration of the follower is proportional to the square of angular velocity, since

$$\frac{d^2u[\theta(t)]}{dt^2} = \omega\frac{d}{dt}\left(\frac{du[\theta(t)]}{d\theta}\right) = \omega^2\frac{d^2u[\theta]}{d\theta^2} \tag{4.30}$$

Because the angular velocity of the cam is a scaling factor in both velocity and acceleration of the follower, the solutions given by Equations 4.12 through 4.20 are general solutions of the kinematics of the cycloid cam. In this respect, the plots of the follower displacement (Figure 4.15), velocity (Figure 4.16), and acceleration (Figure 4.17), except for the specific values of the angles θ_A, θ_B, and θ_C mentioned above, are generic plots characterizing the cycloid cam.

4.4.2 Harmonic Cams

Now it will be shown that Equation 4.10 can also be used to describe the cam profile and it meets the requirement of smoothness. The plot of the function given by Equation 4.10 in normalized, y/r, coordinates over the interval 0 to 2π is shown in Figure 4.18.

As can be seen, a part of this function within the interval 0 to π can be used for the rise part of the displacement diagram, whereas the second part, within the interval π to 2π for the return part of the diagram.

Again, points θ_A, θ_B, and θ_C (see Figure 4.12) will be used as transition points from one continuous function to another on a cam displacement diagram. However, in this case the 0 to θ_A interval will be mapped on 0 to π and θ_B to θ_C on π to 2π of the harmonic function. The corresponding mapping relationships are similar to Equations 4.11 and 4.17 except, instead of 2π, a π is used in both. Thus, the displacements of the follower during the rise and return parts of the cycle are obtained from Equation 4.10 by substituting corresponding expressions for α

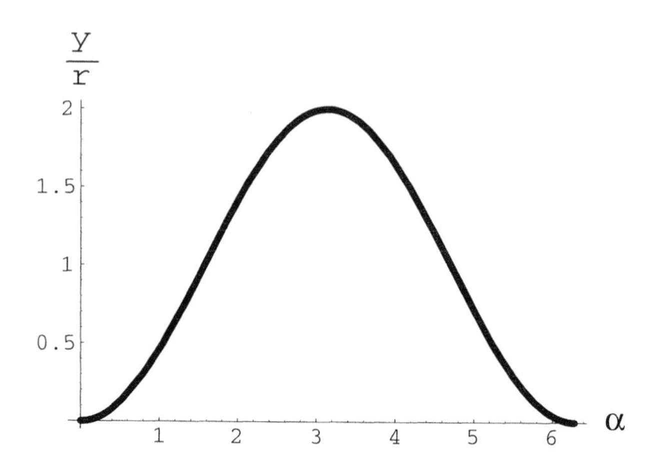

FIGURE 4.18 Function 1 – cos α.

$$\alpha_{\text{rise}} = \pi \frac{\theta}{\theta_A} \tag{4.31}$$

and

$$\alpha_{\text{return}} = \pi \frac{\theta - \theta_C}{\theta_B - \theta_C} \tag{4.32}$$

Now the corresponding displacement formulas are as follows:

$$u_1 = \frac{L}{2}\left(1 - \cos \pi \frac{\theta}{\theta_A}\right) \tag{4.33}$$

and

$$u_3 = \frac{L}{2}\left(1 - \cos \pi \frac{\theta - \theta_C}{\theta_B - \theta_C}\right) \tag{4.34}$$

It is easy to check that at $\theta = \theta_A$ and $\theta = \theta_B$, $u = L$, whereas at $\theta = \theta_C$, $u = 0$. The normalized displacement diagram is shown in Figure 4.19.

Now, one can check whether the smoothness requirements are satisfied. The first and the second derivatives of the function Equation 4.10 are $r \sin \alpha$ and $r \cos \alpha$, respectively. For the rise part, α from Equation 4.31 is substituted, and for the return one, α from Equation 4.32 is substituted for both derivatives. One can see that at both points A and B (see Figure 4.12) the velocities are equal to zero, while accelerations are not. Thus, the harmonic cam does not satisfy all the smoothness requirements. The jump in acceleration while passing through these points means a jump in inertial forces. For high-speed cams when there are design constraints on forces or noise, this cam may not be acceptable.

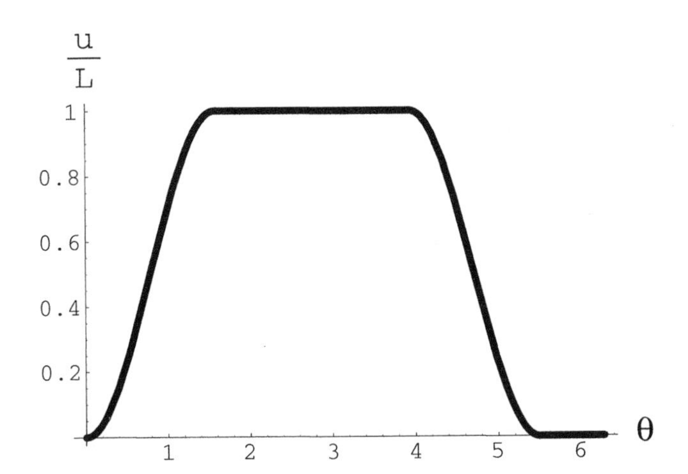

FIGURE 4.19 Normalized displacement diagram for a harmonic cam.

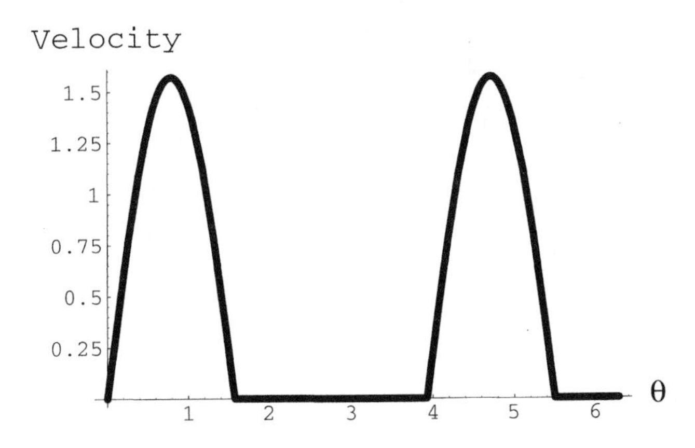

FIGURE 4.20 Normalized velocity plot for a harmonic cam.

In Figure 4.20 the normalized velocity and in Figure 4.21 the normalized acceleration plots are shown. These plots confirm that there is a jump in accelerations at the end of the rise and the beginning of the return. One can also see the amount of this jump in normalized coordinates. The normalization of displacements, velocities, and accelerations in this case is as for a cycloid cam (Equations 4.21 through 4.26).

4.4.3 Comparison of Two Cams: Cycloid vs. Harmonic

Here the kinematic properties of two cams are compared: displacements, velocities, and accelerations in the same normalized coordinates. The displacements are shown in Figure 4.22. One can see that the displacement curves in Figure 4.22 look sufficiently close. However, the differences at specific angles may be significant. One can check the displacements, for example, at $\theta = \theta_A/4$. They are 0.0908451 and 0.146447 for the cycloid and harmonic cams, respectively; i.e., the difference is 38%.

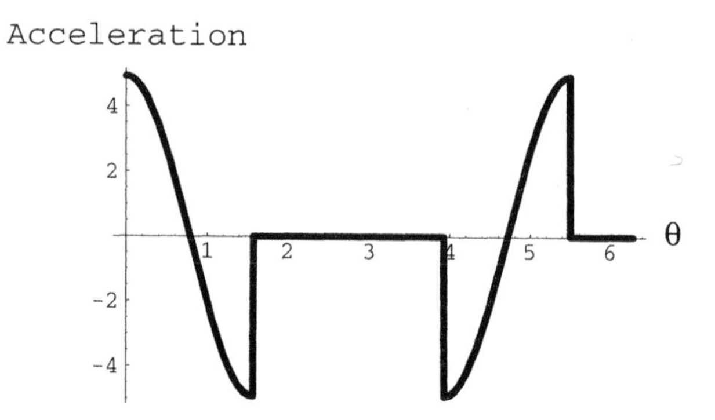

FIGURE 4.21 Normalized acceleration plot for a harmonic cam.

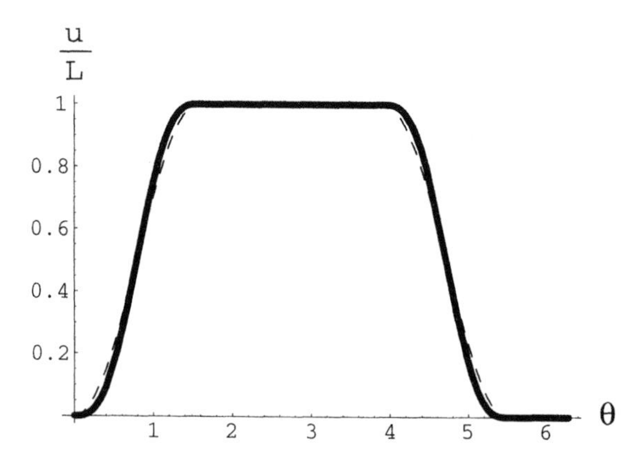

FIGURE 4.22 Comparison of normalized displacement diagrams for two cams: cycloid (solid line) and harmonic (dashed line).

It is also of interest to compare the accelerations of two cams at this point. They are 6.28319 and 3.48943 for the cycloid and harmonic cams, respectively; i.e., the difference is 45%. The velocities differ most significantly at the inflection point $\theta = \theta_A/2$. They are 2 and 1.57 for the cycloid and harmonic cams, respectively.

The comparison of normalized velocities and accelerations for the cycloid and harmonic cams are shown in Figures 4.23 and 4.24.

4.4.4 Cubic Spline Cams

The cubic spline method of designing cams is based on using cubic polynomials to fit a given displacement diagram at a predetermined number of points. This method is used for designing nonstandard cams. In general, the design of cams based on this approach requires a numerical solution of a system of linear algebraic equations. This book will limit itself to a simplified version of the method, which retains all the conceptual elements of it but is more manageable from the analytical point of view.

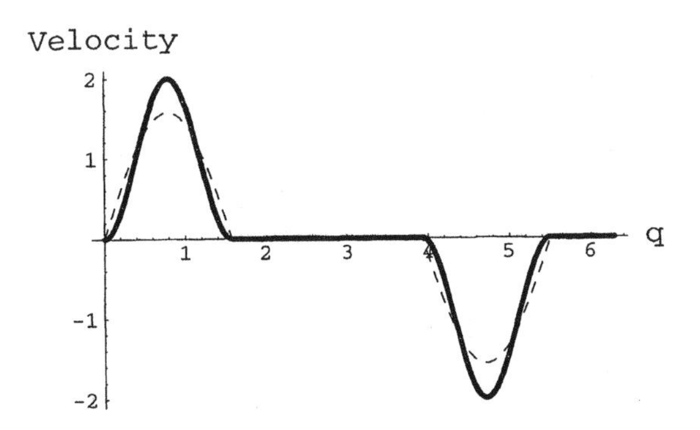

FIGURE 4.23 Comparison of normalized velocity diagrams for two cams: cycloid (solid line) and harmonic (dashed line).

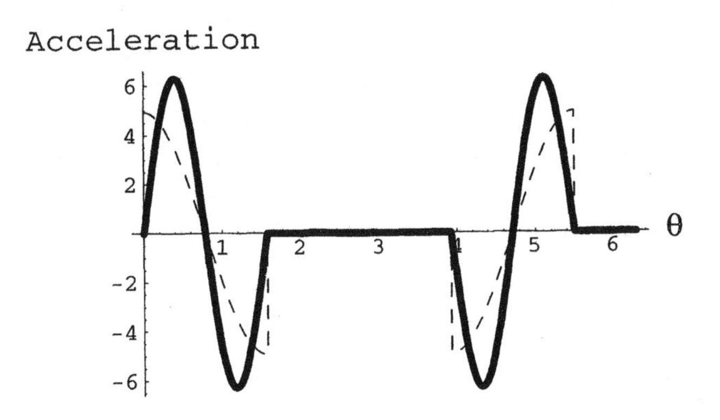

FIGURE 4.24 Comparison of normalized acceleration diagrams for two cams: cycloid (solid line) and harmonic (dashed line).

One can design a displacement diagram comprising six piecewise continuous functions, which are identified in Figure 4.25 by numbers. The first two, 1 and 2, are cubic splines describing the rise, the constant function 3 describes the dwell, the next two, 4 and 5, describe the return, and 6 is again a constant describing the dwell. It is necessary to find such cubic polynomials that meet the smoothness criteria for the cam.

The general form of the cubic polynomial is

$$u_i = a_i\theta_i^3 + b_i\theta_i^2 + c_i\theta_i + h_i \quad i = 1, 2, 4, 5 \tag{4.35}$$

where a_i, b_i, c_i, and h_i are constants to be determined for each spline. In total, for four splines there are 16 unknown constants. For each spline, one may request that it meet the displacement, velocity, and acceleration requirements at its boundaries (in this case at points: 0, $0.5\theta_A$, θ_A, θ_B, and $0.5\,(\theta_B + \theta_C)$). For each boundary (interface of two piecewise functions), one will have three equations defining the smoothness

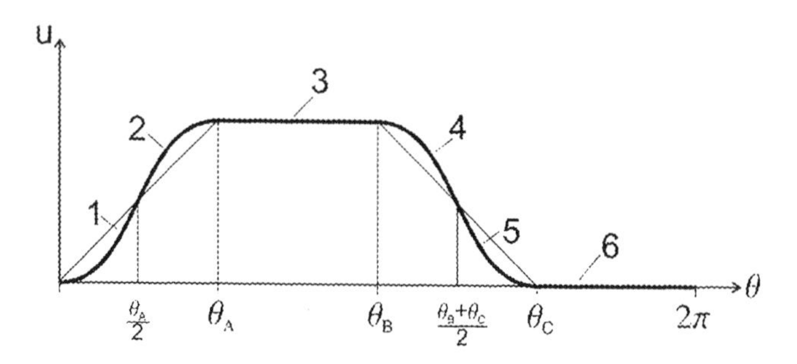

FIGURE 4.25 Piecewise continuous functions describing a displacement diagram.

requirements. In total, for six boundaries, there will be 18 requirements. For 16 unknown constants there is need for only 16 equations, which means that not all smoothness requirements will be satisfied. Thus, there is a freedom to choose which smoothness requirements to satisfy when using the cubic polynomial splines.

First, write down the first and second derivatives for the polynomial given by Equation 4.35. As before, they represent the velocity and acceleration of the follower in normalized (with respect to the angular velocity) coordinates.

$$\dot{u}_i = 3a_i\theta_i^2 + 2b_i\theta_i + c_i \quad \text{i} = 1, 2, 4, 5 \tag{4.36}$$

and

$$\ddot{u}_i = 6a_i\theta_i + 2b_i \quad \text{i} = 1, 2, 4, 5 \tag{4.37}$$

Now, consider spline 1 in Figure 4.25, and assume that at $\theta = 0$, the displacement, velocity, and acceleration are zeros.

$$u_1(0) = 0, \ \frac{du_1(\theta)}{d\theta} = 0, \text{ and } \frac{d^2u_1(\theta)}{d\theta^2} = 0 \tag{4.38}$$

and that at $\theta = 0.5\theta_A$ the displacement is equal to $0.5L$:

$$u_1(\theta) = \frac{L}{2} \tag{4.39}$$

This gives four equations, which is sufficient to find four constants for the first spline.

Similarly, for the second spline, if one declares that at $\theta = \theta_A$ the displacement, velocity, and acceleration are, respectively,

$$u_2(\theta) = L, \ \frac{du_2(\theta)}{d\theta} = 0, \text{ and } \frac{d^2u_2(\theta)}{d\theta^2} = 0 \tag{4.40}$$

and that at $\theta = 0.5\theta_A$ the displacement is equal to $0.5L$:

$$u_2(\theta) = \frac{L}{2} \tag{4.41}$$

there will be four equations needed to find four constants for the second spline.

Before finding the spline constants, it is worth making two comments. First, at the inflection point $\theta = 0.5\theta_A$ the requirement of equal displacements only is satisfied; the equality of velocities and accelerations is not guaranteed. The second point is more technical. Namely, if the rise part of the displacement diagram is approximated with two splines only, the system of equations defining constants can be decoupled. In other words, one can find these constants for each spline independently. The decoupling is achieved by specifying the boundary conditions for the splines as above. If, however, one requested that at $\theta = 0$, $u_1 = 0$ and at $\theta = \theta_A$, $u_2 = L$, and at $\theta = 0.5\theta_A$

$$\frac{du_1(\theta)}{d\theta} = \frac{du_2(\theta)}{d\theta} \tag{4.42}$$

and

$$\frac{d^2u_1(\theta)}{d\theta^2} = \frac{d^2u_2(\theta)}{d\theta^2} \tag{4.43}$$

then the equations for two splines would be coupled. Moreover, now the smoothness requirements would be met at the inflection point, and not guaranteed at $\theta = 0$ and at $\theta = \theta_A$.

Only the first case, when the smoothness is satisfied at $\theta = 0$, $\theta = \theta_A$, $\theta = \theta_B$, and $\theta = \theta_C$, will be discussed.

From Equations 4.38 it follows that $b_1 = c_1 = h_1 = 0$, and from Equation 4.39, the last unknown is found:

$$a_1 = \frac{4L}{\theta_A^3} \tag{4.44}$$

Thus, the first spline is described by the formula:

$$u_1 = 4L\frac{\theta^3}{\theta_A^3} \tag{4.45}$$

For the second spline from Equations 4.40, it follows that

$$b_2 = -3a_2\theta_A,\ c_2 = 3a_2\theta_A^2,\ \text{and}\ h_2 = L - a_2\theta_A^3 \tag{4.46}$$

and from Equation 4.41,

$$a_2 = \frac{4L}{\theta_A^3} \tag{4.47}$$

Thus, the second spline is described by the formula:

$$u_2 = 4L\left[\left(\frac{\theta}{\theta_A}\right)^3 - 3\left(\frac{\theta}{\theta_A}\right)^2 + 3\left(\frac{\theta}{\theta_A}\right)\right] - 3L \tag{4.48}$$

The boundary conditions for the fourth spline at $\theta = \theta_B$ are

$$u_4(\theta) = L, \frac{du_4(\theta)}{d\theta} = 0, \text{ and } \frac{d^2u_4(\theta)}{d\theta^2} = 0 \tag{4.49}$$

and at $\theta = (\theta_B + \theta_C)/2$ is

$$u_4(\theta) = \frac{L}{2} \tag{4.50}$$

Satisfying the first set of boundary conditions, Equations 4.49 gives the expressions for the three constants:

$$b_4 = -3a_4\theta_B,\ c_4 = 3a_4\theta_B^2, \text{ and } h_4 = L - a_4\theta_B^3 \tag{4.51}$$

Satisfying the fourth boundary condition, Equation 4.50 gives the expression for a_4:

$$a_4 = \frac{4L}{(\theta_B - \theta_C)^3} \tag{4.52}$$

Thus the formula for the fourth spline is

$$u_4 = L + \frac{4L\theta^3}{(\theta_B - \theta_C)^3} - \frac{12L\theta_B\theta^2}{(\theta_B - \theta_C)^3} + \frac{12L\theta_B^2\theta}{(\theta_B - \theta_C)^3} - \frac{4L\theta_B^3}{(\theta_B - \theta_C)^3} \tag{4.53}$$

The boundary conditions for the fifth spline at $\theta = \theta_C$ are

$$u_5(\theta) = 0, \frac{du_5(\theta)}{d\theta} = 0, \text{ and } \frac{d^2u_5(\theta)}{d\theta^2} = 0 \tag{4.54}$$

and at $\theta = (\theta_B + \theta_C)/2$ is

$$u_5(\theta) = \frac{L}{2} \tag{4.55}$$

Satisfying the first set of boundary conditions, Equation 4.54 gives the expressions for the three constants:

$$b_5 = -3a_5\theta_C,\; c_5 = 3a_5\theta_C^2, \text{ and } h_5 = -a_5\theta_C^3 \tag{4.56}$$

Satisfying the fourth boundary condition, Equation 4.55 gives the expression for a_5:

$$a_5 = \frac{4L}{(\theta_B - \theta_C)^3} \tag{4.57}$$

As a result, the formula for the fifth spline is

$$u_5 = 4L\left[\frac{\theta^3}{(\theta_B - \theta_C)^3} - \frac{3\theta_C\theta^2}{(\theta_B - \theta_C)^3} + \frac{3\theta_C^2\theta}{(\theta_B - \theta_C)^3} - \frac{\theta_C^3}{(\theta_B - \theta_C)^3}\right] \tag{4.58}$$

The displacement diagram thus is described by the following piecewise function:

$$u(\theta) = \begin{cases} u_1 & \text{if } 0 \le \theta < 0.5\theta_A \\ u_2 & \text{if } 0.5\theta_A \le \theta < \theta_A \\ \mathrm{L} & \text{if } \theta_A \le \theta < \theta_B \\ u_4 & \text{if } \theta_B \le \theta < 0.5(\theta_B + \theta_C) \\ u_5 & \text{if } 0.5(\theta_B + \theta_C) \le \theta < \theta_C \\ 0 & \text{if } \theta_C \le \theta < 2\pi \end{cases} \tag{4.59}$$

The velocities and acceleration functions are obtained by differentiating $u(\theta)$. Thus,

$$\frac{du(\theta)}{d\theta} = \begin{cases} \dot{u}_1 & \text{if } 0 \le \theta < 0.5\theta_A \\ \dot{u}_2 & \text{if } 0.5\theta_A \le \theta < \theta_A \\ 0 & \text{if } \theta_A \le \theta < \theta_B \\ \dot{u}_4 & \text{if } \theta_B \le \theta < 0.5(\theta_B + \theta_C) \\ \dot{u}_5 & \text{if } 0.5(\theta_B + \theta_C) \le \theta < \theta_C \\ 0 & \text{if } \theta_C \le \theta < 2\pi \end{cases} \tag{4.60}$$

and, correspondingly,

$$\frac{d^2u(\theta)}{d\theta^2} = \begin{cases} \ddot{u}_1 & \text{if } 0 \le \theta < 0.5\theta_A \\ \ddot{u}_2 & \text{if } 0.5\theta_A \le \theta < \theta_A \\ 0 & \text{if } \theta_A \le \theta < \theta_B \\ \ddot{u}_4 & \text{if } \theta_B \le \theta < 0.5(\theta_B + \theta_C) \\ \ddot{u}_5 & \text{if } 0.5(\theta_B + \theta_C) \le \theta < \theta_C \\ 0 & \text{if } \theta_C \le \theta < 2\pi \end{cases} \tag{4.61}$$

The normalized displacements for a spline cam are the same as given by Equation 4.21. The normalized velocities and accelerations for a spline cam are, respectively, as follows:

On the rise part:

$$\dot{\bar{u}}_1 = \dot{\bar{u}}_2 = \frac{\dot{u}\theta_A}{2L} \tag{4.62}$$

$$\ddot{\bar{u}}_1 = \ddot{\bar{u}}_2 = \frac{\ddot{u}\theta_A^2}{4L} \tag{4.63}$$

On the return part:

$$\dot{\bar{u}}_4 = \dot{\bar{u}}_5 = \frac{\dot{u}(\theta_B - \theta_C)}{2L} \tag{4.64}$$

$$\ddot{\bar{u}}_4 = \ddot{\bar{u}}_5 = \frac{\ddot{u}(\theta_B - \theta_C)^2}{4L} \tag{4.65}$$

In Figure 4.26 the normalized displacement diagram of a four-spline cam is shown, in Figure 4.27 the normalized velocity diagram of a four-spline cam is shown, and in Figure 4.28 the normalized acceleration diagram of a four-spline cam is shown.

One can see that if the displacements are described by the cubic polynomials (Equation 4.27), the velocities are quadratic functions and accelerations are straight lines (Figure 4.28). The jump in acceleration at the two inflection points is seen in Figure 4.28.

4.4.5 Comparison of Two Cams: Cycloid vs. Four-Spline

Here the kinematic properties of displacement diagrams for two cams in normalized coordinates will be compared. The comparison is shown in Figures 4.29 through 4.31 for the displacement, velocity, and acceleration diagrams.

One can see that the normalized displacement curves in Figure 4.29 look sufficiently close. However, the displacements, for example, at $\theta = \pi/4$ are 0.0908451 and 0.0625 for the cycloid and spline cams, respectively; i.e., the difference is 31%. It is of interest to compare the velocities and accelerations for two cams at the inflection point $\theta = \theta_A/4$. The normalized velocities are 2 and 1.5, and the accelerations are 0 and 6, for the cycloid and four-spline cams, respectively.

There are two lessons one should learn from the comparison of cams: first, that the displacement diagram is not sufficient to assess the cam performance and, second, that relatively small errors in the displacement diagram might result in significant misjudgment of velocities and accelerations.

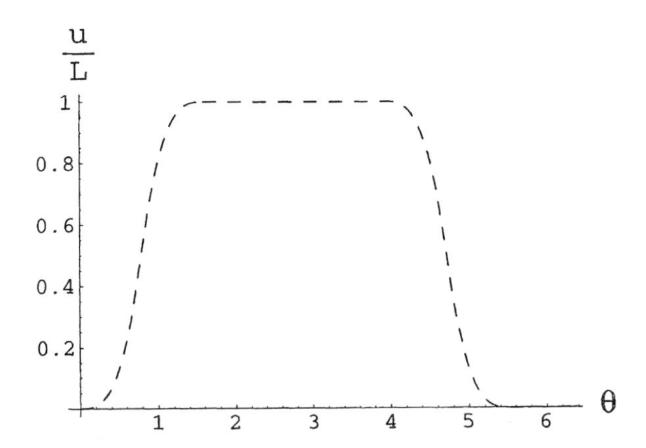

FIGURE 4.26 Normalized displacement diagram for a four-spline cam.

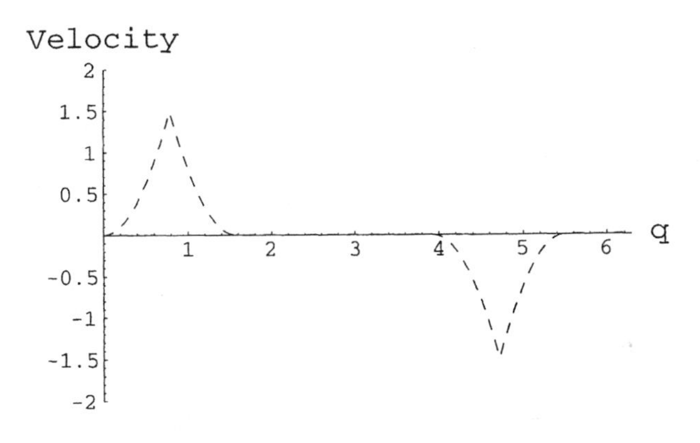

FIGURE 4.27 Normalized velocity diagram for a four-spline cam.

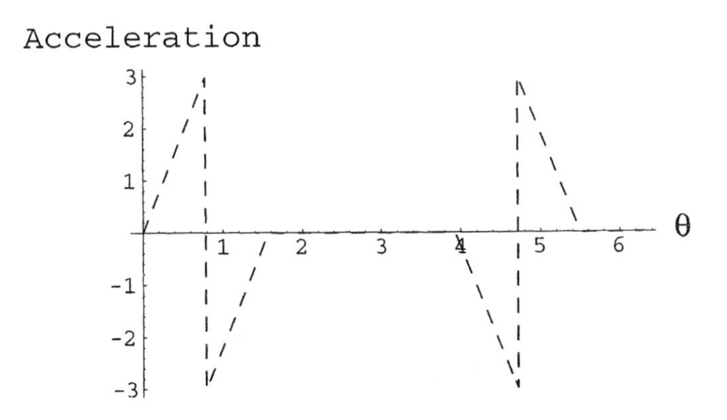

FIGURE 4.28 Normalized acceleration diagram for a four-spline cam.

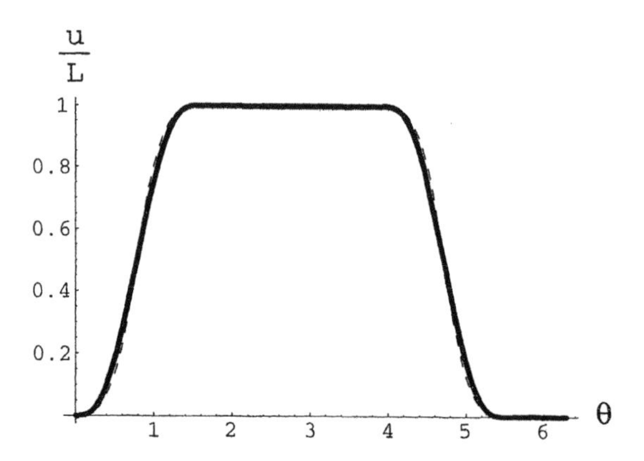

FIGURE 4.29 Comparison of normalized displacement diagrams for two cams: cycloid (solid line) and four-spline (dashed line).

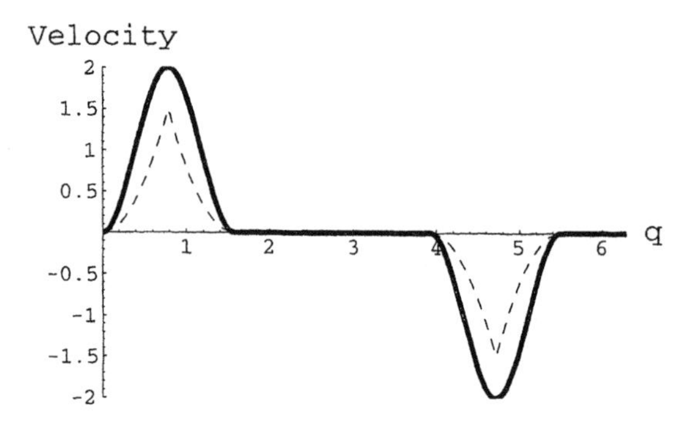

FIGURE 4.30 Comparison of normalized velocity diagrams for two cams: cycloid (solid line) and four-spline (dashed line).

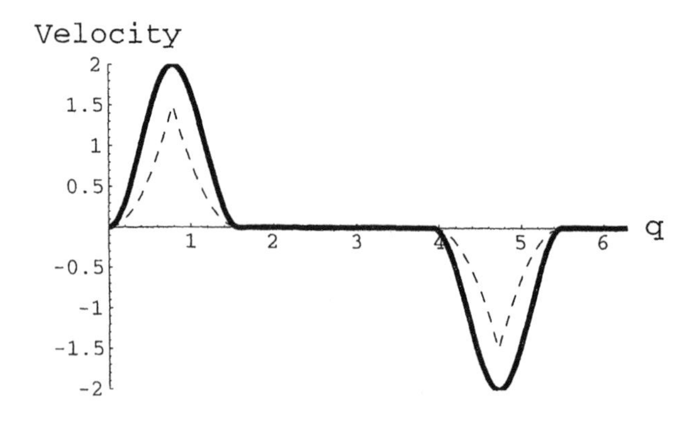

FIGURE 4.31 Comparison of normalized acceleration diagrams for two cams: cycloid (solid line) and four-spline (dashed line).

4.5 EFFECT OF BASE CIRCLE

The expression for the position of the follower:

$$D(\theta) = r_b + u(\theta) \quad (4.66)$$

shows that its velocity and acceleration do not depend on the base radius r_b. However, a corresponding point on the cam does depend on this radius. Indeed, a point on the cam profile is characterized by the vector:

$$\mathbf{r_D} = D(\theta)[\cos\theta, \sin\theta]^T \quad (4.67)$$

and the corresponding velocity vector is (see Chapter 2)

$$\dot{\mathbf{r}}_D = \dot{D}(\theta)[\cos\theta, \sin\theta]^T + D(\theta)[\cos(\theta + \pi/2), \sin(\theta + \pi/2)]^T \quad (4.68)$$

The first term in Equation 4.68 represents the translational component of the velocity vector, and it is equal to the follower velocity. The second term represents the angular (tangential with respect to the follower) component of the cam velocity vector. The follower does not have this component. It means that the tangential velocity is associated with the velocity of sliding of the follower along the cam profile. Although it does not affect the kinematics of motion transfer, it is important in the wear analysis of cams with knife or flat-faced followers since wear is proportional to the coefficient of friction, the normal contact force, and the velocity of sliding.

So, as far as the kinematics is concerned, the base circle is not important. It is important, however, from the geometric point of view, namely, to allow a proper interaction between the cam and the follower, especially a flat-faced one. Look at the profiles of two cams having identical harmonic displacement diagrams wrapped around different base circles: one with $r_b = 0.5L$ (Figure 4.32), and another with $r_b = 2L$ (Figure 4.33). One can see that in Figure 4.32 the curvature of the cam profile at $\theta = 0$ and $\theta = \theta_C$ (see Figure 4.12) is concave, and thus the flat-faced follower, and even the roller follower, may not be able to follow the profile around these points (the follower will ride over the "ditch"). The situation can be improved by increasing the base circle, as shown in Figure 4.33. However, the base circle increase, as is known, leads to the increase of the sliding velocity. In addition, the cam size must conform to dimensional constraints of the design. Thus, a compromise between the kinematic requirements and overall design considerations must be reached.

4.6 PRESSURE ANGLE

The angle between the normal to the cam profile and the axis of the follower is called the *pressure angle* (Figure 4.34). This angle affects the transverse force F_t, creating a bending moment on the follower (note that in Figure 4.34 forces shown are acting on the cam). Thus, it is desirable to keep this angle within an acceptable minimum. The point is that the pressure angle is a function of θ, and so this function should be investigated for its maximum. But, first, the function itself should be derived.

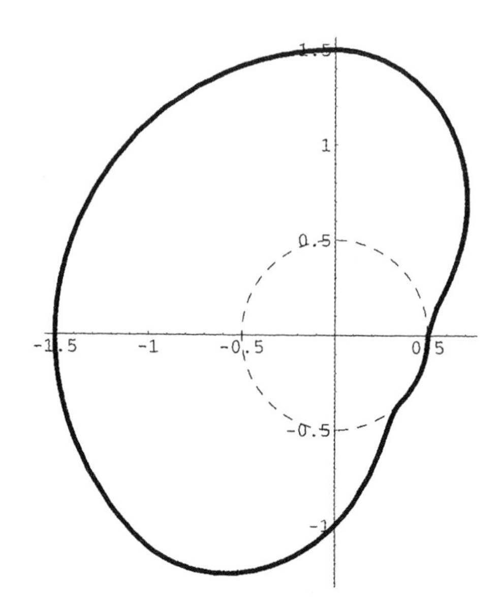

FIGURE 4.32 Harmonic cam profile with base radius $0.5L$.

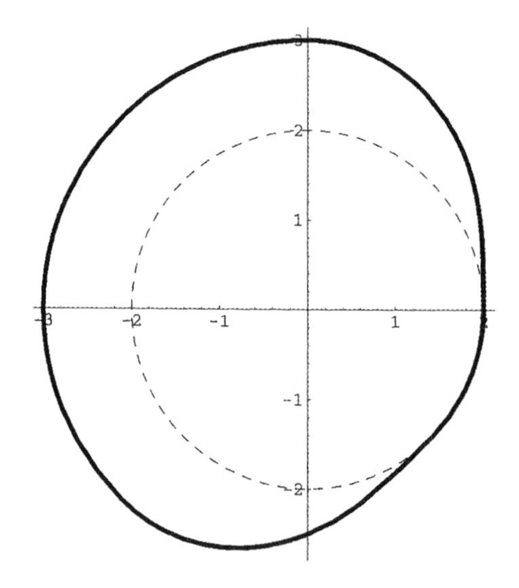

FIGURE 4.33 Harmonic cam profile with base radius $2L$.

The normal is a line perpendicular to the tangent to the cam profile. The tangent is defined by

$$\tan\alpha = \frac{dy(\theta)}{dx(\theta)} \tag{4.69}$$

where $x(\theta)$ and $y(\theta)$ are given by Equation 4.67. To write them explicitly,

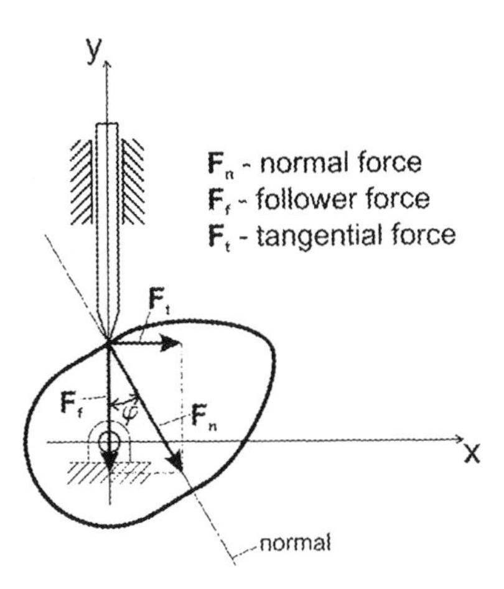

FIGURE 4.34 Pressure angle and forces acting on the cam.

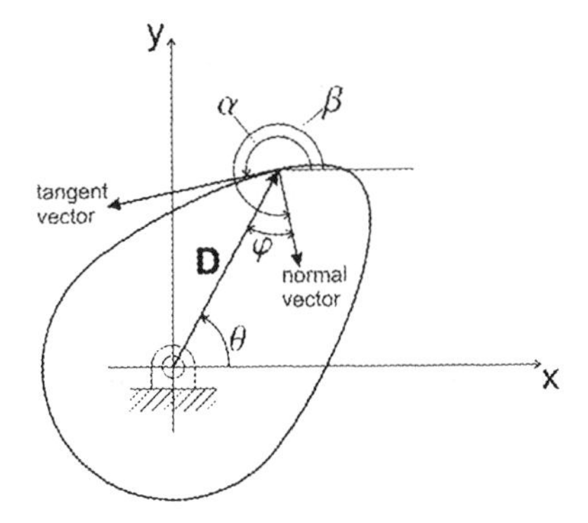

FIGURE 4.35 Tangential and normal to the cam profile.

$$x(\theta) = D(\theta)\cos\theta \tag{4.70}$$

and

$$y(\theta) = D(\theta)\sin\theta \tag{4.71}$$

By differentiating both $x(\theta)$ and $y(\theta)$ with respect to θ, an explicit expression for $\tan\alpha$ is obtained:

$$\tan\alpha = \frac{\dot{D}(\theta)\sin\theta + D(\theta)\cos\theta}{\dot{D}(\theta)\cos\theta - D(\theta)\sin\theta} \tag{4.72}$$

The angle α is measured from the positive direction of the x-axis to the positive direction of the tangent vector. The positive direction of the latter is counterclockwise (see Figure 4.35) in accordance with the right-hand coordinate system. The correct angle α is chosen based on the signs of sin α and cos α which are proportional to dy and dx, respectively; see Equation 4.69. Thus, the expressions for dy and dx are defined by Equation 4.72.

Defining the principal angle α as

$$\alpha^* = \arctan\left[\text{Abs}\left(\frac{\dot{D}(\theta)\sin\theta + D(\theta)\cos\theta}{\dot{D}(\theta)\cos\theta - D(\theta)\sin\theta}\right)\right] \tag{4.73}$$

the correct angle is subject to the signs of sin α and cos α,

$$\alpha = \begin{cases} \alpha^* & \text{if } dx > 0 \text{ and } dy > 0 \\ \pi - \alpha^* & \text{if } dx < 0 \text{ and } dy > 0 \\ \pi + \alpha^* & \text{if } dx < 0 \text{ and } dy < 0 \\ 2\pi - \alpha^* & \text{if } dx > 0 \text{ and } dy < 0 \end{cases} \tag{4.74}$$

The positive direction of the normal vector is then as indicated in Figure 4.35; i.e., the tangent is rotated counterclockwise by $\pi/2$. Thus,

$$\beta = \alpha + \pi/2 \tag{4.75}$$

The pressure angle φ, as is clear from Figure 4.35, is equal to

$$\varphi(\theta) = \beta(\theta) - (\theta + \pi) = \alpha(\theta) - \theta - \pi/2 \tag{4.76}$$

Thus, for every angle θ the angle α is found from Equation 4.72 subject to Equation 4.74, and then the pressure angle φ from Equation 4.76.

Consider now how the pressure angle changes over the cycle of rotation for the cycloid, harmonic, and four-spline cams. The corresponding plots are shown in Figures 4.36 through 4.38 for the case when $r_b/L = 10$, and $\theta_A = \pi/2$, $\theta_B = 5\pi/4$, and $\theta_C = 7\pi/4$ (see Figure 4.13). As can be seen, the pressure angle reaches maximum at the inflection point for all three cams. Also, the maximums of the pressure angles are sufficiently close for the three types of cams designed to the same specifications.

An important point is the effect of the base circle radius on the pressure angle. As is seen from Equations 4.66 and 4.73, the pressure angle depends nonlinearly on this radius. Thus, in each particular case the relationship between the two must be investigated to optimize the design. In Figure 4.39 a comparison of two pressure angle plots is shown for two cases: $r_b/L = 10$ and $r_b/L = 2$. One can see that the effect of the base radius is profound.

In summarizing the effect of the base circle radius on the cam mechanism design, it should be stated that

1. It does not affect the kinematics of motion transfer as long as the cam profile remains convex.

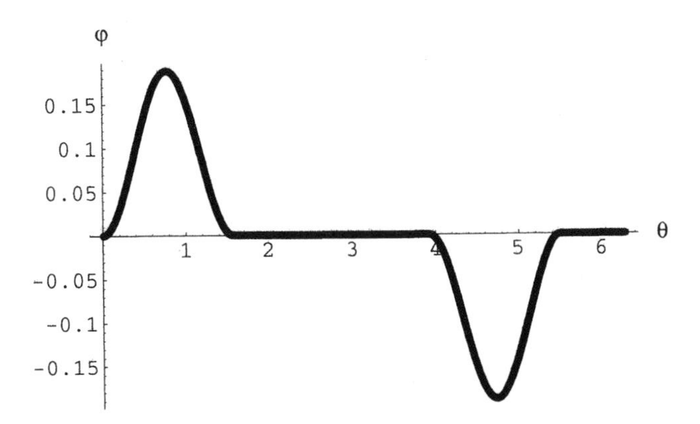

FIGURE 4.36 Variation of the pressure angle φ during the cycle of cycloid cam rotation.

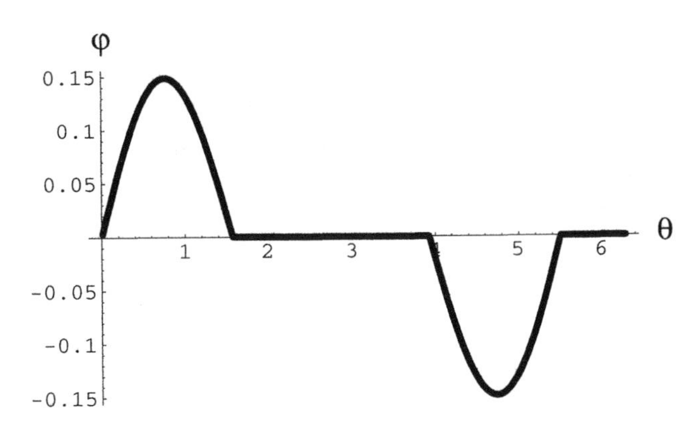

FIGURE 4.37 Variation of the pressure angle φ during the cycle of harmonic cam rotation.

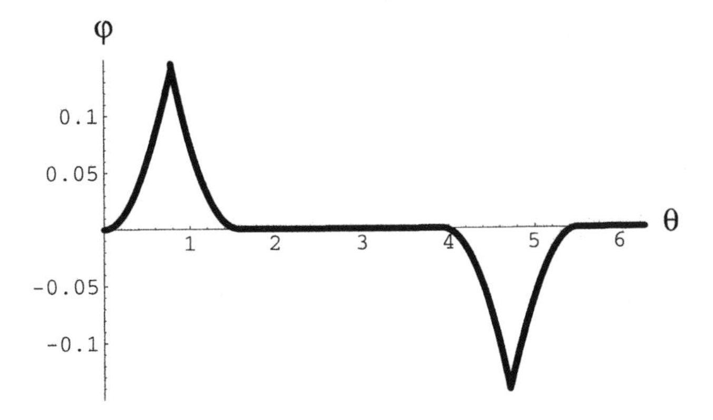

FIGURE 4.38 Variation of the pressure angle φ during the cycle of four-spline cam rotation.

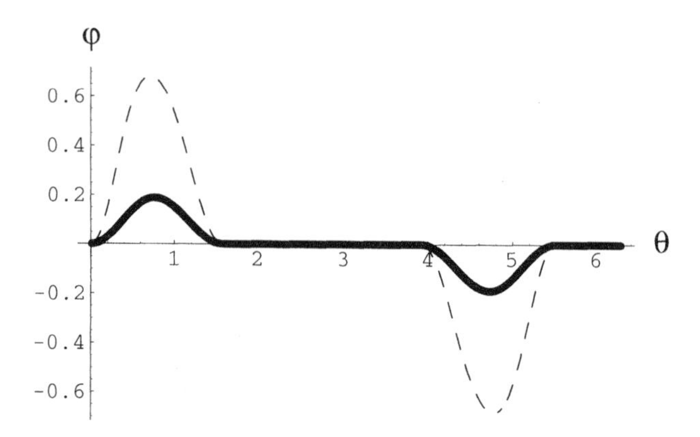

FIGURE 4.39 Variation of the pressure angle φ during the cycle of cycloid cam rotation for two normalized base radii: $r_b/L = 10$ (solid line) and $r_b/L = 2$ (dashed line).

2. Its increase leads to the reduction of the parasitic transverse force on the reciprocating follower, and has no effect on the oscillating one.
3. Its increase leads to the increase of sliding velocity between the cam and knife-edge or flat-faced followers, and to the increase of the angular rotation of the roller in roller-type followers.

As can be seen, the base radius must be optimized based on various design considerations, including size constraints.

PROBLEMS AND EXERCISES

Problems

1. For a given cam, will the choice of the type of follower (knife-edge, flat-faced, roller) affect the displacement diagram?
2. For a circular cam with parameters R and d (Figure 4.8), what are the minimum and maximum displacements of the follower?
3. If in Figure 4.8, instead of a knife-edge follower, a flat-faced follower is to be used, would the displacement diagrams for both designs be identical? And, if not, what would be the difference?
4. Solve Equation 4.1

$$d[\cos\gamma, \sin\gamma]^T + R[\cos\theta, \sin\theta]^T + D\left[\cos\left(\frac{3\pi}{2}\right), \sin\left(\frac{3\pi}{2}\right)\right]^T = 0$$

for the two unknowns: θ and D.

5. The displacement diagram for a cam comprises rise, dwell, and return parts of the cycle. If the transition from one part to another is not smooth (while the continuity of displacements is guaranteed, the continuity of accelerations is not), will it make this cam unacceptable from a functional point of view? Explain.

6. Take a function $x = r(\alpha - \sin\alpha)$ describing the cycloidal cam, and construct a displacement diagram with rise from 0 to $\pi/2$, dwell from $\pi/2$ to π, and return from π to 2π. Take that the lift $L = 1$ cm. Find the maximum accelerations.

7. Take a function $x = r(1 - \cos\alpha)$ describing the harmonic cam and construct a displacement diagram with rise from 0 to $\pi/2$, dwell from $\pi/2$ to π, and return from π to 2π. Assume lift $L = 1$ cm. Find the maximum accelerations and at the transitions from one part to another: $\pi/2$, π, and 2π.

8. The displacement diagram comprises rise, dwell, and return parts. Assuming that the rise and return parts are to be approximated by cubic spline functions, write the smoothness requirements for the transition from one cubic spline function to another. If the number of cubic spline functions to approximate the rise part of the diagram is $2N$, how many equations of smoothness can be formulated?

9. Use two cubic splines to describe the rising part of the displacement diagram (from 0 to $\pi/2$). Find all constants by satisfying smoothness requirements at 0 (transition from dwell) and $\pi/2$ (transition to dwell), and the equality of displacements at the inflection point $\pi/4$.

10. A circular cam in Figure 4.9 rotates with the angular velocity ω.
    a. What is the velocity of the follower for an arbitrary position of the cam?
    b. What is the velocity of a point on the cam interfacing the follower?

11. For the cams in Figures 4.1 and 4.2 show forces acting on cams from the follower, and also the pressure angles.

12. For a circular cam in Figure 4.9 find the expression for the pressure angle.

Exercises (Projects) with Mathematica

1. A circular cam has the d/R ratio 0.2 and rotates with 10 rad/s. Plot the displacement, velocity, and acceleration diagrams of the points on the cam and a knife-edge follower. Assume zero offset.

2. For a cycloidal cam described in Problem 4.6, plot the displacement, velocity, and acceleration diagrams of the points on the knife-edge follower.

3. For a harmonic cam described in Problem 4.7, plot the displacement, velocity, and acceleration diagrams of the points on the knife-edge follower.

4. Construct a displacement diagram, with rise from 0 to $\pi/2$, dwell from $\pi/2$ to π, and return from π to 2π, using two cubic splines to describe each, rise and return, parts of the diagram. Assume lift $L = 1$ cm. Satisfy the smoothness requirements

at the transition points (boundaries), and the equality of displacements at the inflection points. Plot the displacement, velocity, and acceleration diagrams of the points on the knife-edge follower.

5. For the data in Problem 4.6, investigate the effect of the base radius on the pressure angle. Choose two to three values of the base radius.

6. For the data in Problem 4.7, investigate the effect of the base radius on the pressure angle. Choose two to three values of the base radius.

5 Gears

5.1 INTRODUCTION

The kinematic function of gears is to transfer rotational motion from one shaft to another. Since these shafts may be parallel, perpendicular, or at any other angle with respect to each other, gears designed for any of these cases take different forms and have different names: *spur, helical, bevel, worm*, etc.

The fundamental requirement in most applications is that the coefficient of motion transformation (called the gear ratio) remains constant. What is needed to meet this requirement follows from Kennedy's theorem. The important point is that this requirement imposes a constraint on the suitable geometry of gear teeth profiles. Herein only one such profile is considered, called the *involute* profile.

5.2 KENNEDY'S THEOREM

The transformation of motion from one shaft to another involves three bodies: a frame (the position of each shaft is fixed in the frame) and two gears. Consider a general case when two disks with arbitrary profiles (Figure 5.1) represent gears 2 and 3. Also assume that disk 2 rotates with the constant angular velocity ω_2. The motion is transferred through the direct contact at point P (note that P_2 and P_3 are the same point P, but the first is associated with disk 2 whereas the second is associated with disk 3). The question is whether or not the angular velocity ω_3 of disk 3 will also be constant, and, if not, what is needed to make it constant. The answer is given by Kennedy's theorem.

Kennedy's theorem identifies the fundamental property of three rigid bodies in motion.

> *The three instantaneous centers shared by three rigid bodies in relative motion to one another all lie on the same straight line.*

First, recall that the *instantaneous center of velocity* is defined as the instantaneous location of a pair of coincident points of two different rigid bodies for which the absolute velocities of two points are equal. If one considers body 2 and the frame (represented by point O_2) in Figure 5.1, then the instantaneous center of these two bodies is point O_2, which belongs to the frame and to disk 2. The absolute velocities of both bodies at point O_2 are zero. The same is valid for disk 3 and the frame represented by point O_3. For the three bodies in motion there are three instantaneous centers: for all combinations of pairs. Thus, there is an instantaneous center between the two disks.

Look at Figure 5.1 again. Since point P is a common point for two disks, for each disk the velocity component at this point directed along the common normal is the same and equal to V_p. One can move this velocity vector along the common

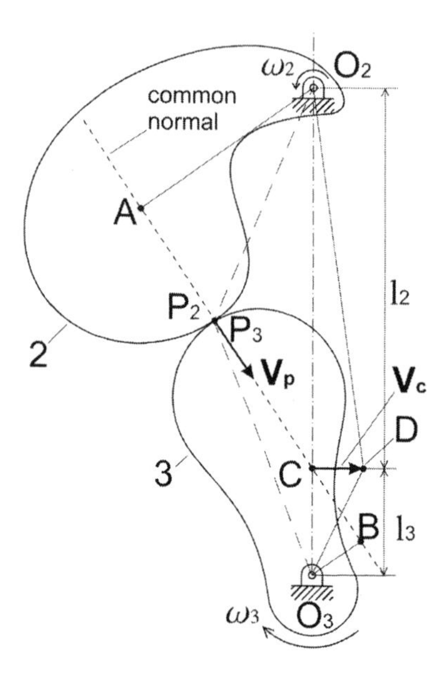

FIGURE 5.1 Illustration of Kennedy's theorem.

normal until it intersects the line connecting the two instantaneous centers O_2 and O_3 at point C. According to Kennedy's theorem, this point is the instantaneous center of velocity for the two disks. Indeed, the velocity V_p, the only instantaneous common velocity for the two disks, is equal to

$$V_p = \omega_2 AO_2 = \omega_3 BO_3 \tag{5.1}$$

From the similarity of the triangles AO_2C and BO_3C it follows that

$$\frac{AO_2}{BO_3} = \frac{CO_2}{CO_3} \tag{5.2}$$

By using the above relationship, the ratio of angular velocities in Equation 5.1 is equal to

$$\frac{\omega_2}{\omega_3} = \frac{l_3}{l_2} \tag{5.3}$$

where l_2 and l_3 denote O_2C and O_3C, respectively (see Figure 5.1).

Thus, velocity V_c in Figure 5.1 is a common velocity for the two disks since

$$V_c = \omega_2 l_2 = \omega_3 l_3 \tag{5.4}$$

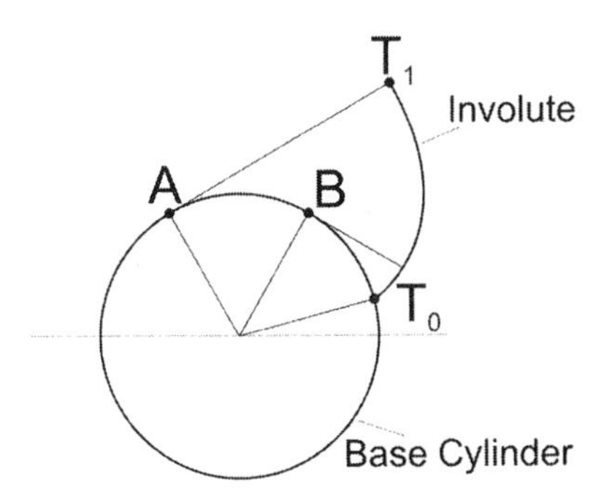

FIGURE 5.2 Illustration of the involute profile generation.

Velocity V_c is also the absolute velocity of each body at this point because there cannot be another velocity component along the line O_2O_3 (since the distance between the frame points O_2 and O_3 is not changing, and bodies 2 and 3 are assumed to be rigid).

Note that the relative velocity of two disks at point C is zero, whereas at point P it is not.

Equation 5.3 gives the transformation of angular velocities from disk 2 to disk 3. As can be seen, in order for the kinematic ratio ω_2/ω_3 to remain constant distances l_2 and l_3 must not change. It is clear from Figure 5.1, however, that for arbitrary profile shapes the common normal changes its direction during the motion, and thus point C moves along line O_2O_3. Thus, the problem of meeting the constant ratio requirement is to find such disk profiles that the kinematic ratio remains constant. It will be shown in the following that if the disk profile is described by the *involute function*, the common normal does not change its direction.

5.3 INVOLUTE PROFILE

An involute is generated by a tracing point on a cord as it is unwrapped from a circle (called a *base circle*) starting at T_0 and ending at T_1 (Figure 5.2). It is seen that points A and B are the instantaneous centers of rotation of the cord. It follows that the cord is normal to the involute at each point. This property is fundamental for the involute profiles to be used in gears. Indeed, consider two circular disks with centers at O_2 and O_3 (Figure 5.3). To each disk a plate with an involute profile is attached. The involutes on these plates are different since they are generated for different base circles, starting at points D_2 and D_3. The two plates have a common point C and a common normal AB. This common normal is tangential to both base circles at any moment during the rotation while the two involutes are engaged. Line AB intersects the line of centers O_2O_3 at point P. It then follows from Kennedy's theorem that this point is the instantaneous center for the two circular disks. And since this point remains the same while the disks rotate, it follows also that the kinematic ratio (called the *transmission ratio*) ω_2/ω_3 remains constant. Note that the common normal AB is called the *line of action* because the force is transmitted from one disk to another along this line.

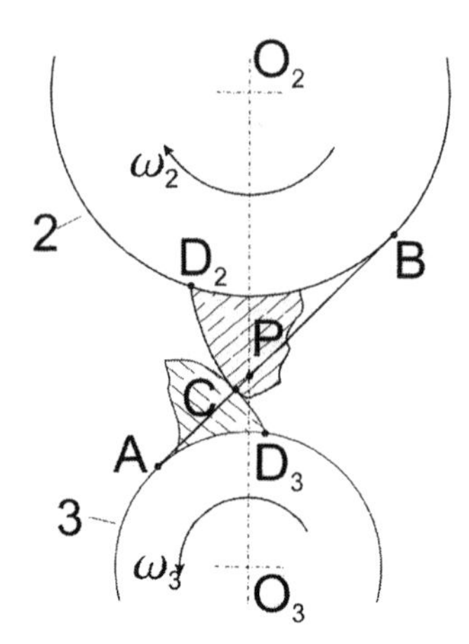

FIGURE 5.3 Illustration of the involute profiles interaction.

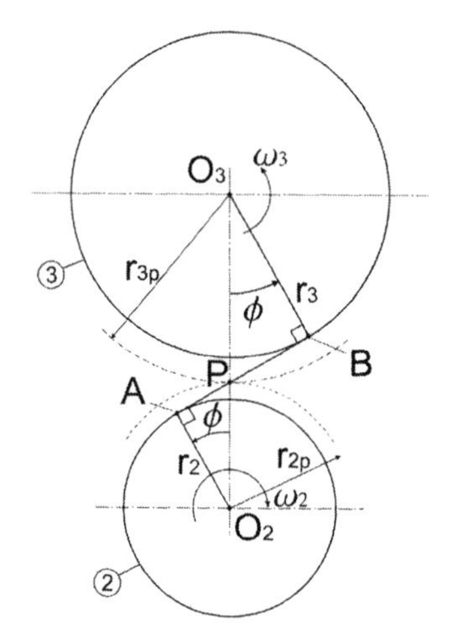

FIGURE 5.4 Illustration of the transmission ratio.

5.4 TRANSMISSION RATIO

The transmission ratio is given by Equation 5.3 (recall that l_2 and l_3 are the distances from the centers of rotation of each body to the instantaneous center for the two bodies). In the case of two gears, the instantaneous center for the two bodies is the intersection of the line of action with the line of centers O_2O_3 (Figure 5.4). This instantaneous center is called the *pitch point*. The corresponding distances from the gear centers to this point are called the *pitch radii* (the corresponding diameters are the *pitch diameters*). Thus, Equation 5.3 in the case of gears is

$$\frac{\omega_2}{\omega_3} = \frac{r_{3p}}{r_{2p}} = \frac{d_{3p}}{d_{2p}} \tag{5.5}$$

where r_{2p}, r_{3p}, and d_{2p}, d_{3p} are the pitch radii and pitch diameters, respectively.

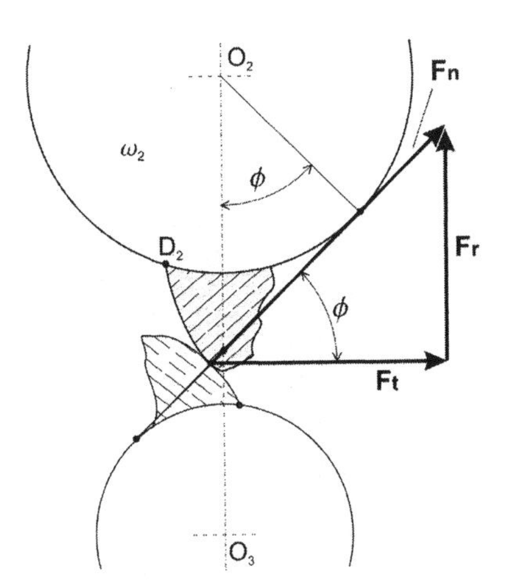

FIGURE 5.5 Illustration of forces at the involute interface.

The distance between any two similar points on the adjacent teeth along the pitch circle is called the *circular pitch*, and it is equal to

$$p = \frac{\pi d_p}{N} \tag{5.6}$$

where N is the number of teeth. Substituting pitch diameters into Equation 5.5 using Equation 5.6, and taking into account that the circular pitch is equal for the two meshing gears, one finds that the transmission ratio can be expressed through the ratio of teeth numbers:

$$\frac{\omega_2}{\omega_3} = \frac{N_3}{N_2} \tag{5.7}$$

It is seen from Equation 5.5 that the transmission ratio of two gears is equivalent to the transmission ratio of two cylindrical disks having a contact at one point, P, and rotating without sliding. Note that the smallest of two intermeshing gears is called a *pinion*.

5.5 PRESSURE ANGLE

The force generated between the two gears acts along the normal to the involute profile, i.e., along the line of action AB in Figure 5.4. This force is called the *normal force,* F_n, and it can be resolved in two components (Figure 5.5): along the line of centers, O_2O_3, and perpendicular to it. The first component is called the *radial force,* F_r, and the second is called the *tangential force,* F_t. The two components of the normal force are

$$F_t = F_n \cos\phi \tag{5.8}$$

and

$$F_r = F_n \sin\phi \tag{5.9}$$

In Equations 5.8 and 5.9 the angle ϕ is called the *pressure angle,* since it characterizes the direction of the normal force between the conjugate teeth at the pitch point P. As follows from the similar triangles, O_3BP and O_2AP, in Figure 5.4, this angle is the same for both gears since

$$\frac{r_3}{r_{3p}} = \frac{r_2}{r_{2p}} = \cos\phi \tag{5.10}$$

The radial forces at the gear interface are parasitic forces since they are not associated with energy transfer, whereas the tangential forces create the transmission torques, T_{2p} and T_{3p}

$$T_{2p} = F_t r_{2p} \tag{5.11}$$

and

$$T_{3p} = F_t r_{3p} \tag{5.12}$$

Since the normal force of interaction is the same for both gears and the angle ϕ is also the same, the radial and tangential forces acting on two meshing gears are the same. The torques, however, as can be seen from Equations 5.11 and 5.12, are not equal. In fact, their ratio equals the transmission ratio:

$$\frac{T_{3p}}{T_{2p}} = \frac{r_{3p}}{r_{2p}} = \frac{\omega_2}{\omega_3} \tag{5.13}$$

The latter equation shows that the energy is conserved during the motion transmission through the gears, since

$$\text{Energy} = T_{3p}\omega_3 = T_{2p}\omega_2 \tag{5.14}$$

In reality the energy is not conserved due to friction losses at the involute interface. It will be seen later that, except at the pitch point, the relative motion of two involutes involves rolling and sliding. The latter results in friction losses.

5.6 INVOLUTOMETRY

A gear can be seen as a disk of radius r_b with teeth attached to it (Figure 5.6). The involute curve for each tooth starts at the base circle with radius r_b and ends on the *addendum circle* with radius r_a. It is clear that the number of teeth must be an integer.

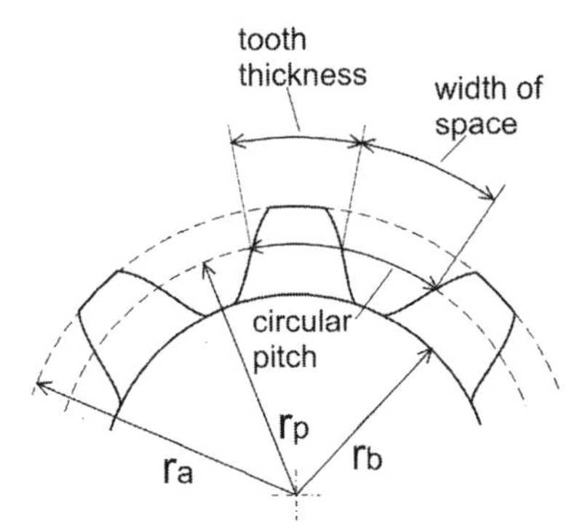

FIGURE 5.6 Teeth geometry.

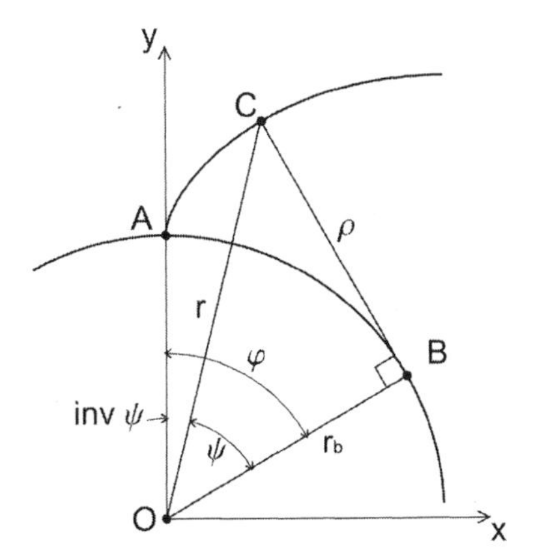

FIGURE 5.7 Involute description.

If this number is chosen, then it defines the *circular pitch*, i.e., the distance between the two similar points on two successive teeth along the pitch circle (see Figure 5.6 and Equation 5.6). Some of this circular distance occupied by the tooth is called the *tooth thickness*, while the rest of it is called the *width of space*. Ideally, the width of space should be equal to the tooth thickness of the engaging gear. Since the circular pitch for the engaging gear must be the same, then the width of space of one gear must be equal to the tooth thickness of another. In other words, the tooth thickness equals half of the circular pitch.

An involute profile can be described analytically. Consider an involute that starts from point A on a base circle (Figure 5.7). Any point C on the involute is described by the radius r and an angle measured from the vertical axis, which is conventionally denoted by inv ψ, and it is equal to $\varphi - \psi$. Since the distance ρ is equal to the arc AB, the angle φ can be expressed through this distance, namely, $\rho = r_b\varphi$. At the same time, from the rectangular triangle OBC it is seen that $\rho = r_b \tan\psi$. Thus, $r_b \tan\psi = r_b\varphi$, or $\tan\psi = \varphi$. As a result, the involute angle inv ψ is equal to

$$\text{inv } \psi = \tan\psi - \psi \tag{5.15}$$

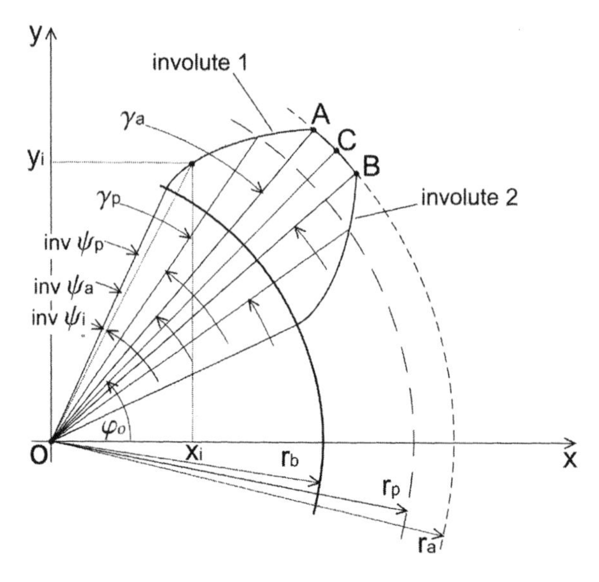

FIGURE 5.8 Description of tooth geometry.

The radius r can also be expressed through the angle ψ from the triangle OBC.

$$r = \frac{r_b}{\cos\psi} \tag{5.16}$$

Having found r and ψ, one can now describe the tooth profile by its x- and y-coordinates. In Figure 5.8 a tooth is shown in the global coordinate system (x,y). Involute 2 is a mirror image of involute 1 with respect to the symmetry line OC. The x- and y-coordinates of the point on involute 1 are given by

$$x_i^1 = r_i\cos(\varphi_0 + \Delta\varphi_i) \tag{5.17}$$

and

$$y_i^1 = r_i\sin(\varphi_0 + \Delta\varphi_i) \tag{5.18}$$

where φ_0 is the angular position of the line of symmetry OC, and the angle $\Delta\varphi_i$ is equal to

$$\Delta\varphi_i = \frac{\gamma_a}{2} + \text{inv}\,\psi_a - \text{inv}\,\psi_i \tag{5.19}$$

where γ_α is the angle corresponding to arc AB on the addendum circle, and inv ψ_α is the involute angle associated with point A on the addendum circle.

Angle ψ_α is found from Equation 5.16 for $r = r_a$.

$$\cos\psi_a = \frac{r_b}{r_a} \tag{5.20}$$

It is seen from Figure 5.8 that the involute angle γ_a is equal to

$$\frac{\gamma_a}{2} = \frac{\gamma_p}{2} + \text{inv}\,\psi_p - \text{inv}\,\psi_a \tag{5.21}$$

where γ_p is the circular pitch angle, and the involute angle ψ_p is given by Equation 5.16 for $r = r_p$.

For involute 2 in Figure 5.8 the x- and y-coordinates are

$$x_i^2 = r_i\cos(\varphi_0 - \Delta\varphi_i) \tag{5.22}$$

$$y_i^2 = r_i\sin(\varphi_0 - \Delta\varphi_i) \tag{5.23}$$

To construct a gear with all teeth, an arbitrary angle φ_0 is chosen, then the above formulas are used in which φ_0 is substituted by $\varphi_0 + j\gamma_p$ ($j = 1, N$), where N is the number of teeth.

5.7 GEAR STANDARDIZATION

The kinematic properties of motion transformation by gears are uniquely determined by the relationships in Equation 5.5, i.e., by the pitch diameters of two meshing gears. However, the ratio given by Equation 5.5 does not uniquely define the gear teeth. Indeed, if the pitch radii expressed through the base radii (see Equation 5.6) are substituted in Equation 5.5, then the transmission ratio becomes

$$\frac{\omega_2}{\omega_3} = \frac{r_3\cos\phi}{r_2\cos\phi} \tag{5.24}$$

What follows from Equation 5.24 is that the same transmission ratio can be achieved for various pressure angles given the pitch radii. The moment the pressure angle is chosen, however, it uniquely defines the base circles. Thus, for the given transmission ratio and center distance O_2O_3, the pitch radii are uniquely defined, whereas the base radii remain uncertain. To determine the latter, some other design requirements should be formulated. These are concerned with *gear interchangeability*, with the *tooth strength*, and with *teeth interference* during rotation (*undercutting*), which are maintainability, design, and assembly considerations.

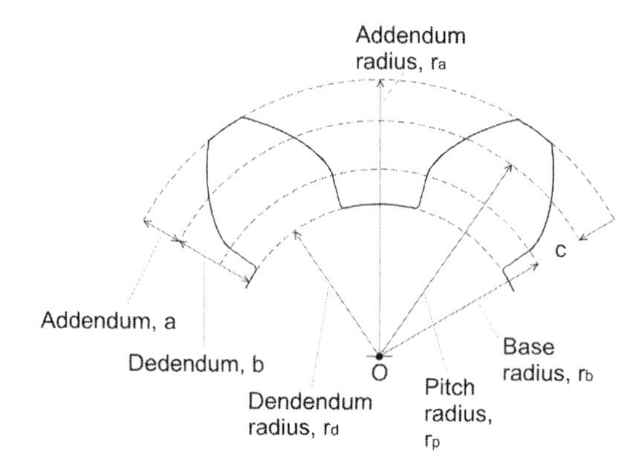

FIGURE 5.9 Teeth terminology.

Recall that the circular pitch is equal to

$$p = \frac{\pi D}{N} \tag{5.25}$$

The two gears are kinematically compatible only if they have the same circular pitch. Thus, to meet the requirement of interchangeability of gears produced by different manufacturers they should agree on a specific set of numbers for the circular pitches. In other words, the circular pitch should be standardized. In practice, instead of circular pitch, the ratio D/N is standardized. This ratio is called the *module* in the metric system.

$$m = \frac{D}{N} \tag{5.26}$$

Since m defines the tooth thickness, it can be used as a normalizing parameter for the tooth height. In other words, the tooth height is made proportional to m. In various countries these coefficients of proportionality, however, might be different.

In the United States the inverse of m is used for standardization purposes, and it is called the *diametral pitch.*

$$P_d = \frac{N}{D} \tag{5.27}$$

The basic elements of gear geometry are shown in Figure 5.9, where the *addendum* and *dedendum* are the radial distances from the pitch circle to the addendum circle and dedendum circles, respectively. These parameters are standardized in terms of the module or the diametrical pitch. For example, the British metric standard gives the following relationships:

Addendum (a)	$1.000m$
Dedendum (b)	$1.250m$
Pressure angle (ϕ)	$20°$

TABLE 5.1
Standard Tooth Systems for Spur Gears

Tooth System	Pressure Angle (°)	Addendum, a	Dedendum, b
Full depth	20	$1/P$ or m	$1.25/P$ or $1.25m$
			$1.35/P$ or $1.35m$
	22.5	$1/P$ or m	$1.25/P$ or $1.25m$
			$1.35/P$ or $1.35m$
	25	$1/P$ or m	$1.25/P$ or $1.25m$
			$1.35/P$ or $1.35m$
Stub	20	$0.8/P$ or $0.8m$	$1/P$ or m

The preferred modules are $m = 1, 1.25, 1.5, 2, 2.5, 3, 4, 5, 6, 8, 10, 12, 16, 20, 25, 32, 40, 50$.

For the coarse pitch the U.S. system has the same relationships. Some standard tooth systems are given in Table 5.1.

As is already known, the two parameters, pressure angle and circular pitch, are fundamentally independent. Thus, since gears having different pressure angles are also incompatible, the pressure angle must also be standardized. For example, the addendum and dedendum values are given together with the pressure angle in the British, or in any other system. However, the same addendum and dedendum can be used with another pressure angle. In addition to 20°, 14° and 25° angles are also used.

It should be noted that this independence has some practical limits, which will be discussed later. At the moment it is worth mentioning that the base circle, which is a function of the pressure angle for a given pitch diameter, also affects the tooth dimension, since the tooth height cannot be larger then the distance between the two base circles (see Figure 5.4).

If the pitch radius of one of the gears becomes infinitely large, then its pitch circle, as well as the base and addendum circles, are transformed into lines (imagine unfolding an infinite cord with respect to an infinite base circle). Such a gear is called an *involute rack*. A rack and pinion pair is shown in Figure 5.10.

FIGURE 5.10 Rack and pinion pair.

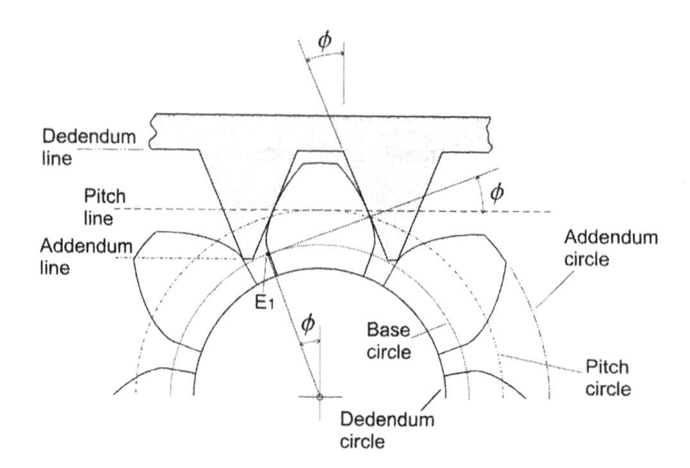

FIGURE 5.11 Illustration of pinion and rack interaction.

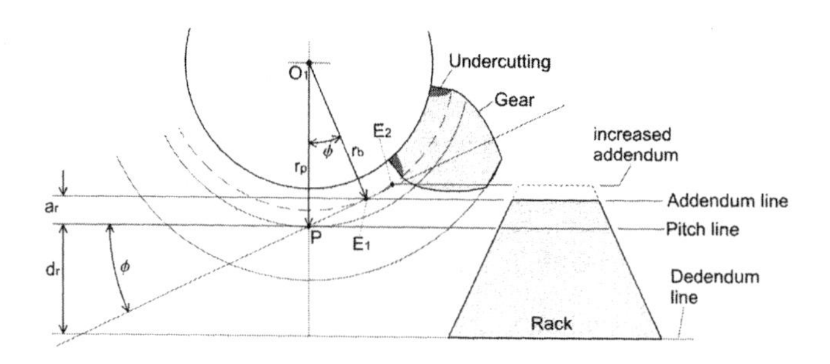

FIGURE 5.12 Illustration of undercutting (the gray lines show the case of interaction with the increased rack addendum).

The involute profiles of the teeth in the rack become straight lines inclined with respect to the vertical in Figure 5.11 by the angle ϕ. The line of action in the pinion–rack pair is a line tangential to the base circle of the pinion at the point E_1 in Figure 5.11.

Recall that the line of action traces points of interaction between the two gears. Thus, if the addendum line of the rack passes through point E_1, the two profiles will interact only along the involutes. If, however, the addendum of the rack, a_r, is increased (as shown by broken lines in Figure 5.12), then its addendum line will cross the line of action at point E_2. This means that some of the interaction between the gears will take place outside of the pinion involute, namely, with a radial tooth profile inside the base disk. The interaction of an involute profile with the noninvolute one violates the constant transmission ratio condition and leads to high contact stresses and, eventually, to wear of the teeth of the pinion inside the base circle. The latter phenomenon is called *undercutting* and is shown in Figure 5.12 by the darkened areas on the pinion tooth profile. The undercutting weakens the tooth against bending forces and may result in tooth failure.

To avoid undercutting, the geometry of the two mating teeth should be compatible; i.e., the addendum line of the rack should not intersect the base circle of the pinion above point E_1. One can use this as a condition for the prevention of undercutting. From Figure 5.11 it is seen that

$$r_b \cos\phi + a_r = r_p \tag{5.28}$$

Taking into account that $r_b = r_p \cos\phi$, Equation 5.28 is reduced to

$$a_r = r_p \sin^2\phi \tag{5.29}$$

The pitch radius can be expressed through module m using Equation 5.26, that is, $r_p = mN/2$. The addendum can also be expressed through m in the form $a_r = km$, where k is some constant. Then Equation 5.29 can be solved for the minimum number of teeth for the addendum of the given rack.

$$N_{\min} = \frac{2k}{\sin^2\phi} \tag{5.30}$$

Indeed, if the number of pinion teeth is larger then $N_{\min}$, then the pinion pitch and base radii will be larger. The point E_1 will now be below the addendum line and thus there will be no undercutting. On the other hand, if the number of teeth is smaller than the minimum, then undercutting will take place.

The constant k in Equation 5.30 is standardized. If $k = 1$, then the tooth is said to have full depth; if $k = 0.8$, the depth is called shortened. In the case of $\phi = 20°$, the minimum number of teeth is 18 for $k = 1$, and 14 for $k = 0.8$.

Another important point is that the minimum number of teeth determined for the pinion–rack transmission is the smallest number. Indeed, if instead of the rack a gear is used with the same addendum, then the point on the line of action where the engagement starts can only be below E_1 in Figure 5.12.

Thus, the standardization of gears allows their interchangeability and guarantees the uniformity of kinematic performance. In this respect, it is important to understand that the kinematic performance of a pair of gears does not depend on small variations of the center distance between the shaft axes. Such a situation is shown in Figure 5.13, where in Figure 5.13a and b the same pair is mounted on shafts with different center distances. It follows from the similarity of the triangles that for both situations the kinematic ratio remains the same, i.e.,

$$\frac{O_3P}{O_2P} = \frac{\omega_2}{\omega_3} = \frac{r_{3b}}{r_{2b}} \tag{5.31}$$

However, aside from the kinematic performance, the change in the center distance may affect the reliability of the gear set. Shortening of this distance may result in jamming of the teeth, while increasing it leads to a clearance between the teeth, which is called *backlash*. The latter may cause knocks, vibrations, and noise.

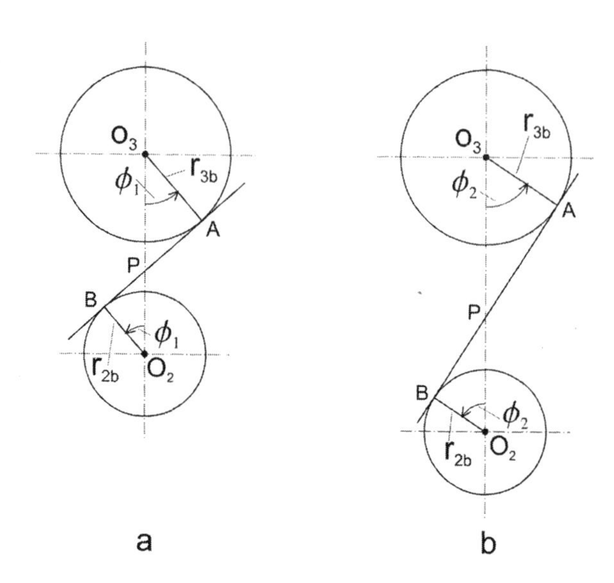

FIGURE 5.13 The effect of center distance on kinematic ratio.

5.8 TYPES OF INVOLUTE GEARS

The concept of an involute profile is used to design various types of gears having different functional and performance properties. Here, *spur, helical, bevel,* and *worm* gears will be briefly discussed.

5.8.1 Spur Gears

This is the most common and most fundamental type of gear. A pair of gears is shown in Figure 5.14. The involute surface of the gear tooth is a cylindrical surface. To visualize it, imagine that one is unwrapping a piece of paper from the base cylinder (Figure 5.15). Thus, in each cross section perpendicular to the cylinder axis the involute profile is identical. The resulting teeth and the corresponding terminology are shown in Figure 5.9. The geometry of spur gears is standardized (see Table 5.1).

FIGURE 5.14 Spur gears.

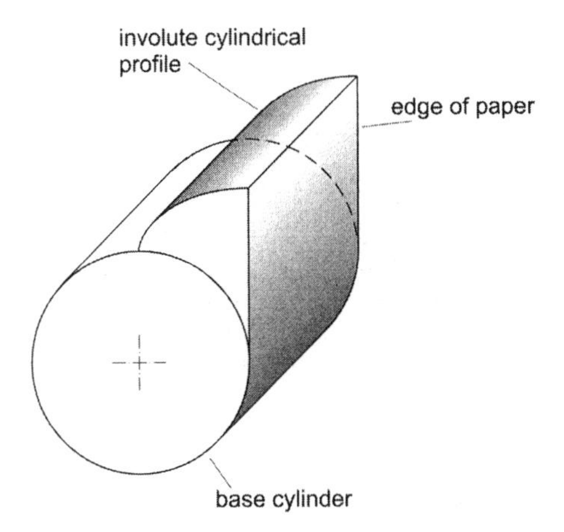

FIGURE 5.15 Illustration of spur gear involute surface generation.

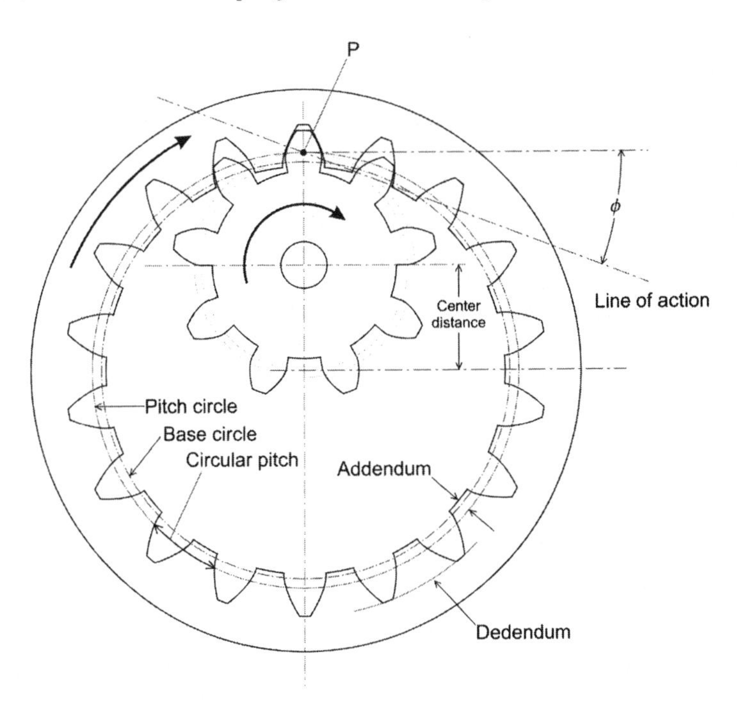

FIGURE 5.16 Annular gear and pinion meshing.

The gears discussed so far are called *external gears*, since their centers are located on both sides of the pitch point. It is possible to design a gear set where a pinion is located inside an *internal* (*annular*) gear. Such a set is shown in Figure 5.16, where P is the pitch point and ϕ is the pressure angle. The kinematic ratio for these gears is given by the same expression as for the external gears, Equations 5.5 and 5.7. However, the sense of rotation is different. Such gears have several advantages over the external gears: they preserve the sense of rotation, they are more compact, and

FIGURE 5.17 Helical gears.

their teeth are subjected to lower contact stresses because the curvature of the involute surfaces is the same. These gears find applications in planetary gear systems (see Section 5.10), and whenever compactness is a critical requirement.

5.8.2 Helical Gears

A pair of helical gears is shown in Figure 5.17. The helical gear is an extension of a spur gear to a more complex involute surface geometry. To understand this extension, one can do a simple physical or imaginary experiment. Take a piece of paper and wrap it around the circular cylinder similar to the way that was done for a spur gear. However, in this case, cut the edge of the paper so that instead of it being parallel to the cylinder axis it will now be inclined to this axis, thus becoming part of a helix line on the cylindrical surface. When the paper is unwrapped, each point on the helix line (on the edge of the paper) will have a trajectory of an involute. Since the generating lines for each point on a helix line are different, they will result in different involutes. The involutes all lie on a surface, which is called an *involute helicoid* (Figure 5.18).

Since the involute helicoid is generated from the base cylinder, the two meshing helical gears will have the transmission ratio defined by the radii of these cylinders, and, correspondingly, by the radii of the pitch cylindrical surfaces, exactly in the same way as in the case of spur gears (see Equations 5.5 and 5.7). Thus, kinematically, from the point of view of motion transfer, helical gears are not different from spur gears. They have, however, other properties that make them attractive. The basic one is the increase in the so-called *contact ratio*. In Figure 5.19a contact lines during the motion of the spur gear are compared with the contact lines in Figure 5.19b of the helical gear. If, in the extreme case, the spur gear has only one tooth in contact at a time, then a comparable helical gear will have, not only more than one tooth in contact, but also a longer contact line. Benefits of this are smaller contact stresses and smoother motion transfer. Also, a helical gear, due to its geometry, is stronger in withstanding bending forces. This leads to one of the drawbacks of helical gears, the generation of parasitic axial forces.

The force generated at the interface of the helical involute surfaces acts along the normal to this surface (assuming that friction forces are ignored). This normal force,

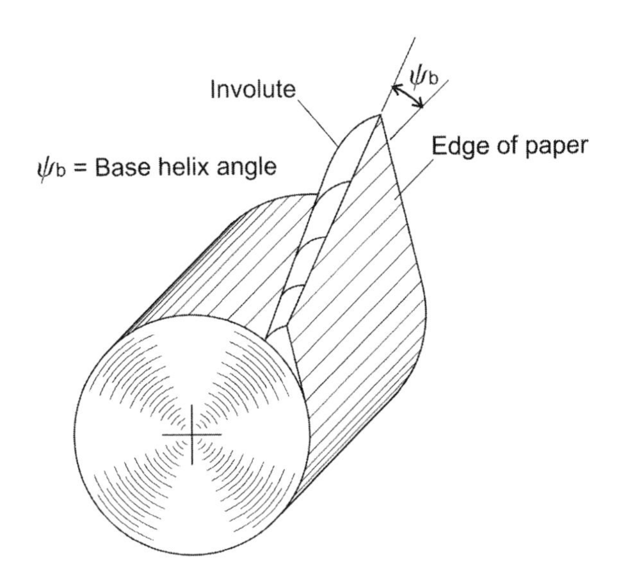

FIGURE 5.18 Illustration of generation of a helical involute surface.

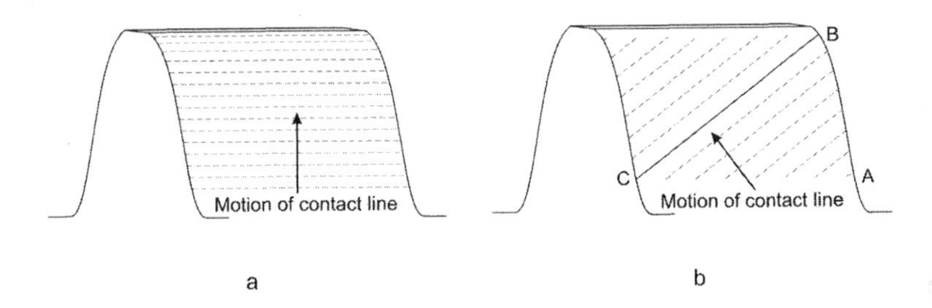

FIGURE 5.19 Comparison of contact lines for spur (a) and helical (b) gears.

$\mathbf{F}_n$, can be resolved into three components, $\mathbf{F}_r$, $\mathbf{F}_a$, and $\mathbf{F}_t$, acting in radial, axial, and tangential (with respect to the surface of the base cylinder) directions, respectively. These forces are shown in Figure 5.20a, where $\mathbf{F}_n$ and $\mathbf{F}_r$ are lying in the plane normal to the helix line (plane *ABC*). The helix line is shown in Figure 5.20b, where it forms an angle ψ with the gear axis. The force $\mathbf{F}_n$ can be resolved in the normal plane *ABC* into two components: $\mathbf{F}_r$ and $\mathbf{F}_{nt}$, where the latter is a component of the normal force lying in the tangential plane *ACE*. The angle ϕ_n between the forces $\mathbf{F}_n$ and $\mathbf{F}_{nt}$ is the pressure angle defined in the same way as for spur gears. Thus, from the resolution of the forces in the normal plane one has the following relationships between the magnitudes of these forces:

$$F_r = F_n \sin \phi_n \tag{5.32}$$

and

$$F_{nt} = F_n \cos \phi_n \tag{5.33}$$

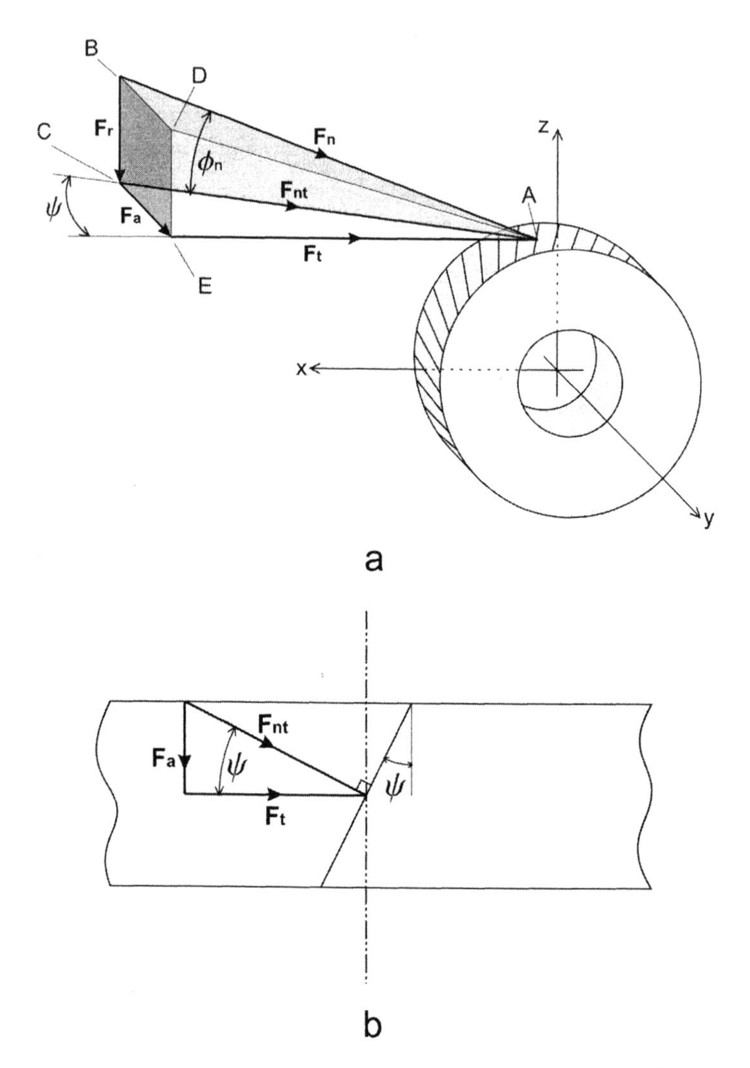

FIGURE 5.20 (a) Forces acting on a helical tooth. (b) Forces acting on a helical tooth in tangential plane.

The force $\mathbf{F_{nt}}$, in turn, can be resolved (see Figure 5.20b) into an axial force $\mathbf{F_a}$ and a tangential force $\mathbf{F_t}$ acting in the plane of rotation *ADE*. Thus, the components of the force $\mathbf{F_{nt}}$ are

$$F_t = F_{nt}\cos\psi \tag{5.34}$$

and

$$F_a = F_{nt}\sin\psi \tag{5.35}$$

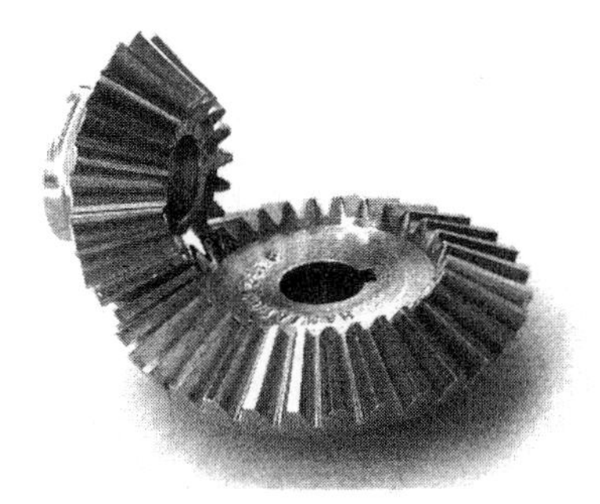

FIGURE 5.21 Straight-tooth bevel gears.

The tangential force is the only useful force, since it is the force transmitted by the gear. This force is known from the transmitted torque:

$$F_t = \frac{T}{r_p} \tag{5.36}$$

where r_p is the pitch radius.

Thus, for a given tooth geometry, ϕ_n and ψ, and given torque, T, the three components of the normal force can be found from the above equations.

5.8.3 Bevel Gears

The spur gears transmit motion only between the parallel shafts, so that the gear planes of rotation are parallel. The helical gears can be used to transmit motion between the shafts that can be either parallel or crossed, and thus the gear planes of rotation can vary from being parallel to perpendicular, but their axes do not intersect. The bevel gears transmit motion between the shafts whose axes intersect while their planes of rotation may vary. In Figure 5.21 a pair of *straight-tooth bevel gears* with perpendicular shafts is shown.

Instead of a cylindrical surface, as in the case of spur and helical gears, serving as a base for an involute surface, in this case the base surface is conical. The corresponding pitch surface is also conical and is called the *pitch cone*. An involute on a cone can be obtained by unfolding a piece of paper from the conical surface while the edge of the paper follows the cone generating line and one point of this edge is attached to the cone apex (Figure 5.22). Each involute T_0T generated in this way is called a *spherical involute* because it lies on a spherical surface. It is important to understand that involutes are obtained in planes perpendicular to the cone surface. Thus, if one takes an infinitesimally thin slice of the involute surface, one can view this slice as if it had been obtained by unfolding a cord from the circle whose plane is perpendicular to the cone surface. Such a slice is shown in Figure 5.23 for a bevel gear. Thus, a very thin bevel gear will have practically the same properties as a spur gear with the equivalent base and pitch circles.

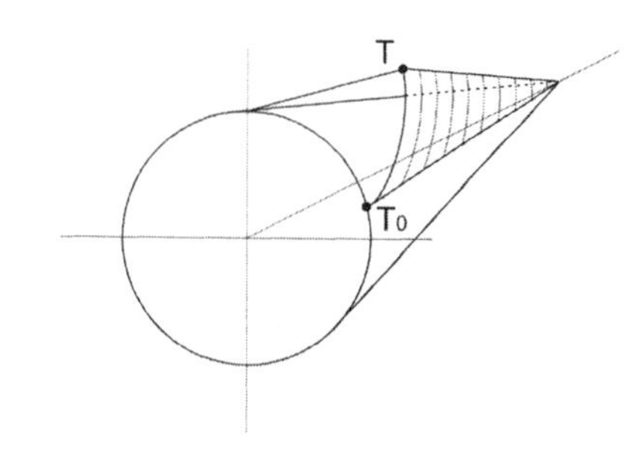

FIGURE 5.22 Generation of the conical involute surface.

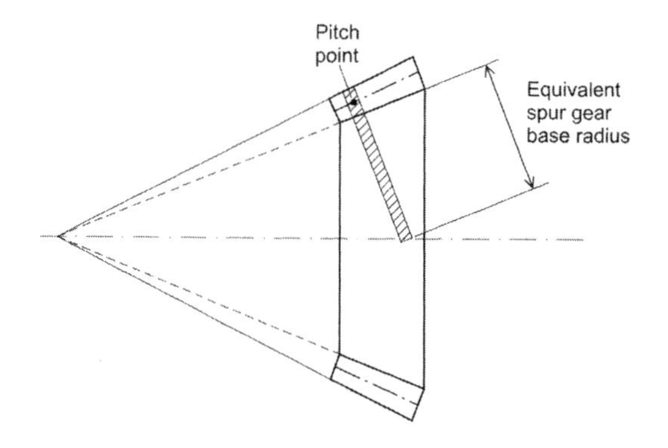

FIGURE 5.23 Illustration of an equivalent spur gear.

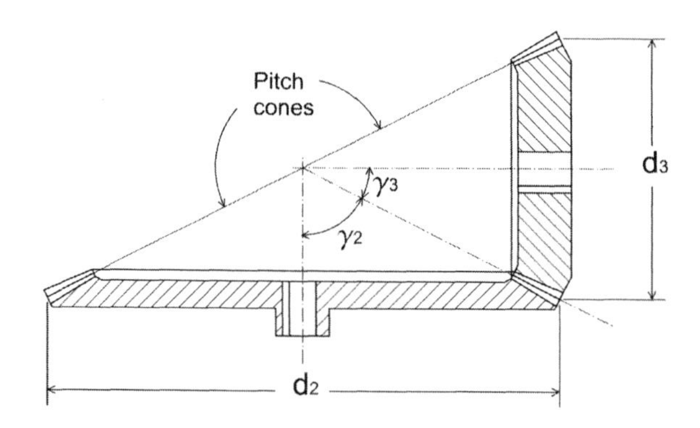

FIGURE 5.24 Bevel gear and pinion.

An example of a bevel pair is shown in Figure 5.24, where a pitch cone is formed by the pitch lines, and γ_2 and γ_3 are the pitch angles characterizing pitch cones.

From the equivalency of bevel and spur gears it follows that the transmission ratio for a pair of bevel gears is given by the same equations as Equations 5.5 and 5.7:

$$\frac{\omega_2}{\omega_3} = \frac{N_3}{N_2} = \frac{d_3}{d_2} \tag{5.37}$$

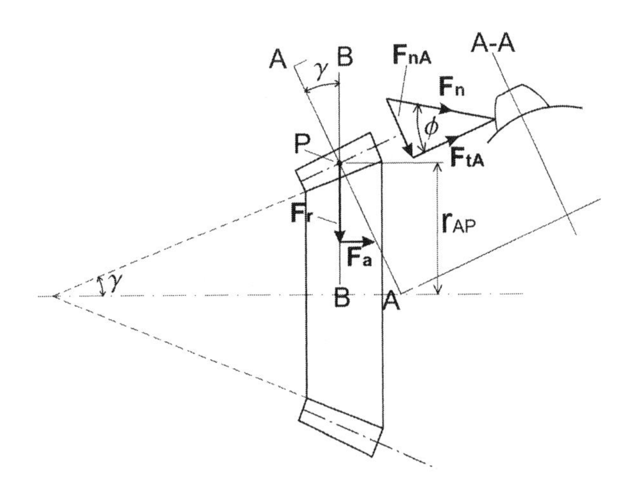

FIGURE 5.25 Forces acting on a tooth of a bevel gear.

where d is the pitch diameter defined on the large end of the tooth (see Figure 5.24).

In Figure 5.25 the forces acting at the interface of a pair of *straight-tooth* bevel gears are shown. Note that in reality the forces are distributed along the tooth length, so those shown in Figure 5.25 should be considered as resultant forces applied in the middle of the tooth, the latter being a simplification. The resultant force is a normal force F_n, shown in the A–A cross section, and it is directed along the line of action, i.e., at angle ϕ with respect to the tangential line. Thus, the resultant normal force has two components in the plane perpendicular to the cone surface: *normal* ($\mathbf{F}_{nA}$) and *tangential* ($\mathbf{F}_{tA}$), which are found from the right-angled triangle to be (Figure 5.25)

$$F_{tA} = F_n \cos\phi \tag{5.38}$$

and

$$F_{nA} = F_n \sin\phi \tag{5.39}$$

The normal component acting in the plane A–A can, in turn, be resolved into two components: one perpendicular to the axis of rotation of the gear (F_r) and another parallel to it (F_a). The former is called the *radial force*, while the latter the *axial force*. These forces are found from the right-angled triangle in Figure 5.25 to be

$$F_r = F_{nA} \cos\gamma \tag{5.40}$$

and

$$F_a = F_{nA} \sin\gamma \tag{5.41}$$

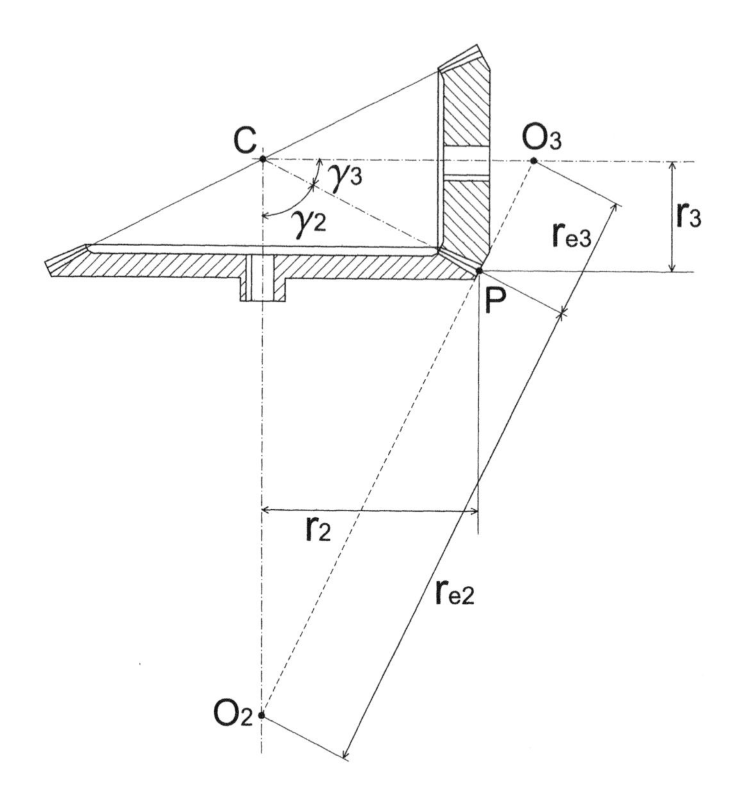

FIGURE 5.26 Illustration of equivalent spur gears.

As is true for all gears, the tangential force is the only useful force and it is found from the known transmitted torque T

$$F_t = F_{tA} = \frac{T}{r_{ap}} \tag{5.42}$$

where r_{ap} is the average pitch radius.

As long as the tangential force is known, the normal resultant force can be found from Equation 5.38, and then the radial and axial forces from Equations 5.39 through 5.41.

In Equations 5.38 and 5.39 the pressure angle ϕ has not been defined. This angle is easier to understand using the concept of equivalent spur gears defined for the large ends of teeth. This concept is illustrated in Figure 5.26, where a normal line to the common pitch cone line CP at point P is shown intersecting the gears axes at points O_2 and O_3. The distances O_2P and O_3P can be viewed as the pitch radii of equivalent spur gears, r_{e2} and r_{e3}. Then the angle ϕ is defined by the relationship between the base radius and pitch radius for the spur gears. For example, for pinion 3,

FIGURE 5.27 Worm gears.

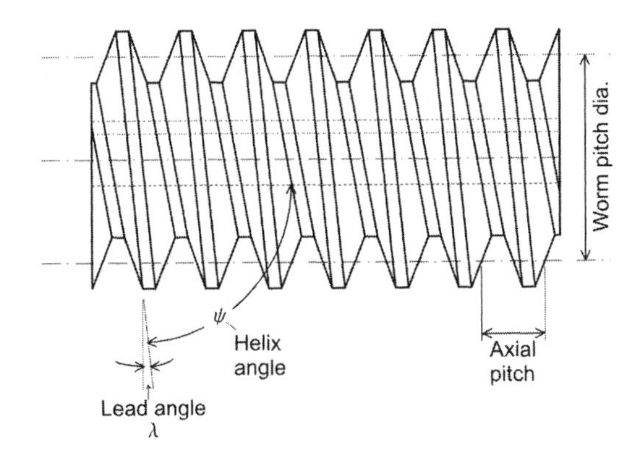

FIGURE 5.28 Worm geometry.

$$\cos\phi = \frac{r_{b3}}{r_3} = \frac{r_{eb3}}{r_{e3}} \tag{5.43}$$

where r_{eb3} is the base radius of an equivalent spur gear 3 in the plane O_2O_3.

It is easy to show from the similarity of triangles that the above relationship is valid for any equivalent spur pair along the tooth length on the pitch cone line *CP* in Figure 5.26. In other words, the angle ϕ is a constant parameter of bevel gears in the same sense as it is in the case of spur gears.

5.8.4 Worm Gears

A worm–worm gear pair is shown in Figure 5.27. Worm gears are usually used when there is a need to transfer motion between perpendicular shafts and a large transmission ratio, from 10:1 to 15:1, is required. The worm has a screwlike thread (Figure 5.28) and is always a driver. The schematic representation of the worm–worm gear pair is shown in Figure 5.29. The teeth profiles are not involutes (since an involute profile for a tooth gives a point contact with the gear). To ensure a line contact of the interfacing teeth, they are cut using the same hob.

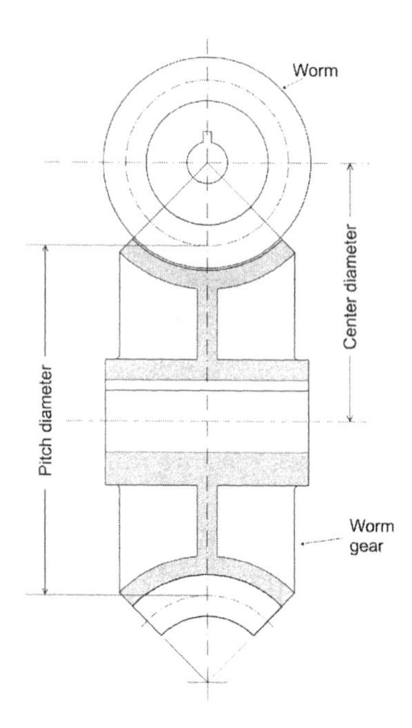

FIGURE 5.29 Worm–worm gear meshing.

From the kinematic point of view it is important to establish the relationship between the pair geometric parameters and the transmission ratio. The worm geometry is illustrated in Figure 5.30a, where ψ is the *helix* angle, while the complementary-to-90° angle λ is called the *lead* angle (see also Figure 5.28). Note that the helix angle of the worm gear equals the lead angle of the worm. If a complete revolution of a thread on a worm is unwrapped, a triangle is obtained (Figure 5.30b). It follows from this triangle that

$$\tan\lambda = \frac{l}{\pi d_2} \tag{5.44}$$

where l is called the *lead*, and it is equal to the axial distance that a point on the helix will move in one revolution of the worm, and d_2 is the worm pitch diameter.

For a single-threaded worm, the lead is equal to its axial pitch p_x (Figure 5.30a). For a multithreaded worm the lead is equal to

$$l = p_x N_2 \tag{5.45}$$

where N_2 is the number of threads (teeth) on the worm.

Taking into account that for the shafts at 90° the axial pitch for the worm is equal to the circular pitch for the gear:

$$p_x = p \tag{5.46}$$

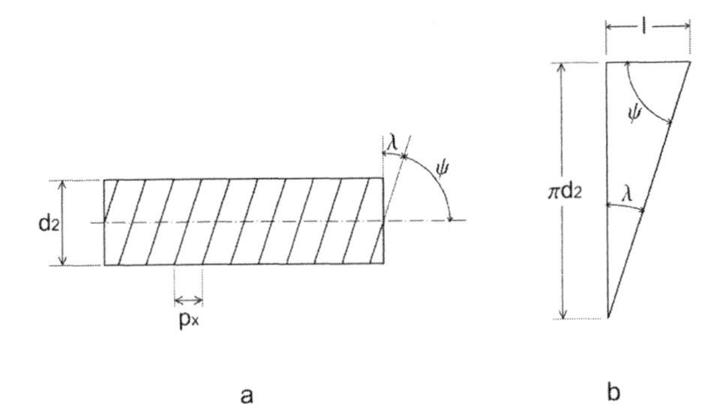

FIGURE 5.30 Illustration of worm geometry.

and that the circular pitch for the gear is $p = \pi d_3/N_3$, then from the equality (Equation 5.46 and Equation 5.45) it follows that

$$\frac{N_3}{N_2} = \frac{\pi d_3}{l} \tag{5.47}$$

For one revolution of a worm with N_2 threads a point on a worm tooth will move by a distance l. Thus, for the angular velocity of the worm ω_2 the axial velocity V_a of the point on a worm tooth is defined by the ratio

$$\frac{2\pi}{l} = \frac{\omega_2}{V_a} \tag{5.48}$$

The axial velocity V_a must be equal to the tangential velocity of the gear:

$$V_t = \frac{d_3\omega_3}{2} \tag{5.49}$$

From the equality $V_a = V_t$ it follows that

$$\frac{\pi d_3}{l} = \frac{\omega_2}{\omega_3} \tag{5.50}$$

and then from the latter and Equation 5.47 that

$$\frac{N_3}{N_2} = \frac{\omega_2}{\omega_3} \tag{5.51}$$

As seen from Equation 5.51, the transmission ratio for the worm–worm gear pair is defined in the same way as for the all other gears, namely, as a ratio between the number of the corresponding teeth. One should recall that the number of teeth for the worm is its number of threads.

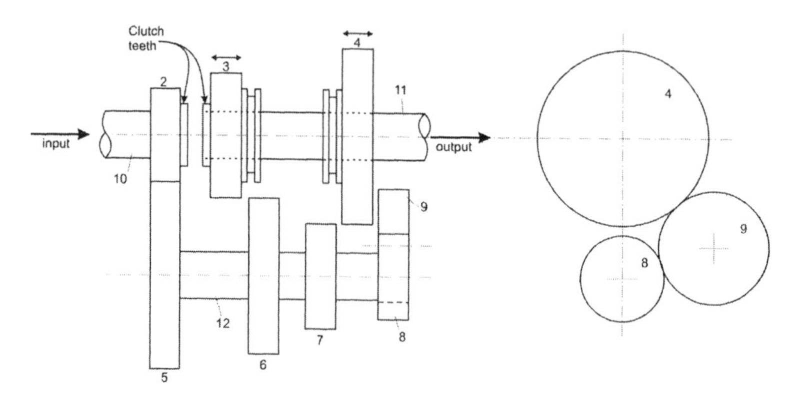

FIGURE 5.31 Automotive transmission.

5.9 PARALLEL-AXIS GEAR TRAINS

Very often a single gear pair transferring motion between two parallel shafts cannot meet the needed transmission ratio requirement. For example, if this requirement is 10:1, then the diameter of the gear must be 10 times larger then the diameter of the pinion. This is usually unacceptable from the design specifications concerning product size. The solution is achieved by arranging a series of gear pairs. Such series is called a *gear train*. A train may comprise different type of gears: spur, helical, bevel, and worm. The gears in a train are *functionally in series* with each other. If a system comprises a few trains, they are *functionally in parallel* with each other. Usually a system of gears arranged physically in one case (box), whether in series or in parallel, is called a *transmission box*.

An example of a transmission box with parallel gear trains is shown in Figure 5.31. This transmission comprises three trains. The input into the box comes from shaft 10, the output shaft is 11, and the auxiliary shaft is 12. To achieve three different paths of motion transmission from the input to the output shafts, the design allows the rearranging of connections (meshing) between the gears by shifting some of them along the shaft axis. In Figure 5.31 gears 3 and 4 can be shifted along the shaft so that the following trains are obtained:

First: 2–5–7–4
Second: 2–5–6–3
Third: 2–5–8–9–4

One should note that there is a fourth mode of motion transmission in this example, when a clutch is engaged and the power is transmitted through the shaft directly from the input to the output without any gears involved.

5.9.1 Train Transmission Ratio

It is customary to define the transmission ratio of a pair as the ratio of driven to driving angular velocities. It is known that if the gears have external meshing, the sense of rotation changes to the opposite, while for gears with internal meshing it remains the same. Thus, for any pair in the train with external meshing the transmission ratio is

$$e = \frac{-\omega_{\text{driven}}}{\omega_{\text{driving}}} \tag{5.52}$$

And thus the train transmission ratio is

$$e_{\text{train}} = e_1 e_2 \ldots e_K \tag{5.53}$$

where K is the number of pairs. It is clear that for gears with external meshing, if K is even, the sense of the rotation of the output is the same as that of the input, and, if K is odd, the sense of rotation changes to the opposite.

If Equation 5.52 is substituted into Equation 5.53, the train transmission ratio is obtained as a ratio of angular velocities of last and first gears in the train, i.e.,

$$e_{\text{train}} = \frac{\omega_{\text{Last}}}{\omega_{\text{First}}} \tag{5.54}$$

Consider the transmission ratio of the first train in Figure 5.31. Recalling that the transmission ratio of a gear pair can be expressed through the ratio of gear teeth numbers, one can write, using Equation 5.53,

$$e_I = (-1)^2 \frac{N_2}{N_5} \frac{N_7}{N_4} \tag{5.55}$$

Similarly, the transmission ratios of the second and third trains are

$$e_{II} = (-1)^2 \frac{N_2}{N_5} \frac{N_6}{N_3} \tag{5.56}$$

$$e_{III} = (-1)^3 \frac{N_2}{N_5} \frac{N_8}{N_9} \frac{N_9}{N_4} \tag{5.57}$$

One can see that in the third transmission ratio the ninth gear does not affect the magnitude of the output angular velocity, but it changes its sign (the sense of rotation). The ninth gear in Figure 5.31 is called the *idler gear*, whose function is to reverse the rotation. The transmission box shown in Figure 5.31 is an example of an automotive transmission with three forward speeds and one reverse.

5.9.2 Design Considerations

In designing a gear box (such as, for example, that shown in Figure 5.31), the input information is the required transmission ratios for each speed. Thus, the problem is, given e_I, e_{II}, and e_{III}, find gears with teeth numbers that will meet other transmission box design requirements, such as, for example, specific center distances between the shafts and the teeth strength, among others. The fact that gear pairs 2–5, 3–6, and 4–7 are mounted on two parallel shafts gives additional equations in the form of the requirement that the center distance for each pair be the same. The center distance equals the sum of two pitch radii, i.e.,

$$C = r_{\text{driven}} + r_{\text{driving}} \tag{5.58}$$

Taking into account that $r = 0.5Nm$, the above equation becomes

$$C = 0.5m(N_{\text{driven}} + N_{\text{driving}}) \tag{5.59}$$

where the tooth module m is another parameter. Thus, for the case of Figure 5.31 three equations of the above type can be written, bringing the total number of equations to six. Since the total number of equations is smaller than the number of unknowns, the solution is not unique. Since the transmission ratio of a train is a product of transmission ratios of gear pairs, there is a freedom of choosing the latter to obtain the same total transmission ratio. In addition, the solution, i.e., the number of teeth for each gear, must be an integer; the type of the gear itself may be different (spur or helical) for each pair. Considering all of the above, the process of gear train (box) design is an iterative one, in which kinematic requirements must be satisfied subject to meeting other (space, strength, etc.) requirements.

5.10 PLANETARY GEAR TRAINS

An elementary planetary gear train is shown in Figure 5.32. It comprises two gears, 2 and 4, each mounted on its own shaft. The new element here is the link 3 connecting these shafts and able to rotate around the fixed axis O_1. This system has two degrees of freedom, which means that if only the input velocity $\omega_{\text{in}} = \omega_2$ is given, the motion of the two other elements cannot be determined. Input gear 2 is called the *sun*

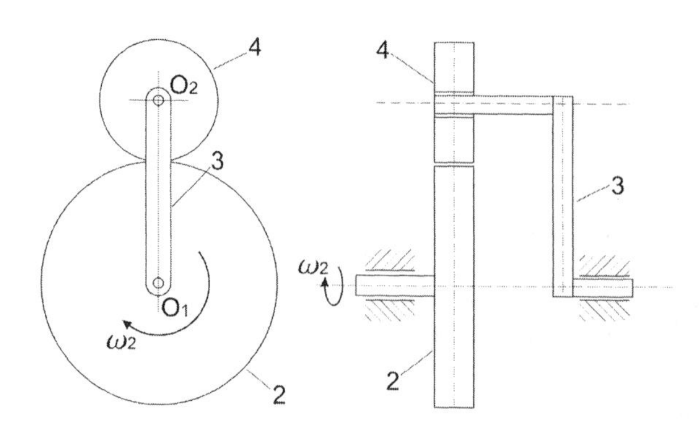

FIGURE 5.32 Planetary gear train.

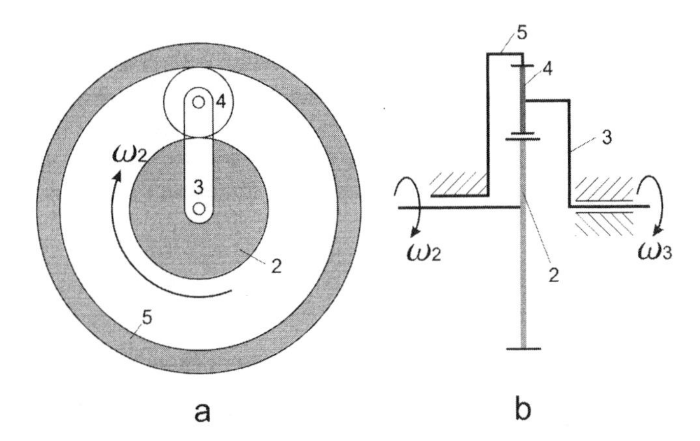

FIGURE 5.33 Planetary gear train with annular gear.

(or *central*) gear, gear 4 is called the *planetary* (or *epicyclic*) gear, and link 3 is called the *planet carrier* (or *crank arm*).

Planetary gear trains allow obtaining high transmission ratios in a compact design, which makes them suitable for applications in, for example, machine tools, hoists, and automatic transmissions. An example of a simple planetary gear box is shown in Figure 5.33, where in addition to the elements in Figure 5.32 an additional annular gear 5 is added, so that the planetary gear 4 is now interfacing both the sun and the annular gears. Note that the annular gear is fixed. The number of degrees of freedom of this system is 1, which means that for a given input $\omega_{in} = \omega_2$ there is a unique output $\omega_{out} = \omega_3$.

5.10.1 Transmission Ratio in Planetary Trains

One can determine the transmission ratio for a planetary train in Figure 5.32. Suppose that arm 3 rotates with angular velocity ω_3. Then, if an observer is sitting on this arm, for this observer the rotation of gears 2 and 4 will not be different from that for a parallel fixed-shaft system, and the corresponding transmission ratio will be ω_4/ω_2. One realizes that an observer on the arm sees rotation in a moving (rotating) coordinate system. Now if the observer is standing on the frame, then the observer will see that the arm rotates with ω_3. The question is what will be the angular velocities of gears 2 and 4 with respect to the observer on a frame.

The answer is given by a general rule for summation of angular velocities in the case when a body rotates with respect to its own axis with velocity ω^1, while the axis itself rotates with respect to another axis with the velocity ω. In the case when the two axes are parallel, the total angular velocity equals the algebraic sum of velocities of two rotations. Thus, the total angular velocity will be

$$\omega_{total} = \omega + \omega^1 \tag{5.60}$$

The above rule is directly applicable to the planetary gear trains. Indeed, the axes of gears 2 and 4 are parallel and one of them rotates with respect to the other with ω_3. Now, the transmission ratio in a coordinate system rotating with ω_3 is known. It is equal to $e_{42} = -N_2/N_4$. Thus, if one applies a counterrotation with $-\omega_3$ to the entire system, then the gear velocities (according to the rule of summation) will be $\omega_4 - \omega_3$ and $\omega_2 - \omega_3$, and link 3 becomes fixed. Thus, the transmission ratio is given by

$$e_{42} = \frac{\omega_4 - \omega_3}{\omega_2 - \omega_3} = -\frac{N_2}{N_4} \tag{5.61}$$

The above equation confirms that the system shown in Figure 5.32 has two degrees of freedom. Indeed, if only ω_2 is given, two unknown velocities remain, ω_3 and ω_4, while the gears are defined.

The example of Figure 5.32 described by Equation 5.61 is equivalent to a one-pair system in conventional gear trains. In this respect, the system in Figure 5.33 is equivalent to a two-pair system where the pairs are functionally in series with each other. Thus, in this case there are two transmission ratios: first, from the sun gear 2 to the planetary gear 4, which is described by Equation 5.61, and, second, from the planetary gear 4 to the annular gear 5. The latter is equal to (in a rotating coordinate system)

$$\frac{\omega_5 - \omega_3}{\omega_4 - \omega_3} = \frac{N_4}{N_5} \tag{5.62}$$

Note that the plus sign on the right-hand side in Equation 5.62 is due to internal gear meshing. If $\omega_5 = 0$ (annular gear is fixed) and ω_2 is known, then ω_3 and ω_4 are found from Equations 5.61 and 5.62.

And, finally, it should be noted that the planetary gear plays the role of an idle gear in Figure 5.33. Indeed, the total transmission ratio for a system in series is equal to the product of transmission ratios of its subsystems (see Equation 5.53). In the case of Figure 5.33 the total transmission ratio is the product of Equations 5.61 and 5.62. The result is

$$e = \frac{\omega_5 - \omega_3}{\omega_2 - \omega_3} = -\frac{N_2}{N_5} \tag{5.63}$$

As one can see, the angular velocity ω_4 of the planetary gear does not affect the transmission ratio.

Given that $\omega_5 = 0$ and ω_2 is known, the unknown output velocity can be immediately found from Equation 5.63:

$$\omega_3 = \omega_2 \frac{n_{25}}{1 + n_{25}} \tag{5.64}$$

where it is denoted $n_{25} = N_2/N_5$.

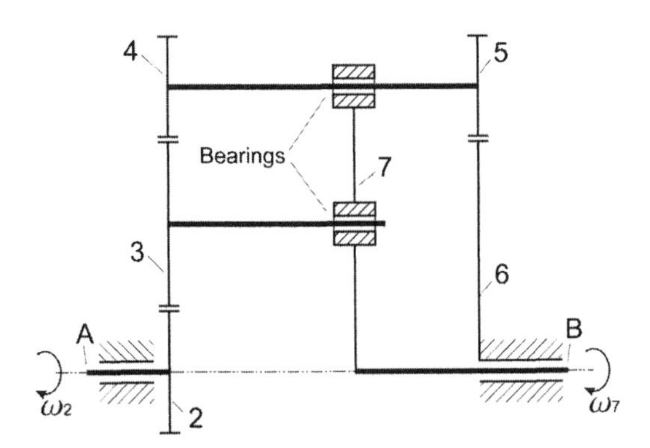

FIGURE 5.34 Skeleton of a complex gear train.

5.10.2 Example of a More Complex Planetary Train

Consider the example shown in Figure 5.34. The sun gear 2 is driven by shaft *A* and the output shaft is *B*, which is driven by the planet carrier 7. The two gears, 4 and 5, are mounted on the same shaft, which means that their angular velocities must be equal. Notice also that gear 6 is fixed. Given all this information, and assuming that the teeth numbers of all gears are known, the problem is to find the velocity of the output shaft given the input velocity ω_2.

The strategy is to follow the chain of gear interfaces 2–3, 3–4, and 5–6, and to write for each interface the equation of the Equation 5.61 type. To shorten this procedure, one can take into account that the planetary gear does not change the kinematics (see Equation 5.63) and thus write a relationship between gears 2 and 4 directly.

$$\frac{\omega_4 - \omega_7}{\omega_2 - \omega_7} = \frac{N_2}{N_4} \tag{5.65}$$

and for gears 5 and 6,

$$\frac{\omega_6 - \omega_7}{\omega_5 - \omega_7} = -\frac{N_5}{N_6} \tag{5.66}$$

Note that in Equation 5.65 the sign is positive because there is an even number of interfaces from gear 2 to gear 4. The system of equations above should be supplemented by the requirement that

$$\omega_4 = \omega_5 \tag{5.67}$$

Given the latter condition and that $\omega_6 = 0$ and ω_2 is known, the two equations (5.65 and 5.66) allow one to find ω_4 and ω_7. The result is

$$\omega_7 = \omega_2 \frac{n_{24}n_{56}}{1 + n_{24}n_{56}} \tag{5.68}$$

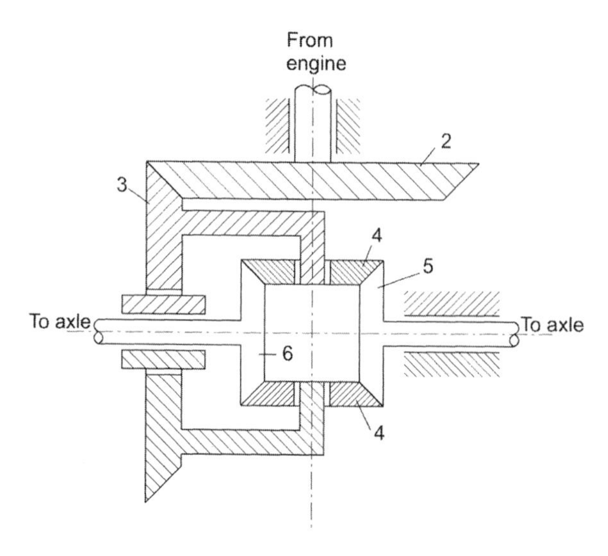

FIGURE 5.35 Automotive differential.

and

$$\omega_4 = \omega_2 \frac{n_{24}(1 + n_{56})}{1 + n_{24}n_{56}} \tag{5.69}$$

where it is denoted

$$\frac{N_2}{N_4} = n_{24} \text{ and } \frac{N_5}{N_6} = n_{56}$$

5.10.3 Differential

Differentials are planetary trains made out of bevel gears and having two degrees of freedom. An example of an automotive differential is shown in Figure 5.35. The rotation from the engine is transferred through bevel gears 2 and 3 to the system of bevel gears 4, 5, and 6. Gears 4 are mounted on the carrier and are the planetary gears, whereas gears 5 and 6 are two independent sun gears. The transmission ratio from 2 to 3 is independent from the rest of the system (in fact, gear 3 could be considered an input gear with known angular velocity). Thus, the transmission through system 3–4–5–6 will be considered.

The important distinction of the bevel planetary mechanism is that the rule of summation of velocities for the planetary and sun gears is not applicable in the sense discussed above since in this case the axes of gears 5 and 4 and gears 6 and 4 are not parallel. Thus, if gear 5 rotates clockwise, when viewed along its axis from the right, gear 6 will be rotating counterclockwise from the same point of view (Figure 5.36). However, the rotation of the gears 3 and 5 and gears 3 and 6 is around parallel axes, and so the rule of summation discussed for planar trains is applicable. Thus, if one applies a counterrotation $-\omega_3$ to the entire system, one will have the system shown in

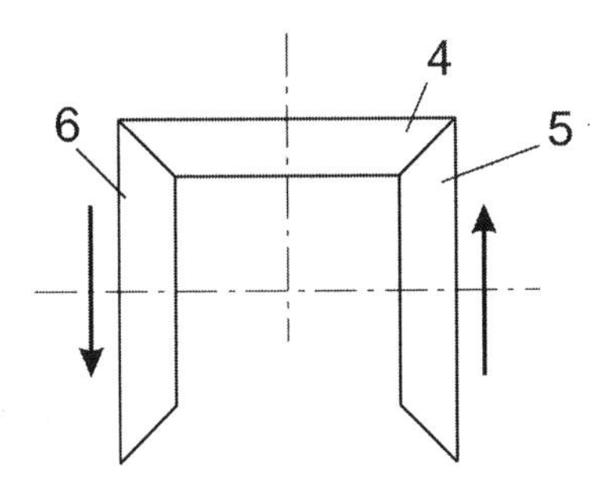

FIGURE 5.36 Illustration of sense of rotations in a differential.

Figure 5.36 in which gear 5 rotates with angular velocity $\omega_5 - \omega_3$ and gear 6 with $\omega_6 - \omega_3$. The transmission ratio between gears 5 and 6 in a rotating coordinate system is equal to

$$\frac{\omega_6 - \omega_3}{\omega_5 - \omega_3} = -\frac{N_5}{N_6} = -1 \tag{5.70}$$

One equation with two unknowns, ω_5 and ω_6, is obtained. Thus, the system has two degrees of freedom. In practical terms this means that the left axle can rotate independently of the right axle. From Equation 5.70 it follows that

$$\omega_5 + \omega_6 = 2\omega_3 \tag{5.71}$$

The latter relationship means that while ω_3 remains constant, the values of ω_5 and ω_6 may change. In the automotive applications ω_5 and ω_6 are the angular velocities of two wheels, and ω_3 can be considered the angular velocity of the engine. So when the vehicle turns, the angular velocities ω_5 and ω_6 become unequal, but it does not affect the engine speed. In other words, the engine maintains its speed during turns. Note also that when the vehicle moves straight, $\omega_5 = \omega_6$, the planetary gear 4 does not rotate because $\omega_5 = \omega_6 = \omega_3$. Otherwise, the planetary gear will be rotating, allowing relative motion between gears 5 and 6.

One more comment concerning Equation 5.71 should be made. In the case of an inverse mechanism, when gears 5 and 6 provide input, while gear 3 is the output, the mechanism performs an operation of summation. If the sign of one of the rotations changes, then it will show the result of subtraction. This property of the differential (the reason it is so named) is used in mechanical calculators.

PROBLEMS

1. Prove Kennedy's theorem that three instantaneous centers shared by three bodies in relative motion to one another all lie on the same straight line.
2. Explain how the involute profile is generated and why the transmission ratio of two disks interacting through the attached plates which have involute profiles is constant.

3. Prove that the transmission ratio of two involute gears does not depend on the center distance between them.
4. Explain why a straight line describes the tooth profile of the rack.
5. For a spur gear,
   a. What is the direction of the force generated at the interface of two meshing teeth?
   b. Is it true that $\omega_2/\omega_3 = T_3/T_2$, where T_2 and T_3 are the torques applied to the corresponding gears? Prove.
6. Spur gears with a module $m = 4$ mm transmit motion between two shafts with center distance $C = 136$ mm. For the given transmission ratio 3:1 find the number of teeth for each gear.
7. Sketch the axial, radial, and tangential components of the force acting on a tooth of a helical gear. How does one find these forces given the torque T transmitted by the gear?
8. What is the involute surface for a straight teeth bevel gear? Draw a sketch explaining how to generate this surface.
9. Explain, using the equivalency of spur and straight teeth bevel gears, that for the latter the transmission ratio is given by $\omega_2/\omega_3 = N_3/N_2 = d_3/d_2$.
10. Prove that the transmission ratio for the worm gears set is $\omega_2/\omega_3 = \pi d_3/l$, where d_3 is the gear pitch diameter and l is the lead of the worm.
11. Prove that the transmission ratio between the sun 2 and the planet gear 4 in Figure 5.32 is

$$e_{42} = \frac{\omega_4 - \omega_3}{\omega_2 - \omega_3} = -\frac{N_2}{N_4}$$

    where ω_3 is the angular velocity of arm 3.
12. Derive the transmission ratio ω_3/ω_2 for the planetary train in Figure 5.33.
13. Derive the transmission ratio ω_7/ω_2 for the planetary train in Figure 5.34.
14. Explain why in an automotive transmission the differential allows two wheels to rotate with different angular velocities while the speed of the engine remains the same.
15. A worm has a double thread with a 10° lead angle and 20-mm pitch diameter. It meshes with a worm gear with a 90° angle between the shafts. For a speed reduction 15:1, determine:
    a. The pitch diameter and the number of teeth of the gear.
    b. The helix angle of the gear.
    c. The center distance.

16. An 18-tooth spur pinion has module $m = 2$ mm. For the transmission ratio 1:2, find:

    a. The number of teeth on the gear.
    b. The center distance.
    c. The pressure angle (take that the base diameter equals pitch diameter minus $2.2m$).

6 Introduction to Linear Vibrations

6.1 INTRODUCTION

A question one may ask is "Why do bodies vibrate?" The explanation is rooted in the *energy conservation principle.* Consider as an example a mass m suspended on a spring having stiffness k (Figure 6.1) and assume that this mass is pushed up (or down) from its static equilibrium position by the amount y_{max} and then released.

By deforming the spring some energy is stored in it, which will be denoted by V_{max}. This energy is called the *potential energy,* and V_{max} represents the maximum energy transferred to the spring by deforming it by y_{max}. After releasing the mass, it will start moving back to its original position. The motion of the mass means that it acquires some *kinetic energy,* which is equal to

$$T = \frac{1}{2} m\dot{y}^2 \tag{6.1}$$

where $\dot{y}$ is the time derivative of mass displacement, i.e., its velocity. At any intermediate position y the potential energy of the deformed spring is equal to

$$V = \frac{1}{2} ky^2 \tag{6.2}$$

If one assumes that the spring is ideal, i.e., its deformation does not lead to any energy losses, then according to the energy conservation principle the sum of the kinetic energy of the mass and the potential energy in the spring must be equal to the original energy introduced into this system, V_{max}. Thus,

$$\frac{1}{2} m\dot{y}^2 + \frac{1}{2} ky^2 = V_{max} = \text{const.} \tag{6.3}$$

The above equation shows that in a spring–mass system an energy transformation takes place, from potential to kinetic and back. More than that, one can see that this process is periodic. Indeed, when $y = \pm y_{max}$ then $V = V_{max}$ and it follows that at these extreme positions $\dot{y} = 0$. Thus, there are two extreme positions of the mass and they are equal in magnitude, but at the opposite sides of the static equilibrium position. On the other hand, when $y = 0$ (or, more correctly, a static displacement), $\dot{y} = \dot{y}_{max}$. This process of a mass moving between two extreme positions in a periodic fashion is called *oscillation.* It is characterized by two parameters: the *amplitude* of oscillation and the *period* of oscillation. The former is the maximum mass displace-

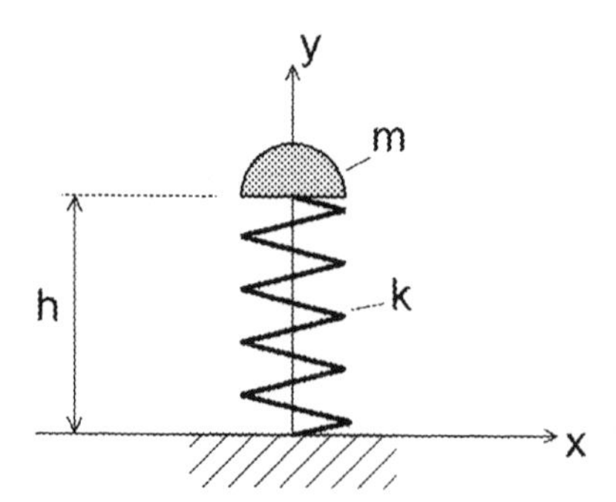

FIGURE 6.1 Spring–mass system.

ment from the position of static equilibrium, y_{max}, whereas the latter is the time between two consecutive maximum displacements.

The relationship given by Equation 6.3 is a differential one. Since it is known that the motion of the mass is periodic with the amplitude y_{max}, one can assume it to be described by a function:

$$y = y_{max} \sin\left(2\pi \frac{t}{T} + \alpha\right) \tag{6.4}$$

where t is the time, T is the period of oscillation, and α is some constant which is defined by the requirement that at $t = 0$, $y = y_{max}$.

In Equation 6.4 the maximum displacement y_{max} is known, and T is the only unknown (besides α) to be found in order to describe the motion of the mass in time. The expression for T follows from the requirement that the function (Equation 6.4) must satisfy Equation 6.3. Substitute y and $\ddot{y}$ into Equation 6.3. After simple derivations, the differential Equation 6.3 is reduced to an algebraic relationship:

$$\left[\frac{m(2\pi)^2}{T^2} - k\right] y_{max}^2 \cos^2\left(2\pi \frac{t}{T} + \alpha\right) = 0 \tag{6.5}$$

For the above equality to be true at any time, the term in the brackets must be equal to zero, i.e.,

$$\left(\frac{m(2\pi)^2}{T^2} - k\right) = 0 \tag{6.6}$$

The latter equation defines the period of oscillation as a function of mass and spring stiffness:

$$T = 2\pi\sqrt{\frac{m}{k}} \tag{6.7}$$

The important point is that the period of oscillation is an intrinsic property of the mass–spring system and does not depend on the initial disturbance (or on the initial energy pumped into the system). The inverse of T gives the number of oscillations per unit time and it is called the *frequency* of oscillation, f. If the time is measured in seconds, then one oscillation per second is called a *hertz*.

$$f = \frac{1}{T} = \frac{1}{2\pi}\sqrt{\frac{k}{m}} \tag{6.8}$$

In Equation 6.4 $2\pi/T$ gives the angular displacement per unit time; i.e., it gives the angular velocity. This angular velocity is called the *circular frequency* of oscillation, ω, and it is equal to

$$\omega = \frac{2\pi}{T} = \sqrt{\frac{k}{m}} \tag{6.9}$$

The mass–spring system shown in Figure 6.1 represents a single-degree-of-freedom (SDOF) system. The motion of the mass discussed above was caused by its initial displacement and Equation 6.3 describes this motion. One can simplify this equation by differentiating both sides of it with respect to time. The result is

$$(m\ddot{y} + ky)\dot{y} = 0 \tag{6.10}$$

Since $\dot{y} \neq 0$ at *all* times, Equation 6.10 can be satisfied only if the first term equals zero, i.e.,

$$m\ddot{y} + ky = 0 \tag{6.11}$$

Now it can be seen that the obtained equation expresses Newton's second law, where ky is the external spring force acting on the mass treated as a free body. In general, if a periodic force $P(t) = P_0 \sin(\omega t)$ acts upon the mass, then its motion equation can be obtained by considering the equilibrium of forces, including inertial force, acting on this mass (Figure 6.2). Taking into account that forces ky and $P(t)$ are opposite, the motion equation becomes

$$m\ddot{y} + ky = P_0 \sin(\omega t) \tag{6.12}$$

Equations 6.11 and 6.12 have been derived assuming that the system is not experiencing any energy losses during oscillations. If this assumption cannot be made, then a force associated with the energy losses should be added to the resultant force acting on the mass in Figure 6.2. The causes of energy losses may be many, such as friction between the moving parts, oil or air resistance, internal losses in materials. These different causes may entail different mathematical models to describe them. In general, unless there is a dominating factor, all of them contribute to the total energy loss. The most convenient way to model energy losses is to assume that they are

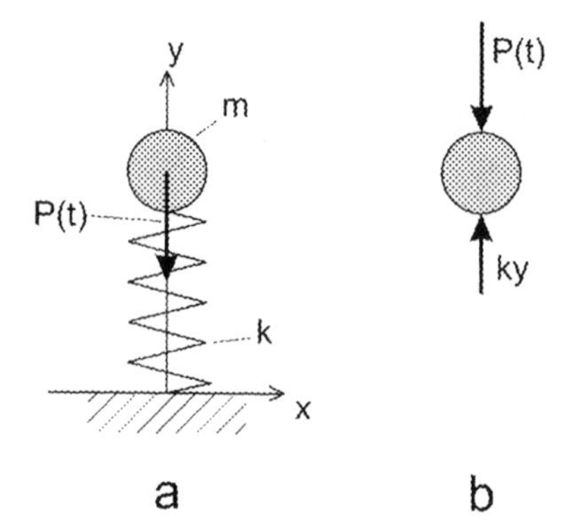

FIGURE 6.2 External force applied to a mass–spring system (a), and a free-body diagram of the mass (b).

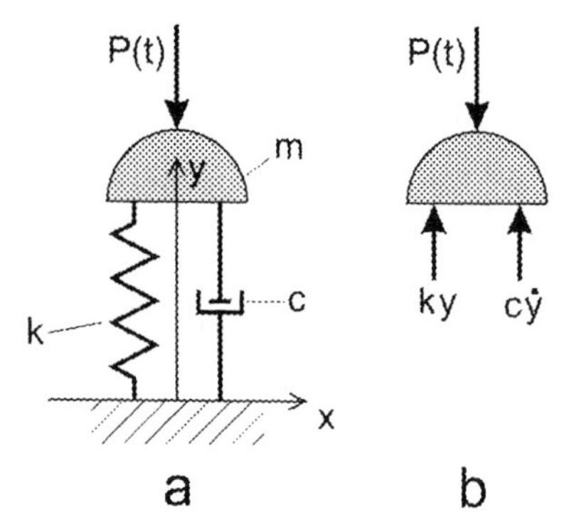

FIGURE 6.3 Mass–spring–damper system (a), and free-body diagram of mass (b).

caused by forces proportional to the body velocity, similar to the resistance experienced by a body moving through a viscous liquid. Accordingly, these forces are called *viscous forces* and are taken in the form:

$$F_r(t) = c\dot{y} \tag{6.13}$$

where c is called the *damping coefficient.*

In Figure 6.3a in addition to the spring a *damping element* is shown, which is conventionally used to identify the presence of viscous resistance force having damping coefficient c. This additional force acts in the same direction as the spring resistance force (Figure 6.3b). The dynamic equilibrium of forces shown in Figure 6.3b gives the following differential equation:

$$m\ddot{y} + c\dot{y} + ky = P_0\sin(\omega t) \tag{6.14}$$

Equations 6.11, 6.12, and 6.14 have one important property; they are *linear* differential equations. Their linearity is due to the assumptions that resistance forces are proportional to either displacement or velocity. If any of these assumptions is not adequate, then the equation becomes *nonlinear*. There are well-developed mathematical techniques for solving linear equations, whereas nonlinear equations very often require an individual analytical approach or a numerical solution. This book will be limited to linear equations.

Equation 6.14 is a *second-order nonhomogeneous linear equation with constant coefficients*. The next section considers solutions of Equation 6.14.

6.2 SOLUTION OF SECOND-ORDER NONHOMOGENEOUS EQUATIONS WITH CONSTANT COEFFICIENTS

Linear equations have a very important property; they allow linear combination of solutions to form new solutions. Namely, if $f_1(t)$ and $f_2(t)$ are two solutions, then their combination with some constant coefficients, $f(t) = a_1 f_1(t) + a_2 f_2(t)$, is also a solution of this equation. This is called the *superposition principle*. The following will use this principle in various situations, but its first application is to split the solution of Equation 6.14 into two parts: one describing the behavior of the system without the external forces (the corresponding differential equation is called *homogeneous* and its solution the *general solution*) and the other describing the behavior of the system subjected to external forces (the corresponding differential equation is called *nonhomogeneous* and its solution the *particular solution*). The solution of the homogeneous equation will first be considered.

6.2.1 Solution of the Homogeneous Equation

The homogeneous equation to be solved is

$$m\ddot{y} + c\dot{y} + ky = 0 \tag{6.15}$$

It is more convenient for the following to transform this equation into one with nondimensional constants by dividing each term by m. The transformed equation is as follows:

$$\ddot{y} + 2\xi\omega_n\dot{y} + \omega_n^2 y = 0 \tag{6.16}$$

where it is denoted

$$\xi = \frac{c}{2m\omega_n} \quad \text{and} \quad \omega_n^2 = \frac{k}{m} \tag{6.17}$$

In Equation 6.17 ξ is called the *nondimensional damping coefficient*, and ω_n is called the *natural frequency* (see Section 6.3).

It is known that an exponential function

$$y(t) = e^{\lambda t} \tag{6.18}$$

is a possible solution of this equation, where λ is some constant. If this function is substituted into Equation 6.16, the following algebraic equation is obtained:

$$(\lambda^2 + 2\xi\omega_n\lambda + \omega_n^2)e^{\lambda t} = 0 \tag{6.19}$$

Since $e^{\lambda t}$ in the above equation cannot be equal to zero, the nontrivial solution of Equation 6.19 exists only if the quadratic polynomial equals zero:

$$\lambda^2 + 2\xi\omega_n\lambda + \omega_n^2 = 0 \tag{6.20}$$

The above equation is called the *characteristic equation*, since it defines two unique roots and thus two possible solutions of the differential equation. The two roots are

$$\lambda_{1,2} = (-\xi \pm \sqrt{\xi^2 - 1})\omega_n \tag{6.21}$$

If λ_1 and λ_2 are two distinct roots, then it means that there are two possible solutions of Equation 6.16. The *general solution* is a combination (again, the principle of superposition is used) of these two solutions with some constants:

$$y(t) = c_1 e^{\lambda_1 t} + c_2 e^{\lambda_2 t} \tag{6.22}$$

The specific form of the solution given by Equation 6.22 depends on the type of roots in Equation 6.21: *real and distinct*, *complex*, or *real and equal*. Each of these cases will be considered separately since, as will be seen later, the type of roots reflects the type of system behavior.

- *Roots are real and distinct*

This is possible if $\xi > 1$ in Equation 6.21; then the solution given by Equation 6.22 is an exponentially decreasing in time function since both roots are negative.

- *Roots are complex*

This is possible if $\xi < 1$ in Equation 6.21; then they can be written in the form $\lambda_{1,2} = \alpha \pm i\beta$, where $\alpha = -\xi\omega_n$ and $\beta = \sqrt{1 - \xi^2}\omega_n$ are real numbers and i is the imaginary unit (square root of –1). Now the solution Equation 6.22 takes the following form:

$$y(t) = c_1 e^{(\alpha + i\beta)t} + c_2 e^{(\alpha - i\beta)t} \tag{6.23}$$

By using the Euler formula

$$e^{\pm i\beta t} = \cos\beta t \pm i \sin\beta t \tag{6.24}$$

the solution Equation 6.23 can be transformed to

$$\begin{aligned} y(t) &= c_1 e^{\alpha t}(\cos\beta t + i\sin\beta t) + c_2 e^{\alpha t}(\cos\beta t - i\sin\beta t) \\ &= e^{\alpha t}[(c_1 + c_2)\cos\beta t + i(c_1 - c_2)\sin\beta t] \\ &= e^{\alpha t}[a_1\cos\beta t + a_2\sin\beta t] \end{aligned} \tag{6.25}$$

where $a_1 = c_1 + c_2$ and $a_2 = i(c_1 - c_2)$ are two new arbitrary constants.

- *Roots are real and equal*

This is possible if $\xi = 1$ in Equation 6.21 so that $\lambda_{1,2} = \lambda = -\xi\omega_n$. In this case, in addition to the solution given by Equation 6.18, there is another one given by the function $te^{\lambda t}$, and thus, the general solution becomes

$$y(t) = c_1 e^{\lambda t} + c_2 t e^{\lambda t} \tag{6.26}$$

6.2.2 Particular Solution of the Nonhomogeneous Equation

The nonhomogeneous equation with nondimensional coefficients is

$$\ddot{y} + 2\xi\omega_n\dot{y} + \omega_n^2 y = p_o\sin\omega t \tag{6.27}$$

where $p_o = P_o/m$. The particular solution caused by the harmonic forcing function is also harmonic and can be taken in either of the following two forms:

$$y_p(t) = d_1\cos\omega t + d_2\sin\omega t = D\sin(\omega t - \phi) \tag{6.28}$$

where d_1, d_2, D, and ϕ are constants.

Taking the second form of the solution in Equation 6.28 and substituting it into Equation 6.27 obtains

$$D[(\omega_n^2 - \omega^2)\sin(\omega t - \phi) + 2\xi\omega_n\omega\cos(\omega t - \phi)] = p_o\sin\omega t \tag{6.29}$$

By using the trigonometric relations

$$\cos(\omega t - \phi) = \cos\omega t\cos\phi + \sin\omega t\sin\phi \tag{6.30}$$

and

$$\sin(\omega t - \phi) = \sin\omega t\cos\phi - \cos\omega t\sin\phi \tag{6.31}$$

Equation 6.29 is transformed into the following:

$$D[-(\omega_n^2 - \omega^2)\sin\phi + 2\xi\omega_n\omega\cos\phi]\cos\omega t + D[(\omega_n^2 - \omega^2)\cos\phi + 2\xi\omega_n\omega\sin\phi]\sin\omega t = p_o\sin\omega t \tag{6.32}$$

The above equation can be satisfied at any moment in time if the harmonic functions on the left and on the right have equal amplitudes. This requirement leads to two equations:

$$D[-(\omega_n^2 - \omega^2)\sin\phi + 2\xi\omega_n\omega\cos\phi] = 0 \tag{6.33}$$

and

$$D[(\omega_n^2 - \omega^2)\cos\phi + 2\xi\omega_n\omega\sin\phi] = p_o \tag{6.34}$$

To solve the above system for the unknowns D and ϕ, transform it into a simpler system. This is done by first multiplying Equation 6.33 by $-\sin\phi$ and Equation 6.34 by $\cos\phi$ and adding them. The result is

$$D(\omega_n^2 - \omega^2) = p_o\cos\phi \tag{6.35}$$

Now, multiply Equation 6.33 by $\cos\phi$ and Equation 6.34 by $\sin\phi$ and add them. The result is

$$D2\xi\omega_n\omega = p_o\sin\phi \tag{6.36}$$

The new system, Equations 6.35 and 6.36, can be easily solved. Square both sides in Equations 6.35 and 6.36 and add the equations. This gives the expression for the unknown D

$$D = \frac{p_o}{\sqrt{(\omega_n^2 - \omega^2)^2 + (2\xi\omega_n\omega)^2}} \tag{6.37}$$

Now if Equation 6.36 is divided by Equation 6.35, the expression for the angle ϕ is obtained:

$$\phi = \arctan\left(\frac{2\xi\omega_n\omega}{\omega_n^2 - \omega^2}\right) \tag{6.38}$$

Note that the constants d_1, d_2 in Equation 6.28 are equal to

$$d_1 = -D\sin\phi \text{ and } d_2 = D\cos\phi \tag{6.39}$$

Using Equations 6.35 and 6.36 to express sin ϕ and cos ϕ through D gives the following expressions for these coefficients:

$$d_1 = -\frac{2D^2\xi\omega_n\omega}{p_o} \tag{6.40}$$

and

$$d_2 = \frac{D^2(\omega_n^2 - \omega^2)}{p_o} \tag{6.41}$$

where D is given by Equation 6.37.

It is of interest to consider the case when, instead of forcing function $p_0 \sin \omega t$, a function $p_0 \cos \omega t$ is used in Equation 6.27. It can be checked that the solutions for the amplitude, Equation 6.37, and for the phase angle, Equation 6.38, remain the same.

6.2.3 Complete Solution of the Nonhomogeneous Equation

The complete solution is a superposition of the general solution of the homogeneous equation (Equation 6.22) and the particular solution of the nonhomogeneous equation (Equation 6.28). Thus,

$$y_c(t) = y_g(t) + y_p(t) = c_1 e^{\lambda_1 t} + c_2 e^{\lambda_2 t} + d_1 \cos \omega t + d_2 \sin \omega t \tag{6.42}$$

where c_1 and c_2 are unknown constants, and d_1 and d_2 are given by Equations 6.40 and 6.41.

As was discussed in Section 6.1, the vibration of the mass on a spring is caused by the initial displacement of this mass. By displacing and releasing the mass, one introduces into the mass–spring system some initial potential energy. It was tacitly assumed that this initial displacement was slow enough so that the corresponding kinetic energy of motion could be ignored. However, if this is not the case, then the total energy transferred to the system is a sum of both potential and kinetic energies. The former is associated with the initial displacement of the mass, while the second is associated with its initial velocity. Thus, in general, the motion of a body from the undisturbed position starts with some *initial displacement* and with some *initial velocity*. These are called *initial conditions*. For a single body in a uniaxial motion, there are two initial conditions, which are stated as follows

$$y_c(0) = Y_0 \text{ and } \left.\frac{dy_c t}{dt}\right|_{t=0} = V_0 \tag{6.43}$$

These two initial conditions define the constants c_1 and c_2 in Equation 6.42. Satisfying the initial displacement condition gives

$$c_1 + c_2 + d_1 = Y_0 \tag{6.44}$$

Then taking the derivative of $y_c(t)$ and then satisfying the initial velocity requirement gives

$$c_1\lambda_1 + c_2\lambda_2 + d_2\omega = V_0 \tag{6.45}$$

Solving Equations 6.44 and 6.45 for the unknowns c_1 and c_2 gives

$$c_1 = \frac{(Y_0 - d_1)\lambda_2 - V_0 + d_2\omega}{\lambda_2 - \lambda_1} \tag{6.46}$$

and

$$c_2 = \frac{(Y_0 - d_1)\lambda_1 - V_0 + d_2\omega}{\lambda_1 - \lambda_2} \tag{6.47}$$

where λ_1 and λ_2 are given by Equation 6.21.

The type of roots (real or complex) in Equation 6.21 leads to different forms of complete solution. Each case will be looked into separately.

- *Roots are real and distinct* ($\xi > 1$)

In this case the complete solution is

$$y_c(t) = c_1 e^{-|\lambda_1|t} + c_2 e^{-|\lambda_2|t} + d_1 \cos\omega t + d_2 \sin\omega t \tag{6.48}$$

One can see that the first two terms in Equation 6.48, associated with the initial disturbance, tend to zero exponentially and so after some time, practically, only the periodic terms caused by the external load will remain. A system in which the nondimensional damping coefficient $\xi > 1$ is said to be *overdamped.*

- *Roots are complex* ($\xi < 1$)

In this case the complete solution is (see Equation 6.25 for the general part of the solution)

$$y_c(t) = e^{-\xi\omega_n t}(a_1 \cos\beta t + a_2 \sin\beta t) + d_1 \cos\omega t + d_2 \sin\omega t \tag{6.49}$$

One can see that in this case the amplitude of the initial disturbance is also decreasing exponentially while oscillating. A system in which the nondimensional damping coefficient $\xi < 1$ is said to be *underdamped.* However, in this case, like in the previous one, the effect of initial disturbance disappears after some time and only the periodic oscillation remains.

- *Roots are real and equal* ($\xi = 1$)

In this case the complete solution is

$$y_c(t) = e^{-\xi\omega_n t}(c_1 + c_2 t) + d_1 \cos\omega t + d_2 \sin\omega t \tag{6.50}$$

One can see that the initial disturbance tends to zero and after some time only the oscillation caused by the periodic external force remain. This is the boundary case between the oscillating and nonoscillating initial disturbance, and the corresponding damping is called *critical damping.*

The next section considers various applications of the above solution to an SDOF system.

6.3 FREE VIBRATIONS OF AN SDOF SYSTEM WITH NO DAMPING ($\xi = 0, p_o = 0$)

The characteristic equation, Equation 6.20, in this case has imaginary roots (see Equation 6.21) $\lambda_{1,2} = \pm i(\omega_n)$. This is a particular case of the complete solution obtained in Section 6.2.3 and the corresponding solution can be obtained from Equation 6.49 by taking, $\xi = 0, d_1 = 0, d_2 = 0,$ and $\beta = \omega_n$. As a result, one obtains

$$y_g(t) = a_1 \cos\omega_n t + a_2 \sin\omega_n t \tag{6.51}$$

where a_1 and a_2 are constants defined by the initial conditions:

$$y_g(0) = a_1 = Y_0 \tag{6.52}$$

and

$$\dot{y}_g(0) = a_2\omega_n = V_0 \tag{6.53}$$

Thus, the final form of the solution for an SDOF system without damping is

$$y_g(t) = Y_0 \cos\omega_n t + \frac{V_0}{\omega_n}\sin\omega_n t \tag{6.54}$$

This equation once more shows that any disturbance causes the system to oscillate with the circular frequency ω_n, which is defined by the system properties. This frequency is called the *natural frequency* of the system. One can also see that the effects of initial displacement and initial velocity are uncoupled, which is another manifestation of the *principle of superposition* in linear systems. In other words, one can solve first for initial displacement, second for initial velocity, and then combine the results.

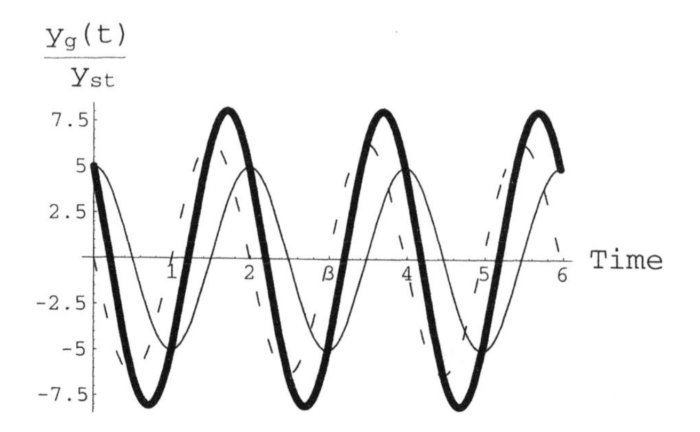

FIGURE 6.4 Normalized amplitude of free vibrations of an SDOF system without damping (y_{st} is the static displacement). Dashed line = initial velocity; solid thin line = initial displacement; solid thick line = both.

In Figure 6.4 oscillations of an SDOF system are shown for the case of initial displacement only, initial velocity only, and for both, displacement and velocity, conditions. The time between two consecutive peaks is the *period of oscillation.* One can see that the period is the same for any type of initial disturbance. Recall that this period is equal to $T = 2\pi/\omega_n$ and ω_n does not depend on the type of disturbance.

6.4 FORCED VIBRATIONS OF AN SDOF SYSTEM WITH NO DAMPING ($\xi = 0$)

It is assumed that forced vibrations start at time zero from the undisturbed state of the system. Thus, the initial conditions are $Y_0 = 0$ and $V_0 = 0$ at $t = 0$. The forcing function is $p_0 \sin(\omega t)$. The complete solution is given by Equation 6.49 in which the first two terms on the right describing the vibrations caused by the initial conditions contain two unknown (a_1 and a_2) constants, whereas the constants in the last two terms (d_1 and d_2) are given by Equations 6.40 and 6.41 and Equation 6.37. Taking into account that, for the case of $\xi = 0$, $\beta = \omega_n$ and $d_1 = 0$, the complete solution is

$$y_c(t) = a_1 \cos \omega_n t + a_2 \sin \omega_n t + d_2 \sin \omega t \tag{6.55}$$

where the constants a_1 and a_2 are found by satisfying the initial conditions

$$y_c(0) = a_1 = 0 \tag{6.56}$$

and

$$\left.\frac{dy_c(t)}{dt}\right|_{t=0} = a_2\omega_n + d_2\omega = 0 \tag{6.57}$$

It follows from the latter equation that

$$a_2 = -d_2\frac{\omega}{\omega_n} \tag{6.58}$$

Substituting the above constants into Equation 6.55 yields

$$y_c(t) = d_2\left(\sin\omega t - \frac{\omega}{\omega_n}\sin\omega_n t\right) \tag{6.59}$$

where (see Equation 6.41)

$$d_2 = \frac{D^2(\omega_n^2 - \omega^2)}{p_0} \tag{6.60}$$

and (see Equation 6.37)

$$D = \frac{p_0}{\omega_n^2 - \omega^2} \tag{6.61}$$

Thus, the final form of the solution of forced vibrations starting from zero initial conditions is

$$y_c(t) = \frac{p_0}{\omega_n^2 - \omega^2}\left(\sin\omega t - \frac{\omega}{\omega_n}\sin\omega_n t\right) \tag{6.62}$$

The obtained result shows that the forced vibration in this case is a superposition of two motions: one with the frequency of the external force ω and the other with the natural frequency ω_n. The resultant amplitude of vibrations is a function of the frequency ω, and at $\omega = \omega_n$ it becomes undetermined since both numerator and denominator equal zero at this frequency. One can use L'Hopital's rule to resolve the uncertainty by taking the ratio of derivatives of the numerator and denominator with respect to ω and then setting $\omega = \omega_n$. The result is

$$y_c(\omega = \omega_n) = \frac{-\sin\omega_n t + \omega_n t\cos\omega_n t}{-2\omega_n} \tag{6.63}$$

One can see that the amplitude of vibrations in this case is proportional to the time t and thus grows to infinity. This phenomenon of unlimited amplitude growth is called *resonance*, and the corresponding frequency is called the *resonance frequency*. As is clear, the resonance frequency is the natural frequency of the system. It will be seen later that in real systems with damping these frequencies are close but not equal.

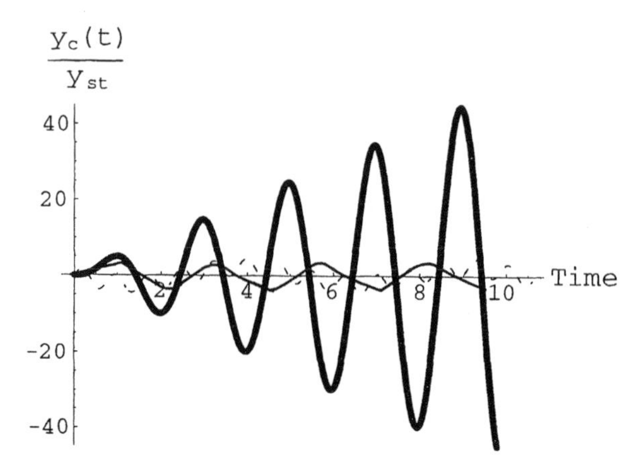

FIGURE 6.5 Dynamic response in time of an SDOF system without damping (y_{st} is the static displacement). Dashed line = $\omega/\omega_n = 0.5$; solid thin line = $\omega/\omega_n = 0.75$; solid thick line = $\omega/\omega_n = 1$.

In Figure 6.5 the displacement of the mass in a system without damping is shown at different frequencies of external force. One can see that at the resonance frequency the amplitude of displacement grows linearly with time.

6.5 STEADY-STATE FORCED VIBRATIONS OF AN SDOF SYSTEM WITH NO DAMPING ($\xi = 0$)

In this case the effect of the initial conditions is neglected and thus only the particular solution is considered. The latter is given by Equation 6.55 in which both a_1 and a_2 equal zero. Thus, Equation 6.62 in which the second term in the parentheses is absent gives the solution

$$y_c(t) = \frac{p_0}{\omega_n^2 - \omega^2} \sin \omega t \tag{6.64}$$

This solution shows that the frequency of vibrations in this case is equal to the frequency of the forcing function, and the amplitude is a function of this frequency. One can see now that when $\omega = \omega_n$ the amplitude becomes infinite. Now look more closely at the amplitude as a function of forcing frequency ω.

$$D = \frac{p_0}{\omega_n^2 - \omega^2} \tag{6.65}$$

From the latter equation it follows that, when $\omega = 0$, then

$$D = \frac{p_0}{\omega_n^2} = \frac{P_0 m}{mk} = \frac{P_0}{k} = D_{st}$$

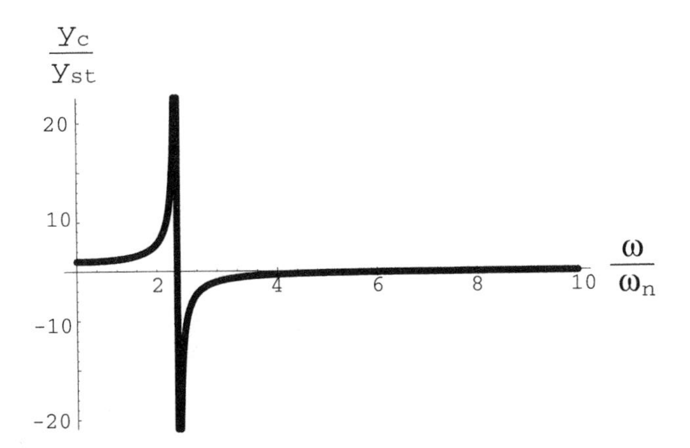

FIGURE 6.6 Amplitude–frequency diagram of an SDOF system without damping (y_{st} is the static displacement).

where D_{st} is the static displacement of the mass caused by the force P_0. Furthermore, when ω approaches ω_n from the left ($\omega < \omega_n$), then the amplitude tends to infinity while remaining positive, whereas when ω approaches ω_n from the right ($\omega > \omega_n$), then the amplitude tends to infinity while remaining negative. Also, when ω tends to infinity, then the amplitude tends to zero while being negative. This dependence of the amplitude on the frequency is shown in the diagram of Figure 6.6 in normalized coordinates. This diagram is called the *amplitude–frequency diagram.*

The function in the diagram depicted in Figure 6.6 comprises two continuous functions. Such a function is called a *piecewise continuous function* and can be described as follows:

$$y_c(t) = \begin{cases} \left|\dfrac{p_0}{\omega_n^2 - \omega^2}\right| \sin(\omega t) & \text{if } \omega < \omega_n \\ \left|\dfrac{p_0}{\omega_n^2 - \omega^2}\right| \sin(\omega t - \pi) & \text{if } \omega > \omega_n \end{cases} \tag{6.66}$$

In Equation 6.66 the amplitude is now always positive, whereas the sign change after the resonance is controlled by the shift in the angle by π. This angle π is called the *phase angle.* Thus, up to the resonance frequency the phase angle is equal to 0, and after the resonance frequency it becomes $-\pi$. Below it will be seen that for systems with damping this change in the phase angle is continuous.

6.6 FREE VIBRATIONS OF AN SDOF SYSTEM WITH DAMPING ($\xi \neq 0, p_0 = 0$)

As was discussed in Section 6.2.1, the form of the general part of the solution depends on the degree of damping. Consider each case separately.

- *Overdamped system* ($\xi > 1$)

In this case Equation 6.42, in which d_1 and d_2 equal zero, gives the complete solution:

$$y_c(t) = c_1 e^{\lambda_1 t} + c_2 e^{\lambda_2 t} \tag{6.67}$$

and c_1 and c_2 are determined by the initial conditions:

$$y_c(0) = c_1 + c_2 = Y_0 \tag{6.68}$$

and

$$\dot{y}_c(0) = c_1\lambda_1 + c_2\lambda_2 = V_0 \tag{6.69}$$

Taking into account that $\lambda_1 = (-\xi + \sqrt{\xi^2 - 1})\omega_n$, $\lambda_2 = (-\xi - \sqrt{\xi^2 - 1})\omega_n$ (see Equation 6.21), the expressions for the constants c_1 and c_2 become

$$c_1 = -\frac{Y_0\lambda_2 - V_0}{2\sqrt{\xi^2 - 1}\,\omega_n} \tag{6.70}$$

and

$$c_2 = \frac{Y_0\lambda_1 - V_0}{2\sqrt{\xi^2 - 1}\,\omega_n} \tag{6.71}$$

One can see that the two initial conditions are independent of each other. In Figure 6.7 the response of the overdamped system to initial conditions is shown for the case of $\xi = 1.2$, $\omega_n = 10$ rad/s, $Y_0 = 1$ cm, and $V_0 = 5$ cm/s.

- *Underdamped system* ($\xi < 1$)

In this case the complete solution is given by Equation 6.49 in which d_1 and d_2 equal zero.

$$y_c(t) = e^{-\xi\omega_n t}(a_1 \cos\beta t + a_2 \sin\beta t) \tag{6.72}$$

where a_1 and a_2 are obtained from the initial conditions requirements

$$y_c(0) = a_1 = Y_0 \tag{6.73}$$

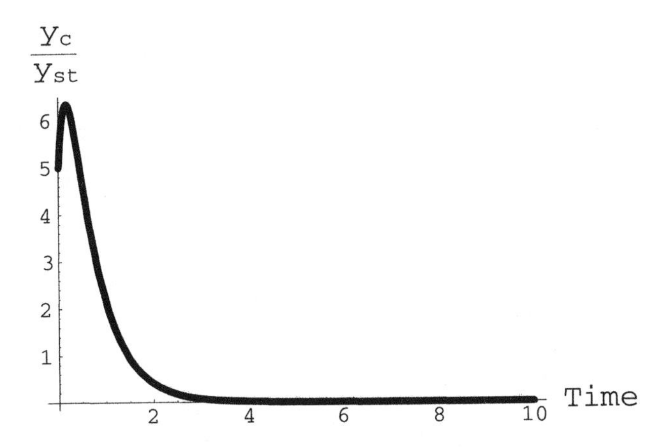

FIGURE 6.7 Response of the overdamped ($\xi = 1.2$) SDOF system to initial conditions.

and

$$\left.\frac{dy_c(t)}{dt}\right|_{t=0} = -\xi\omega_n a_1 + a_2\beta = V_0 \tag{6.74}$$

Taking into account Equation 6.73, the second constant is

$$a_2 = \frac{V_0 + \xi\omega_n Y_0}{\beta} \tag{6.75}$$

where $\beta = \sqrt{1-\xi^2}\,\omega_n$ (see Section 6.2.1).

Recall that the undamped system oscillates with natural frequency ω_n when subjected to initial disturbance. In the case of a damped system, the frequency of oscillation becomes

$$\beta = \omega_d = \sqrt{1-\xi^2}\,\omega_n \tag{6.76}$$

One can see that this *frequency of free damped vibrations* is always smaller than the natural frequency, $\omega_d < \omega_n$.

The displacement of the mass in time is shown in Figure 6.8. As is seen from Equation 6.72 the amplitude of the oscillating motion is decreasing exponentially. This property of the declining amplitude is used to determine the damping coefficient experimentally. This will be discussed in Section 6.8.

It is important to point out that the coefficient of damping is not the property of the material only, since it depends also on the system mass and stiffness. Note that the material properties are not present explicitly in motion equations. It was only assumed in Section 6.2 that the material possesses some damping properties represented by the damping coefficient.

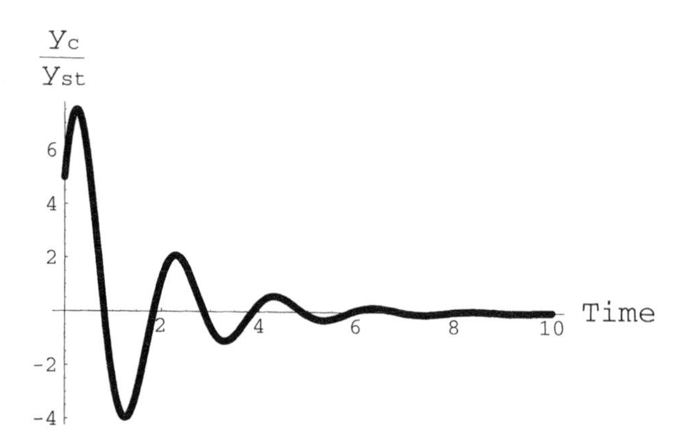

FIGURE 6.8 Response of the underdamped ($\xi = 0.2$) SDOF system to initial conditions.

- *Critical damping* ($\xi = 1$)

The solution is given by Equation 6.50 in which d_1 and d_2 equal zero.

$$y_c(t) = e^{-\omega_n t}(c_1 + c_2 t) \tag{6.77}$$

Satisfying the initial conditions yields

$$y_c(0) = c_1 = Y_0 \tag{6.78}$$

and

$$\left.\frac{dy_c(t)}{dt}\right|_{t=0} = -\omega_n c_1 + c_2 = V_0 \tag{6.79}$$

From Equations 6.78 and 6.79 the coefficient c_2 follows

$$c_2 = V_0 + \omega_n Y_0 \tag{6.80}$$

In Figure 6.9 the system response to initial conditions in the case of critical damping is shown. In Figure 6.10 the graphs from Figures 6.7 through 6.9 are shown together for the sake of comparison.

6.7 FORCED VIBRATIONS OF A DAMPED ($\xi < 1$) SDOF SYSTEM WITH INITIAL CONDITIONS

Equation 6.49, repeated here for the sake of convenience, gives the complete solution:

$$y_c(t) = e^{-\xi\omega_n t}(a_1\cos\beta t + a_2\sin\beta t) + d_1\cos\omega t + d_2\sin\omega t \tag{6.81}$$

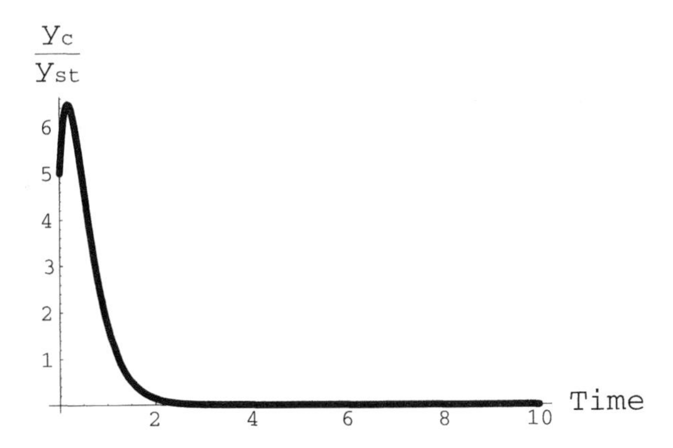

FIGURE 6.9 Response of the critically damped ($\xi = 1$) SDOF system to initial conditions.

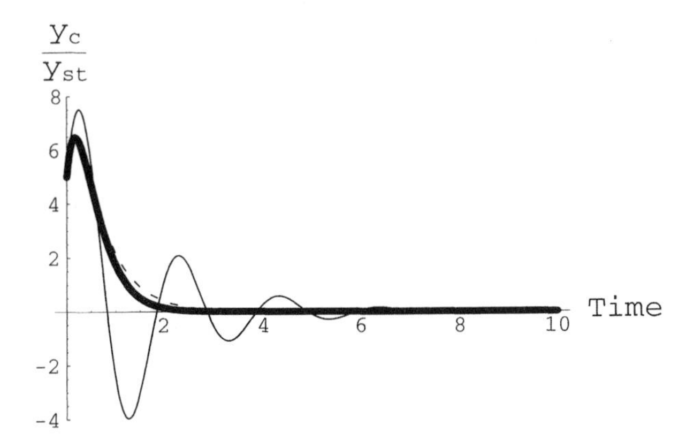

FIGURE 6.10 Comparison of an SDOF system response to initial conditions. Dashed line = overdamping ($\xi = 1.2$); solid thin line = underdamping ($\xi = 0.2$); solid thick line = critical damping ($\xi = 1$).

Take the same initial conditions as in the case of no damping in Section 6.4, namely, that $Y_0 = 0$ and $V_0 = 0$ at $t = 0$, and the forcing function is also the same, $p_0 \, sin(\omega t)$. The equations defining constants are given by Equations 6.46 and 6.47:

$$c_1 = \frac{-d_1\lambda_2 + d_2\omega}{\lambda_2 - \lambda_1} \tag{6.82}$$

and

$$c_2 = \frac{-d_1\lambda_1 + d_2\omega}{\lambda_1 - \lambda_2} \tag{6.83}$$

where λ terms are given by Equation 6.21, $\lambda_{1,2} = (-\xi \pm \sqrt{\xi^2 - 1})\omega_n$, and d_1 and d_2 by Equations 6.40 and 6.41, $d_1 = -2D^2\xi\omega_n\omega/p_0$ and $d_2 = D^2(\omega_n^2 - \omega^2)/p_0$. One can take into account that the roots of the characteristic equation in this case are complex and use simplified expressions for the λ, $\lambda_{1,2} = \alpha \pm i\beta$, where $\alpha = -\xi\omega_n$ and $\beta = \sqrt{1 - \xi^2}\,\omega_n$. Then, the constants c_1 and c_2 become complex conjugate numbers:

$$c_1 = -0.5\left[d_1 - i\left(-d_1\frac{\alpha}{\beta} + d_2\frac{\omega}{\beta}\right)\right] \tag{6.84}$$

and

$$c_2 = -0.5\left[d_1 + i\left(-d_1\frac{\alpha}{\beta} + d_2\frac{\omega}{\beta}\right)\right] \tag{6.85}$$

Thus, the constants a_1 and a_2 in Equation 6.81 become [recall that $a_1 = c_1 + c_2$ and $a_2 = i(c_1 - c_2)$]

$$a_1 = c_1 + c_2 = -d_1 \tag{6.86}$$

and

$$a_2 = i(c_1 - c_2) = d_1\frac{\alpha}{\beta} - d_2\frac{\omega}{\beta} \tag{6.87}$$

In Figures 6.11 and 6.12 the vibrations of an SDOF initially undisturbed system are shown for two forcing frequencies and $\xi = 0.4$, $p_0 = 1$. It is seen that for a frequency close enough to the resonance frequency, $\omega = 0.75\omega_n$, the effect of initial conditions is visible only for the first cycle of motion (Figure 6.11), while for a frequency far enough from the resonance frequency, $\omega = 0.25\omega_n$, the effect of initial conditions is not visible at all (Figure 6.12).

In general, the effect of initial conditions diminishes very quickly. This allows one to neglect it all together in applications and thus to consider forced vibrations of systems with damping as a steady-state process.

6.8 FORCED VIBRATIONS OF AN SDOF SYSTEM WITH DAMPING ($\xi < 1$) AS A STEADY-STATE PROCESS

In this case the coefficients a_1 and a_2 in Equation 6.81 are set to zero. Thus, the motion is described by the particular solution of the nonhomogeneous equation, Equation 6.28:

$$y_p(t) = D\sin(\omega t - \phi) \tag{6.88}$$

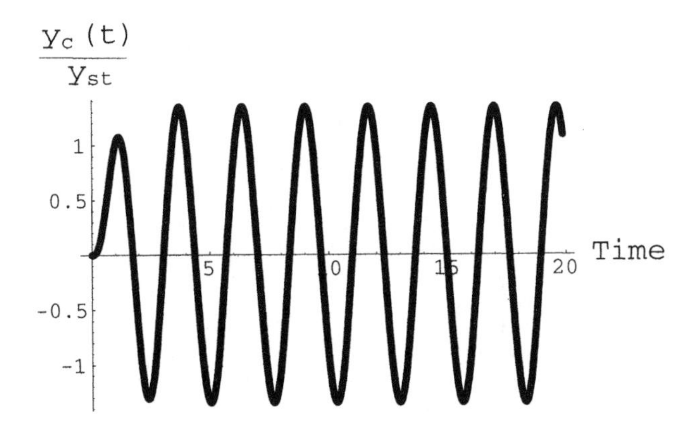

FIGURE 6.11 Forced vibrations of an SDOF system with damping from undisturbed state ($\omega = 0.75\omega_n$).

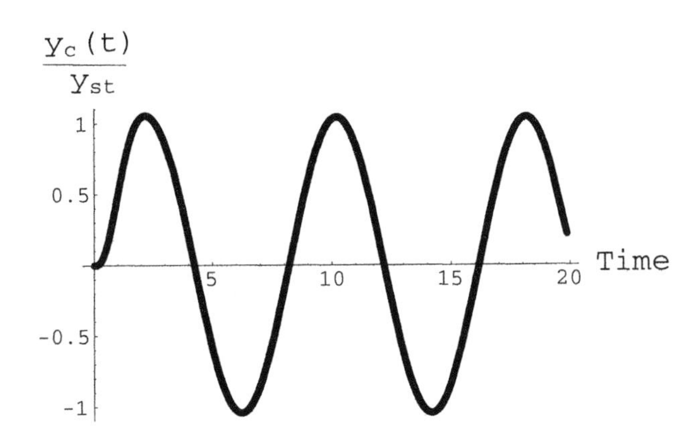

FIGURE 6.12 Forced vibrations of an SDOF system with damping from undisturbed state ($\omega = 0.25\omega_n$).

where

$$D = \frac{p_0}{\sqrt{(\omega_n^2 - \omega^2)^2 + (2\xi\omega_n\omega)^2}} \tag{6.89}$$

and the angle ϕ

$$\phi = \arctan\left(\frac{2\xi\omega_n\omega}{\omega_n^2 - \omega^2}\right) \tag{6.90}$$

The maximum amplitude is achieved when $\omega = \omega_n\sqrt{1 - 2\xi^2}$ and it is equal to

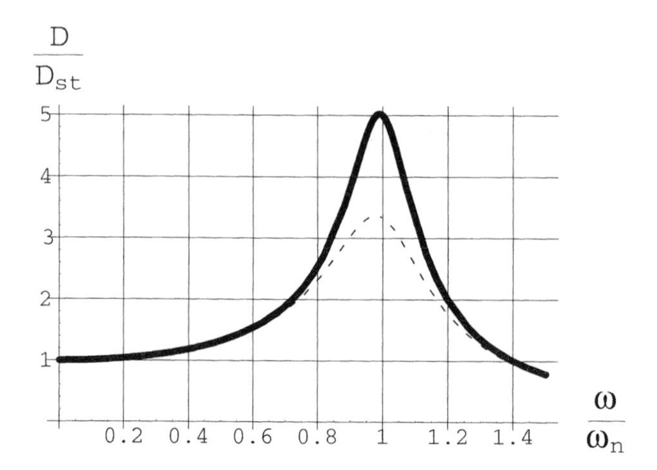

FIGURE 6.13 Amplitude–frequency diagrams for an SDOF system with damping; $\xi = 0.1$ (solid line), $\xi = 0.15$ (dashed line).

$$D_{\max} = \frac{p_0}{2\xi\omega_n^2\sqrt{1-2\xi^2}} \tag{6.91}$$

where $p_0 = P_0/m$ and $\xi = c/2m\omega_n$.

Equation 6.91 can be used to measure damping properties experimentally by finding the amplitude of vibrations at the resonance for the given amplitude of the forcing function.

In Figure 6.13 the normalized (with respect to the static displacement) amplitude–frequency diagram for a system with damping are shown for two damping properties. One can see that the maximum amplitude takes place at $\omega < \omega_n$.

In Figure 6.14 the change of the phase angle with frequency is shown. For comparison, a phase angle in the system without damping is also given. One can see that the higher the damping, the more it affects the phase angle. The change in sign of the phase angle means that below the resonance frequency the response lags behind the forcing function, whereas above the resonance frequency it forestalls this function.

It was assumed above that the forcing function has the form $p_0 \sin \omega t$. Assume a more general complex forcing function, $P(t) = P_0 e^{i\omega t}$. Then Equation 6.27 becomes

$$\ddot{y} + 2\xi\omega_n\dot{y} + \omega_n^2 y = P_0 e^{i\omega t} \tag{6.92}$$

Here a particular solution of Equation 6.92 is of interest. One can see that the solution will be a complex number as well. By separating it into real and imaginary parts, one can find solutions to periodic excitations by $p_0 \sin \omega t$ and $p_0 \cos \omega t$. The particular solution of Equation 6.92 has the form:

$$y(t) = De^{i\omega t} \tag{6.93}$$

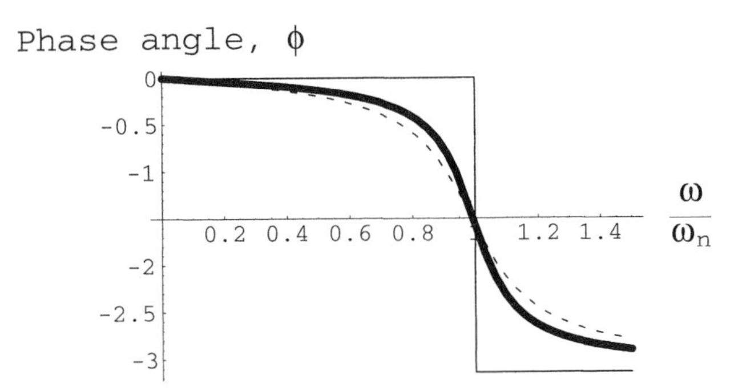

FIGURE 6.14 Phase angle vs. normalized frequency for an SDOF system: $\xi = 0.1$ (solid thick line), $\xi = 0.15$ (dashed line), $\xi = 0$ (solid thin line).

After substitution of the above form of the solution into Equation 6.92 and canceling $e^{i\omega t}$, one obtains $D = D(i\omega)$

$$D(i\omega) = \frac{p_0}{\omega_n^2 - \omega^2 + i2\xi\omega_n\omega} \tag{6.94}$$

The latter can be transformed into a conventional complex number by multiplying the numerator and denominator by $\omega_n^2 - \omega^2 - i2\xi\omega_n\omega$. As a result,

$$D(i\omega) = p_0\left[\frac{\omega_n^2 - \omega^2}{(\omega_n^2 - \omega^2)^2 + (2\xi\omega_n\omega)^2} - i\frac{2\xi\omega_n\omega}{(\omega_n^2 - \omega^2)^2 + (2\xi\omega_n\omega)^2}\right] \tag{6.95}$$

Any complex number can be represented as $a + ib = |a + ib|e^{i\phi}$, where $|a + ib| = \sqrt{a^2 + b^2}$, and $\phi = \arctan(b/a)$. Applying this to Equation 6.95 yields

$$|D(i\omega)| = \frac{p_0}{\sqrt{(\omega_n^2 - \omega^2)^2 + (2\xi\omega_n\omega)^2}} \tag{6.96}$$

and

$$\phi = -\arctan\left[\frac{2\xi\omega_n\omega}{\omega_n^2 - \omega^2}\right] \tag{6.97}$$

The results are the same as in Equations 6.89 and 6.90, except for the sign of the phase angle. Now one can present the particular solution with complex forcing function in the form:

$$y_p(t) = p_0|H(i\omega)|e^{i(\omega t - \phi)} \tag{6.98}$$

where $|H(i\omega)| = |D(i\omega)|/p_0$, and $H(i\omega)$ is called the *complex frequency response function,* which does not depend on the external force but only on the properties of the system. From Equation 6.98, it can be seen that

$$H(i\omega) = |H(i\omega)|e^{-i\phi} \tag{6.99}$$

The frequency response function is a convenient way to describe the dynamic properties of systems and it will be considered again later for the case of a 2DOF system.

6.9 COEFFICIENT OF DAMPING, LOGARITHMIC DECREMENT, AND ENERGY LOSSES

In the previous section it was indicated that the coefficient of damping c can be determined from vibration tests at resonance frequency (Equation 6.91). At the same time, Figure 6.8 showed that in a free vibrating system the amplitude decreases exponentially according to Equation 6.72. If the curve shown in Figure 6.8 is found experimentally, then it provides enough information to find the damping properties of the system. Identify two consecutive maximums on a curve in Figure 6.8. Apparently, the time difference between these two points is the period of free damped oscillation T_d. Now, consider the ratio of these two amplitudes:

$$\frac{y_c(t)}{y_c(t+T_d)} = \left[\frac{e^{-\xi\omega_n t}}{e^{-\xi\omega_n(t+T_d)}}\right]\left[\frac{a_1\cos\beta t + a_2\sin\beta t}{a_1\cos\beta(t+T_d) + a_2\sin\beta(t+T_d)}\right] \tag{6.100}$$

In the above equation, the second ratio on the right-hand side is equal to 1, since T_d is the period of the trigonometric functions. Thus, the above ratio is reduced to

$$\frac{y_c(t)}{y_c(t+T_d)} = e^{\xi\omega_n T_d} \tag{6.101}$$

The logarithm of this ratio is called the *logarithmic decrement*, δ, and it is equal to

$$\delta = \ln\frac{y_c(t)}{y_c(t+T_d)} = \xi\omega_n T_d \tag{6.102}$$

Taking into account that

$$T_d = \frac{2\pi}{\beta} = \frac{2\pi}{\omega_n\sqrt{1-\xi^2}} \tag{6.103}$$

the nondimensional damping coefficient can be expressed through the logarithmic decrement:

$$\xi = \frac{\delta}{\sqrt{4\pi^2 - \delta^2}} \tag{6.104}$$

By recalling that $\xi = c/(2m\omega_n)$, the damping coefficient c can be found for a given logarithmic decrement:

$$c = \frac{2m\omega_n\delta}{\sqrt{4\pi^2 - \delta^2}} \tag{6.105}$$

If $\delta^2/4\pi^2 << 1$, then the above relationships can be simplified:

$$\xi = \frac{\delta}{2\pi} \tag{6.106}$$

and

$$c = \frac{m\omega_n\delta}{\pi} \tag{6.107}$$

Once again it is worth noting that δ found experimentally reflects the properties of a system tested, and so does the coefficient of damping c.

The presence of damping, and associated viscous forces, results in energy losses during vibrations. In Section 6.1 the case of energy conservation in a mass–spring system was considered. In general, systems in which there are no losses are called *conservative*, whereas systems with losses are *nonconservative*. Since mechanical losses are transformed into thermal energy, it is important to know how much energy is lost during one cycle (or the *dissipation energy* of the system).

The lost energy is equal to the product of the viscous force $F_d = c\dot{y}$, and the velocity of moving mass, $\dot{y}$

$$V_d = c\dot{y}^2 \tag{6.108}$$

Here, a case of forced vibrations of a system with damping in a steady-state regime is of interest. As is known (see Section 6.8), the motion is described by Equation 6.88. Assume that the forcing frequency is equal to the frequency of free vibrations of a damped system, i.e., $\omega = \omega_d = \sqrt{1-\xi^2}\omega_n$.

$$y_p(t) = D\sin(\omega_d t - \phi) \tag{6.109}$$

The work done by the viscous force during one cycle is equal to

$$\Delta W_d = \int_0^{T_d} c\dot{y}_p^2 dt = cD^2\omega_d^2\int_0^{T_d}\cos^2(\omega_d t - \phi)dt \tag{6.110}$$

Using the trigonometric identity $\cos^2\gamma = 1/2(1 + \cos 2\gamma)$ and then taking into account that the integral from a periodic function $\cos 2(\omega_d t - \phi)$ within the limits 0 to T_d is zero, the above expression after integration becomes

$$\Delta W_d = \frac{1}{2} c D^2 \omega_d^2 T_d = \pi c D^2 \omega_d \tag{6.111}$$

It is of interest to know the ratio of the energy dissipated during one cycle to the maximum energy in the system during this cycle. The latter is equal to either maximum kinetic energy $(W_{max} = (1/2) m D^2 \omega_d^2)$ or maximum potential energy $(W_{max} = (1/2) k D^2)$. The ratio then is equal to

$$\frac{\Delta W}{W_{max}} = \frac{2\pi\omega_d c}{k} \tag{6.112}$$

The latter, taking into account Equation 6.107 and that $\omega_d = \sqrt{1 - \xi^2}\,\omega_n$, is reduced to

$$\frac{\Delta W}{W_{max}} = 2\delta\sqrt{1 - \xi^2} \tag{6.113}$$

In the case of $\xi \ll 1$, the above ratio can be simplified:

$$\frac{\Delta W}{W_{max}} = 2\delta \tag{6.114}$$

The ratio $\Delta W / W_{max}$ characterizes the damping capacity of the system.

6.10 KINEMATIC EXCITATION

Thus far, energy transferred to the system was either by initial disturbance or by external force. An alternative way is to shake the frame on which the mass–spring–dashpot system is mounted, in the same way a building is shaken by an earthquake or a car is shaken when going over a bump. This type of excitation is called *kinematic excitation.*

Consider an SDOF system, as shown in Figure 6.15, in which the base oscillates with some frequency ω. The absolute displacement of the mass is equal to the sum of the displacement of the base, $y_b(t)$, and the displacement of the mass with respect to the base, $y_1(t)$:

$$y(t) = y_b(t) + y_1(t) \tag{6.115}$$

The forces in the spring and the dashpot depend on the differences of displacements, $y - y_b$, and velocities, $\dot{y} - \dot{y}_b$ (see Figure 6.15b), whereas the inertial force depends on the absolute acceleration of the mass. Thus, the following motion equation exists:

$$m\ddot{y} + c\dot{y}_1 + k y_1 = 0 \tag{6.116}$$

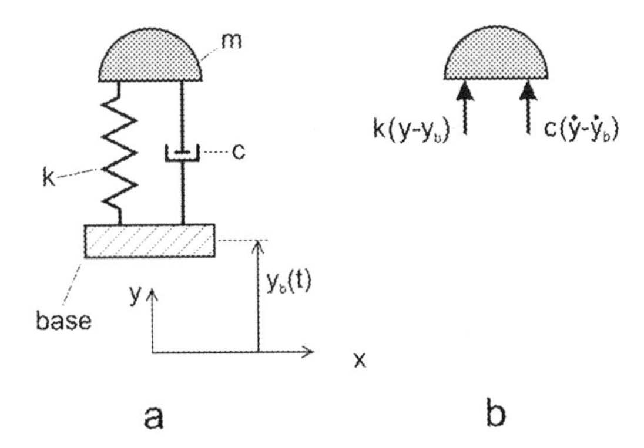

FIGURE 6.15 Kinematic excitation of an SDOF system.

Considering the periodic motion of the base,

$$y_b(t) = Y_b \sin \omega t \tag{6.117}$$

Equation 6.116 is reduced to

$$m\ddot{y}_1 + c\dot{y}_1 + ky_1 = -mY_b\omega^2 \sin \omega t \tag{6.118}$$

Thus, the problem of kinematic excitation of a damped SDOF system is reduced to the problem of forced excitation of this system. The force in this case is the inertial force of the mass created by a moving frame. This equation can be rewritten with nondimensional coefficients:

$$\ddot{y}_1 + 2\xi\omega_n c\dot{y}_1 + \omega_n^2 y_1 = p_0 \sin \omega t \tag{6.119}$$

which is the same as Equation 6.27, except in this case $p_0 = Y_b\omega^2$. Thus, the results given in Sections 6.2.1 through 6.2.3 can be used in this case.

6.11 GENERAL PERIODIC EXCITATION

There are many situations when the forcing function is a sum of a few periodic functions. For example, a gear generates a periodic force with the so-called teeth frequency. Thus, if two or more gears are mounted on the same shaft, the shaft will be subjected to periodic forces with different frequencies. Since the sum of two periodic functions is also a periodic function, this case can be analyzed using all the previous results.

In general, if the external force $F(t)$ has a period T, then this function can be represented as a Fourier series:

$$F(t) = \frac{a_0}{2} + \sum_{j=1}^{\infty} a_j \cos j\omega t + \sum_{j=1}^{\infty} b_j \sin j\omega t \tag{6.120}$$

where

$$a_j = \frac{2}{T}\int_0^T F(t)\cos j\omega t dt, \quad j = 0, 1, 2\ldots \tag{6.121}$$

and

$$b_j = \frac{2}{T}\int_0^T F(t)\sin j\omega t dt, \quad j = 0, 1, 2\ldots \tag{6.122}$$

The equation of motion of an SDOF system subjected to force $F(t)$ is

$$\ddot{y} + 2\xi\omega_n\dot{y} + \omega_n^2 y = \frac{a_0}{2m} + \frac{1}{m}\sum_{j=1}^{\infty} a_j\cos j\omega t + \frac{1}{m}\sum_{j=1}^{\infty} b_j\sin j\omega t \tag{6.123}$$

The forcing function on the right-hand side is a sum of periodic functions. One can use the *principle of superposition* to find the solution as a sum of solutions for each periodic function. Then the problem is reduced to finding the solution for the first term on the right and then for the terms with the index j.

$$\ddot{y}_0 + 2\xi\omega_n\dot{y}_0 + \omega_n^2 y_0 = \frac{a_0}{2m} \tag{6.124}$$

$$\ddot{y}_{j1} + 2\xi\omega_n\dot{y}_{j1} + \omega_n^2 y_{j1} = \frac{1}{m}a_j\cos j\omega t \tag{6.125}$$

$$\ddot{y}_{j2} + 2\xi\omega_n\dot{y}_{j2} + \omega_n^2 y_{j2} = \frac{1}{m}b_j\sin j\omega t \tag{6.126}$$

The particular solution of the first equation (Equation 6.124) is a constant:

$$y_0 = \frac{a_0}{2m\omega_n^2} = \frac{a_0}{2k} \tag{6.127}$$

The particular solutions of the next two equations (Equations 6.125 and 6.126) are given in Section 6.2, where it was indicated that the solutions for both cosine and sine functions remain the same. However, in this case the value of p_0 is different and it should be taken into account in expressions for the amplitudes. Thus,

$$y_{j1}(t) = D_{j1}\sin(j\omega t - \phi_{j1}) \tag{6.128}$$

and

$$y_{j2}(t) = D_{j2}\sin(j\omega t - \phi_{j2}) \tag{6.129}$$

where

$$D_{j1} = \frac{a_j}{m\sqrt{(\omega_n^2 - j\omega^2)^2 + (2\xi\omega_n j\omega)^2}} \tag{6.130}$$

$$D_{j2} = \frac{b_j}{m\sqrt{(\omega_n^2 - j\omega^2)^2 + (2\xi\omega_n j\omega)^2}} \tag{6.131}$$

$$\phi_{j1} = \phi_{j2} = \arctan\left(\frac{2\xi\omega_n j\omega}{\omega_n^2 - j\omega^2}\right) \tag{6.132}$$

Thus, the solution of Equation 6.123 is

$$y(t) = \frac{a_0}{2k} + \sum_{j1=1}^{\infty} D_{j1}\sin(j\omega t - \phi_{j1}) + \sum_{j2=1}^{\infty} D_{j2}\sin(j\omega t - \phi_{j2}) \tag{6.133}$$

6.12 TORSIONAL VIBRATIONS

Vibrations of a mass on a spring take place along the axis (axis y in Figure 6.1). In Figure 6.16 a system is shown in which one end of the shaft with a diameter d is attached to a frame, while at the other end a disk is mounted on it. If the disk is twisted and then released, the disk will be oscillating by rotating around the shaft axis in a periodic fashion. By twisting the shaft, some resistance moment, $k_t\theta$, is generated, caused by the angular stiffness of the shaft, k_t. If the shaft material has some damping properties, then another resistance moment, $c_t\dot{\theta}$, will be generated proportional to the angular velocity of the disk, $\dot{\theta}$. Taking into account that the inertial moment is equal to $J\ddot{\theta}$, the equilibrium of the disk subjected to all these forces (Figure 6.1b) is

$$J\ddot{\theta} + c_t\dot{\theta} + k_t\theta = 0 \tag{6.134}$$

where J is the moment of inertia of the disk, c_t is the torsional damping coefficient, and k_t is the torsional stiffness.

If an external periodic moment is applied to the disk, then the disk oscillations are described by a nonhomogeneous equation:

$$J\ddot{\theta} + c_t\dot{\theta} + k_t\theta = M_0\sin\omega t \tag{6.135}$$

One can transform this equation into one with nondimensional coefficients, similar to the transformation in Section 6.2.1. As a result,

$$\ddot{\theta} + 2\xi_t\omega_n\dot{\theta} + \omega_n^2\theta = m_0\sin\omega t \tag{6.136}$$

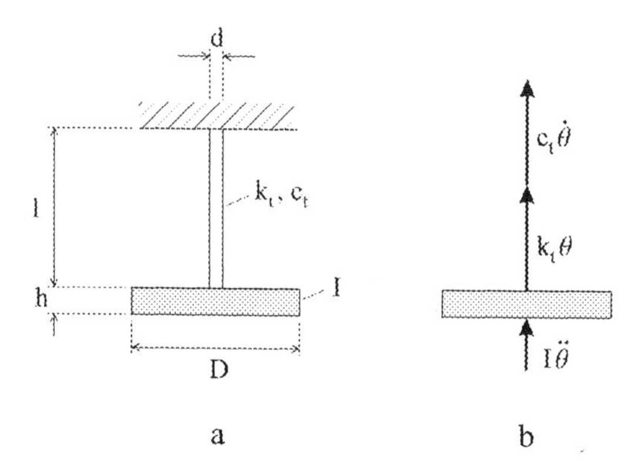

FIGURE 6.16 Torsional vibrations of a disk (a), and free-body diagram of the disk (b).

where it is denoted

$$\xi_t = \frac{c_t}{2J\omega_n} \text{ and } \omega_n^2 = \frac{k}{J} \tag{6.137}$$

One can see that Equation 6.136 differs from Equation 6.27 only in notations, and thus, all solutions developed for the latter equation are applicable to the former.

6.13 MULTIDEGREE-OF-FREEDOM SYSTEMS

An N-degree-of-freedom system is shown in Figure 6.17a. One can write the equation of motion for an intermediate mass m_j. Forces acting on this mass depend on the difference in displacement and velocities of this mass and its neighbors (Figure 6.17b). Thus, Newton's law written for this mass gives the system of equations for all masses:

$$\begin{aligned} &m_j\ddot{y}_j + c_{j+1}(\dot{y}_j - \dot{y}_{j+1}) + c_j(\dot{y}_j - \dot{y}_{j-1}) \\ &+ k_{j+1}(y_j - y_{j+1}) + k_j(y_j - y_{j-1}) = 0, \quad j = 1,2,\ldots N \end{aligned} \tag{6.138}$$

where it should be taken into account that

$$y_0 = \dot{y}_0 = y_{N+1} = \dot{y}_{N+1} = c_{N+1} = k_{N+1} = 0.$$

If there are no external forces, then Equation 6.138 describes free vibrations of an N-degree-of-freedom system. Since Equation 6.138 comprises second-order differential equations, it will have $2N$ integration constants. To determine these constants, $2N$ initial conditions, two for each mass, must be specified. For example, one may displace the Nth mass only, while assigning zero initial conditions for all other masses.

It is more convenient to present Equation 6.138 in a matrix form:

$$\mathbf{M}\ddot{\mathbf{y}} + \mathbf{C}\dot{\mathbf{y}} + \mathbf{K}\mathbf{y} = \mathbf{0} \tag{6.139}$$

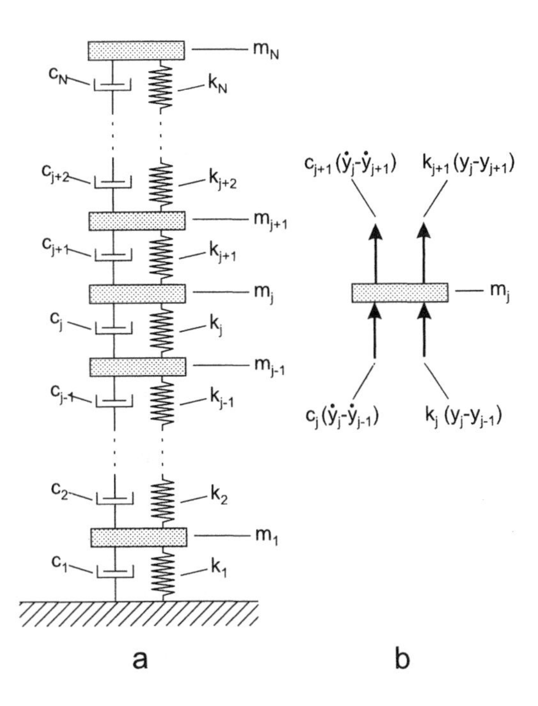

FIGURE 6.17 Multidegree-of-freedom system (a), and free-body diagram of the *j*th mass (b).

where

$$\mathbf{M} = \begin{vmatrix} m_1 & & & & \\ & \cdots & & & \\ & & m_j & & \\ & & & \cdots & \\ & & & & m_N \end{vmatrix} \tag{6.140}$$

$$\mathbf{C} = \begin{vmatrix} c_1 + c_2 & -c_2 & & & \\ & \cdots & & & \\ & -c_j & c_j + c_{j+1} & (-c_{j+1}) & \\ & & & \cdots & \\ & & & -c_N & c_N \end{vmatrix} \tag{6.141}$$

$$\mathbf{K} = \begin{vmatrix} k_1 + k_2 & -k_2 & & & \\ & \cdots & & & \\ & -k_j & k_j + k_{j+1} & -k_{j+1} & \\ & & & \cdots & \\ & & & -k_N & k_N \end{vmatrix} \tag{6.142}$$

and

$$\mathbf{Y} = (y_1, \ldots, y_j, \ldots, y_N)^T \tag{6.143}$$

In the case the external forces are applied to masses, the force vector is

$$\mathbf{P} = (P_1, \ldots, P_j, \ldots, P_N)^T \tag{6.144}$$

and Equation 6.139 then becomes

$$\mathbf{M\ddot{y}} + \mathbf{C\dot{y}} + \mathbf{Ky} = \mathbf{P} \tag{6.145}$$

6.13.1 Free Vibrations of a 2DOF System without Damping

Consider free oscillations of an undamped system having two masses, Figure 6.18a. The equations of motion are a particular case of the system Equation 6.139:

$$\begin{vmatrix} m_1 & \\ & m_2 \end{vmatrix} \begin{vmatrix} \ddot{y}_1 \\ \ddot{y}_2 \end{vmatrix} + \begin{vmatrix} k_1 + k_2 & -k_2 \\ -k_2 & k_2 \end{vmatrix} \begin{vmatrix} y_1 \\ y_2 \end{vmatrix} = \begin{vmatrix} 0 \\ 0 \end{vmatrix} \tag{6.146}$$

Since the above is the system of linear equations with constant coefficients, the solution is an amplitude vector multiplied by an exponential function (see Equation 6.18):

$$\mathbf{Y} = \mathbf{A}e^{\lambda t} \text{ or } \begin{vmatrix} y_1 \\ y_2 \end{vmatrix} = \begin{vmatrix} D_1 \\ D_2 \end{vmatrix} e^{\lambda t} \tag{6.147}$$

If this form of the solution is substituted into Equation 6.146, then

$$\left[\begin{vmatrix} m_1 & \\ & m_2 \end{vmatrix} \lambda^2 + \begin{vmatrix} k_1 + k_2 & -k_2 \\ -k_2 & k_2 \end{vmatrix} \right] \begin{vmatrix} D_1 \\ D_2 \end{vmatrix} e^{\lambda t} = \begin{vmatrix} 0 \\ 0 \end{vmatrix} \tag{6.148}$$

Since $e^{\lambda t} \neq 0$, the latter equation is reduced to a requirement:

$$\left[\begin{vmatrix} m_1 & \\ & m_2 \end{vmatrix} \lambda^2 + \begin{vmatrix} k_1 + k_2 & -k_2 \\ -k_2 & k_2 \end{vmatrix} \right] \begin{vmatrix} D_1 \\ D_2 \end{vmatrix} = \begin{vmatrix} 0 \\ 0 \end{vmatrix} \tag{6.149}$$

Since the amplitude vector $\mathbf{D} = (D_1, D_2)^T$ is arbitrary, the matrix–vector product is equal to zero only if the *determinant* of the matrix is equal to zero. For a matrix of rank 2 the second-order determinant is defined as follows:

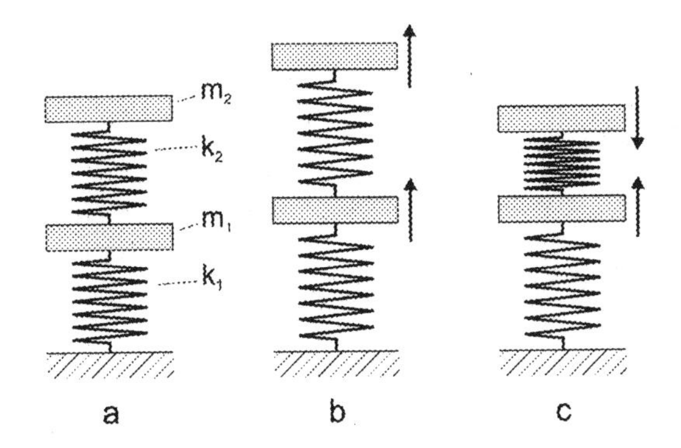

FIGURE 6.18 A 2DOF system (a), first mode of oscillation (b), and second mode of oscillation (c).

$$\det\begin{bmatrix} b_1 & d_1 \\ b_2 & d_2 \end{bmatrix} = b_1 d_2 - b_2 d_1 \tag{6.150}$$

Using the above definition, one can write the determinant of the matrix in Equation 6.149:

$$\det\begin{bmatrix} m_1\lambda^2 + k_1 + k_2 & -k_2 \\ -k_2 & m_2\lambda^2 + k_2 \end{bmatrix} = (m_1\lambda^2 + k_1 + k_2)(m_2\lambda^2 + k_2) - k_2^2 = 0 \tag{6.151}$$

The latter equation is a fourth-order polynomial with respect to λ:

$$m_1 m_2 \lambda^4 + [m_2(k_1 + k_2) + m_1 k_2]\lambda^2 + k_1 k_2 = 0 \tag{6.152}$$

The above equation is called the *characteristic equation* (see Equation 6.20) and it defines such values of λ that the nontrivial solutions of Equation 6.146 exist. These values of λ are called *eigenvalues*. Thus, the eigenvalues are the roots of Equation 6.152. Solving Equation 6.152 for λ^2 yields

$$\lambda_{1,2}^2 = -\frac{1}{2}\left(\frac{k_1 + k_2}{m_1} + \frac{k_2}{m_2}\right) \pm \sqrt{\frac{1}{4}\left(\frac{k_1 + k_2}{m_1} + \frac{k_2}{m_2}\right)^2 - \frac{k_1}{m_1}\frac{k_2}{m_2}} \tag{6.153}$$

If the parameters of the system are such that the expression under the square root is positive, then the two roots are negative, which means that in this case λ is an imaginary number. The two imaginary numbers are

$$\lambda_1 = i\omega_1 \quad \text{and} \quad \lambda_2 = i\omega_2 \tag{6.154}$$

The circular frequencies ω_1 and ω_2 are the *natural frequencies* of a two-mass system. For each natural frequency the amplitudes D_1 and D_2 remain undetermined since the determinant of the matrix in Equation 6.149 is equal to zero. However, the ratio of these amplitudes can be found from any of the two scalar equations for each natural frequency. For example, for $\lambda_1 = i\omega_1$,

$$[-m_1\omega_1^2 + (k_1 + k_2)]D_1^1 - k_2 D_2^1 = 0 \tag{6.155}$$

or

$$-k_2 D_1^1 + (-m_2\omega_1^2 + k_2)D_2^1 = 0 \tag{6.156}$$

Thus, the ratio between the amplitudes for the first natural frequency is

$$r_1 = \frac{D_2^1}{D_1^1} = \frac{-m_1\omega_1^2 + (k_1 + k_2)}{k_2} = \frac{k_2}{-m_2\omega_1^2 + k_2} \tag{6.157}$$

Similarly, for the second natural frequency,

$$r_2 = \frac{D_2^2}{D_1^2} = \frac{-m_1\omega_2^2 + (k_1 + k_2)}{k_2} = \frac{k_2}{-m_2\omega_2^2 + k_2} \tag{6.158}$$

Thus, for each natural frequency there is an amplitude vector **D** (see Equation 6.147)

$$\mathbf{D}^1 = \begin{vmatrix} D_1^1 \\ r_1 D_1^1 \end{vmatrix} \text{ and } \mathbf{D}^2 = \begin{vmatrix} D_1^2 \\ r_2 D_1^2 \end{vmatrix} \tag{6.159}$$

Vectors $\mathbf{D}^1$ and $\mathbf{D}^2$ characterize the *mode* of vibrations. The vector corresponding to the lowest natural frequency characterizes the *first mode*, and then the other characterizes the *second mode*. The meaning of modes will be made clear in the example below.

It is very important to underscore that neither the natural frequencies nor the modes of oscillation depend on the initial conditions. As was the case in SDOF systems, both reflect the properties of the system: masses and stiffnesses.

Recalling again Euler's formula, $e^{i\omega} = \cos\omega + i\sin\omega$, the solution $e^{i\lambda t}$ entails two functions as possible solutions: $\cos\omega t$ and $\sin\omega t$. Take, for example, the first root, $\lambda_1 = i\omega_1$, and substitute it in Equation 6.147. Then,

$$\begin{vmatrix} y_1^1 \\ y_2^1 \end{vmatrix} = \begin{vmatrix} D_1^1 \\ D_2^1 \end{vmatrix}(\cos\omega_1 t + i\sin\omega_1 t) = \begin{vmatrix} D_1^1 \\ r_1 D_1^1 \end{vmatrix}\cos\omega_1 t + i\begin{vmatrix} D_1^1 \\ r_1 D_1^1 \end{vmatrix}\sin\omega_1 t \tag{6.160}$$

Similarly, for the second root $\lambda_2 = i\omega_2$,

$$\begin{vmatrix} y_1^2 \\ y_2^2 \end{vmatrix} = \begin{vmatrix} D_1^2 \\ D_2^2 \end{vmatrix}(\cos\omega_2 t + i\sin\omega_2 t) = \begin{vmatrix} D_1^2 \\ r_2 D_1^2 \end{vmatrix}\cos\omega_2 t + i\begin{vmatrix} D_1^2 \\ r_2 D_1^2 \end{vmatrix}\sin\omega_2 t \qquad (6.161)$$

The latter expressions mean that any constant vector multiplied by either $\cos\omega_1 t$, $\sin\omega_1 t$, $\cos\omega_2 t$, or $\sin\omega_2 t$ is a solution of Equation 6.146. These functions, $\cos\omega_1 t$, $\sin\omega_1 t$, $\cos\omega_2 t$, and $\sin\omega_2 t$ are called *fundamental solutions.*

The general solution of the homogeneous equation Equation 6.146 is a combination of two solutions corresponding to two roots (again, the *superposition principle* is used)

$$\mathbf{Y} = b_1\mathbf{Y}^1 + b_2\mathbf{Y}^2 \qquad (6.162)$$

where b_1 and b_2 are arbitrary constants.

For the two roots in Equation 6.154 one thus has four functions, $\cos\omega_1 t$, $\sin\omega_1 t$, $\cos\omega_2 t$, and $\sin\omega_2 t$, as possible solutions. It means that the general solution is a combination of these four fundamental solutions. From Equation 6.162 and Equations 6.160 and 6.161,

$$y_1(t) = c_{11}^1\cos\omega_1 t + c_{12}^1\sin\omega_1 t + c_{11}^2\cos\omega_2 t + c_{12}^2\sin\omega_2 t \qquad (6.163)$$

and

$$y_2(t) = r_1 c_{11}^1\cos\omega_1 t + r_1 c_{12}^1\sin\omega_1 t + r_2 c_{11}^2\cos\omega_2 t + r_2 c_{12}^2\sin\omega_2 t \qquad (6.164)$$

where c_{11}^1, c_{12}^1, c_{11}^2, and c_{12}^2 are constants, and r_1, r_2 are given by Equations 6.157 and 6.158.

Alternatively, Equations 6.163 and 6.164 can be written in the form:

$$y_1(t) = D_1\cos(\omega_1 t + \phi_1) + D_2\cos(\omega_2 t + \phi_2) \qquad (6.165)$$

and

$$y_2(t) = r_1 D_1\cos(\omega_1 t + \phi_1) + r_2 D_2\cos(\omega_2 t + \phi_2) \qquad (6.166)$$

where D_1 and D_2 are the amplitudes of displacement of mass 1 and 2, and ϕ_1 and ϕ_2 are the corresponding phase angles.

Define the initial conditions, namely, that at $t = 0$

$$y_1(0) = D_{10},\ \dot{y}_1(0) = V_{10},\ y_2(0) = D_{20},\ \text{and}\ \dot{y}_2(0) = V_{20} \qquad (6.167)$$

Subjecting the solutions (Equations 6.165 and 6.166) to the above initial conditions, one obtains four equations for four unknowns:

$$y_1(0) = D_1\cos\phi_1 + D_2\cos\phi_2 = D_{10} \tag{6.168}$$

$$\dot{y}_1(0) = -\omega_1 D_1\sin\phi_1 - \omega_2 D_2\sin\phi_2 = V_{10} \tag{6.169}$$

$$y_2(0) = r_1 D_1\cos\phi_1 + r_2 D_2\cos\phi_2 = D_{20} \tag{6.170}$$

$$\dot{y}_2(0) = -\omega_1 r_1 D_1\sin\phi_1 - \omega_2 r_2 D_2\sin\phi_2 = V_{20} \tag{6.171}$$

This system of four equations can be solved for four unknowns: $D_1\cos\phi_1$, $D_2\cos\phi_2$, $D_1\sin\phi_1$, and $D_2\sin\phi_2$. Notice that this system is uncoupled, namely, Equations 6.168 and 6.170 form one independent subsystem, whereas Equations 6.169 and 6.171 form another. These smaller systems can be easily solved, and, as a result,

$$D_1\cos\phi_1 = \frac{r_2 D_{10} - D_{20}}{r_2 - r_1} \tag{6.172}$$

$$D_2\cos\phi_2 = \frac{r_1 D_{10} - D_{20}}{r_1 - r_2} \tag{6.173}$$

$$D_1\sin\phi_1 = \frac{-r_2 V_{10} + V_{20}}{\omega_1(r_2 - r_1)} \tag{6.174}$$

$$D_2\sin\phi_2 = \frac{-r_1 V_{10} + V_{20}}{\omega_2(r_1 - r_2)} \tag{6.175}$$

Now, from Equations 6.172 and 6.174, D_1 and ϕ_1 are found:

$$D_1 = \frac{1}{r_2 - r_1}\sqrt{(r_2 D_{10} - D_{20})^2 + \frac{(-r_2 V_{10} + V_{20})^2}{\omega_1^2}} \tag{6.176}$$

$$\phi_1 = \arctan\left[\frac{-r_2 V_{10} + V_{20}}{\omega_1(r_2 D_{10} - D_{20})}\right] \tag{6.177}$$

And, similarly, from Equations 6.173 and 6.175, D_2 and ϕ_2 are found:

$$D_2 = \frac{1}{r_1 - r_2}\sqrt{(r_1 D_{10} - D_{20})^2 + \frac{(-r_1 V_{10} + V_{20})^2}{\omega_2^2}} \tag{6.178}$$

$$\phi_2 = \arctan\left[\frac{-r_1 V_{10} + V_{20}}{\omega_2(r_1 D_{10} - D_{20})}\right] \tag{6.179}$$

Thus, Equations 6.165 and 6.166 together with Equations 6.176 through 6.179 give the general solution of vibrations of a 2DOF system without damping subjected to the initial disturbance.

Consider a numerical example for the case when $m_1 = m_2 = m$ and $k_1 = k_2 = k$. Then Equation 6.153 becomes

$$\lambda_{1,2}^2 = \frac{3}{2}\frac{k}{m}\left(-1 \pm \frac{\sqrt{5}}{3}\right) \tag{6.180}$$

and the natural frequencies ω_1 and ω_2 in Equation 6.154 are

$$\omega_1^2 = \frac{k}{m}\left(\frac{3-\sqrt{5}}{2}\right) \text{ and } \omega_2^2 = \frac{k}{m}\left(\frac{3+\sqrt{5}}{2}\right) \tag{6.181}$$

The ratio between the amplitudes for the first natural frequency, Equation 6.157, is

$$r_1 = \frac{a_2^1}{a_1^1} = \frac{-m\omega_1^2 + 2k}{k} = -\frac{3-\sqrt{5}}{2} + 2 = \frac{1+\sqrt{5}}{2} \tag{6.182}$$

and for the second frequency, Equation 6.158, is

$$r_2 = \frac{a_2^2}{a_1^2} = \frac{-m\omega_2^2 + 2k}{k} = -\frac{3+\sqrt{5}}{2} + 2 = \frac{1-\sqrt{5}}{2} \tag{6.183}$$

Recall that the amplitude ratio reflects the mode of vibrations, Equation 6.159. Now the meaning of the mode becomes clear from Equations 6.182 and 6.183. In the *first mode*, when the system oscillates with the low frequency ω_1, the ratio of amplitudes is positive, which means that the two masses move synchronously. In the *second mode*, when the system oscillates with the high frequency ω_2, the ratio of amplitudes is negative, which means that the two masses move asynchronously. Note also that in the first mode the absolute value of the ratio of amplitude is larger.

In Figure 6.19 vibrations of the first mass are shown, caused by its initial displacement. One can see that it oscillates with two frequencies, low frequency ω_1, and high frequency ω_2. The resulting motion is the superposition of motions with these two frequencies. In Figure 6.20 the motions of the first and second masses are superimposed. One can see that the two masses move synchronously at low frequency

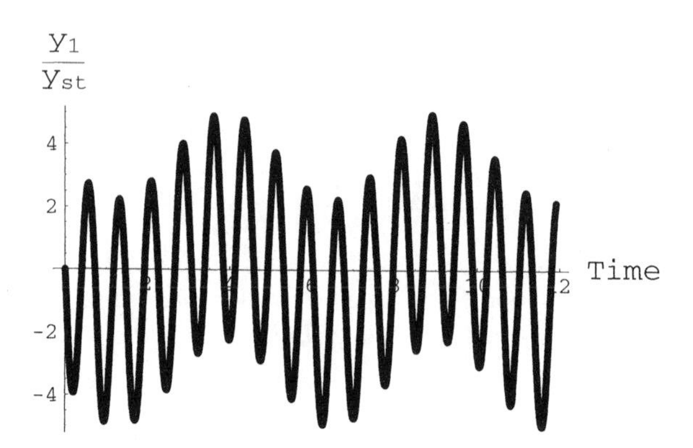

FIGURE 6.19 Vibrations of the first mass.

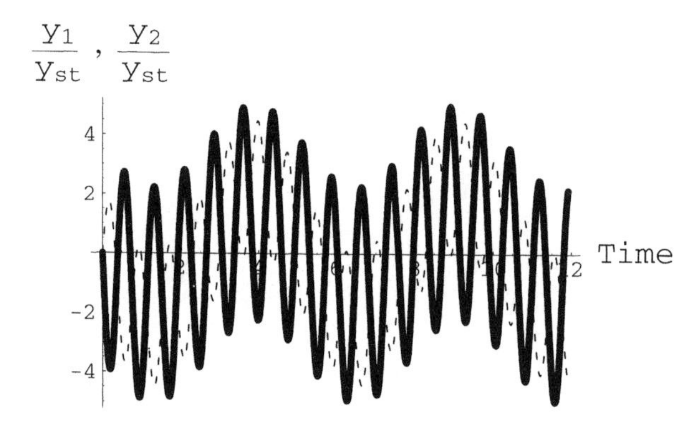

FIGURE 6.20 Vibrations of a two-mass system. Solid line = mass 1; dashed line = mass 2.

and asynchronously (in other words, either apart or toward each other) at the high frequency.

These two modes of vibrations are shown schematically in Figure 6.18b and c.

6.13.2 Free Vibrations of a 2DOF System with Damping

Equation 6.145 adds viscous forces defined by the matrix **C** (Equation 6.141)

$$\begin{vmatrix} m_1 & \\ & m_2 \end{vmatrix} \begin{vmatrix} \ddot{y}_1 \\ \ddot{y}_2 \end{vmatrix} + \begin{vmatrix} c_1 + c_2 & -c_2 \\ -c_2 & c_2 \end{vmatrix} \begin{vmatrix} \dot{y}_1 \\ \dot{y}_2 \end{vmatrix} + \begin{vmatrix} k_1 + k_2 & -k_2 \\ -k_2 & k_2 \end{vmatrix} \begin{vmatrix} y_1 \\ y_2 \end{vmatrix} = \begin{vmatrix} 0 \\ 0 \end{vmatrix} \quad (6.184)$$

The general solution of Equation 6.184 has the form:

$$\mathbf{Y} = \begin{vmatrix} D_1 \\ D_2 \end{vmatrix} e^{\lambda t} \quad (6.185)$$

Substituting the latter into Equation 6.184 yields an algebraic equation similar to Equation 6.149:

$$\left[\begin{vmatrix} m_1 & \\ & m_2 \end{vmatrix}\lambda^2 + \begin{vmatrix} c_1+c_2 & -c_2 \\ -c_2 & c_2 \end{vmatrix}\lambda + \begin{vmatrix} k_1+k_2 & -k_2 \\ -k_2 & k_2 \end{vmatrix}\right]\begin{vmatrix} D_1 \\ D_2 \end{vmatrix} = \begin{vmatrix} 0 \\ 0 \end{vmatrix} \tag{6.186}$$

The determinant of the matrix must equal zero in order for the nontrivial solution to exist. This requirement gives the equation to determine λ:

$$\det\begin{bmatrix} m_1\lambda^2 + (c_1+c_2)\lambda + k_1 + k_2 & -c_2\lambda - k_2 \\ -c_2\lambda - k_2 & m_2\lambda^2 + c_2\lambda + k_2 \end{bmatrix} \tag{6.187}$$
$$= (m_1\lambda^2 + (c_1+c_2)\lambda + k_1 + k_2)(m_2\lambda^2 + c_2\lambda + k_2) - (c_2\lambda + k_2)^2 = 0$$

The above is a characteristic equation (polynomial) with respect to λ:

$$m_1m_2\lambda^4 + [m_1c_2 + m_2(c_1+c_2)]\lambda^3 + [m_1k_2 + m_2(k_1+k_2) + c_1c_2]\lambda^2 + (c_1k_2 + c_2k_1)\lambda + k_1k_2 = 0 \tag{6.188}$$

The solution of the above equation can be obtained with the help of Mathematica, but the expressions are complicated. Instead, consider a simplified case when $m_1 = m_2 = m$, $k_1 = k_2 = k$, and $c_1 = c_2 = c$. In this case Equation 6.188 becomes

$$m^2\lambda^4 + 3mc\lambda^3 + (3mk + c^2)\lambda^2 + 2ck\lambda + k^2 = 0 \tag{6.189}$$

The roots of this equation are

$$\lambda_{1,2} = \frac{(3+\sqrt{5})c}{4m} \pm \frac{\sqrt{2}}{4}\sqrt{\frac{c^2}{m^2}(7-3\sqrt{5}) + 4\frac{k}{m}(-3+\sqrt{5})} \tag{6.190}$$

and

$$\lambda_{3,4} = \frac{-(3+\sqrt{5})c}{4m} \pm \frac{\sqrt{2}}{4}\sqrt{\frac{c^2}{m^2}(7+3\sqrt{5}) - 4\frac{k}{m}(3+\sqrt{5})} \tag{6.191}$$

To introduce the same notations as for an SDOF system, Equation 6.17,

$$\xi = \frac{c}{2m\omega_n} \text{ and } \omega_n^2 = \frac{k}{m} \tag{6.192}$$

Then, Equations 6.190 and 6.191 become

$$\lambda_{1,2} = \left(\frac{(3+\sqrt{5})}{2}\xi \pm \frac{\sqrt{2}}{4}\sqrt{4\xi^2(7-3\sqrt{5}) + (12-4\sqrt{5})}\right)\omega_n \tag{6.193}$$

and

$$\lambda_{3,4} = \left(\frac{-(3+\sqrt{5})}{2}\xi \pm \frac{\sqrt{2}}{4}\sqrt{4\xi^2(7+3\sqrt{5})-(12+4\sqrt{5})}\right)\omega_n \tag{6.194}$$

Similar to the case of the system without damping, for each root the amplitudes in Equation 6.186 are undetermined, but their ratio can be found from Equation 6.186. For any root λ_i one has, taking, for example, the first scalar equation in Equation 6.186,

$$[m_1\lambda_i^2 + (c_1+c_2)\lambda_i + k_1 + k_2]D_1^i + [-c_2\lambda_i - k_2]D_2^i = 0 \tag{6.195}$$

and

$$r_i = \frac{D_2^i}{D_1^i} = \frac{m_1\lambda_i^2 + (c_1+c_2)\lambda_i + k_1 + k_2}{c_2\lambda_i + k_2} = \frac{m\lambda_i^2 + 2c\lambda_i + 2k}{c\lambda_i + k} \tag{6.196}$$

The general solution is a superposition of four fundamental solutions of the Equation 6.185 type

$$y_1(t) = D_1^1 e^{\lambda_1 t} + D_1^2 e^{\lambda_2 t} + D_1^3 e^{\lambda_3 t} + D_1^4 e^{\lambda_4 t} \tag{6.197}$$

and

$$y_2(t) = r_1 D_1^1 e^{\lambda_1 t} + r_2 D_1^2 e^{\lambda_2 t} + r_3 D_1^3 e^{\lambda_3 t} + r_4 D_1^4 e^{\lambda_4 t} \tag{6.198}$$

As in the case of an SDOF system, the type of response depends on the amount of damping the system has, in other words, on the type of roots given by Equations 6.193 and 6.194. The damping is determined by the coefficient c or a nondimensional coefficient ξ. In the case of an 2DOF system there are more combinations of responses since there are more combinations of roots. First of all, find the critical values of ξ that identify the boundary between complex and real roots in Equations 6.193 and 6.194. These values are found by equating the expressions under the radical to zero. From Equation 6.193,

$$\xi_1^{cr} = \sqrt{\frac{3-\sqrt{5}}{7-3\sqrt{5}}} = 1.61803 \tag{6.199}$$

and

$$\xi_2^{cr} = \sqrt{\frac{3+\sqrt{5}}{7+3\sqrt{5}}} = 0.618034 \tag{6.200}$$

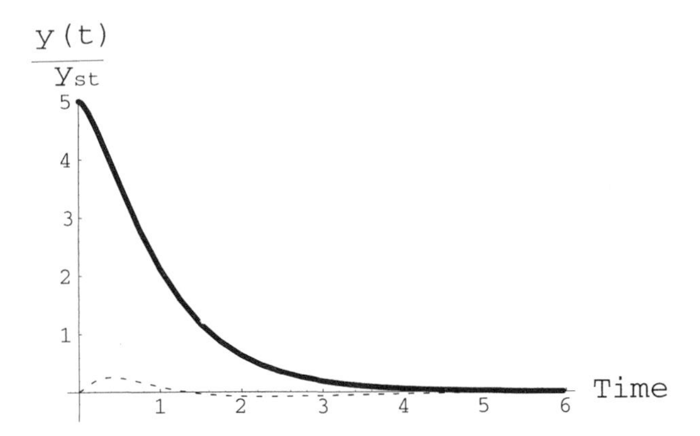

FIGURE 6.21 Normalized displacements of m_2 (solid line) and m_1 (dashed line) caused by initial displacement $y_2(0)/y_{st}$ =5 of the second mass ($\xi = 1.7 > \xi_1^{cr}$).

One can now see various possibilities:

- If $\xi > \xi_1^{cr}$, then both roots are real and it means that both masses are over-damped. There are no vibrations in this case due to initial disturbance.
- If $\xi_2^{cr} < \xi < \xi_1^{cr}$, then roots in Equation 6.194 are real, while in Equation 6.193 they are complex. In this case both masses will oscillate with one frequency (Equations 6.197 and 6.198).
- If $\xi < \xi_2^{cr}$, then roots in both Equations 6.194 and 6.195 are complex, and the system will oscillate with two frequencies.

Consider now each case, taking, for example, that $\omega_n^2 = 10$.

1. $\xi = 1.7 > \xi_1^{cr}$. The displacements of both bodies in time, caused by the initial displacement of the second body, are shown in Figure 6.21. It is seen that after the second body is released the two of them move to the position of static equilibrium exponentially. Note that the first body slightly overshoots the position of static equilibrium (which is zero).
2. $\xi_2^{cr} < \xi = 1.0 < \xi_1^{cr}$. The displacements of both bodies in time, caused by the initial displacement of the second body, are shown in Figure 6.22. It is seen that after the second body is released, the two of them cross the position of static equilibrium and then move to this position exponentially. There is a periodic term in the solutions, the amplitude of which becomes very small after the first half-cycle that it is not seen on the graph.
3. $\xi = 0.2 < \xi_2^{cr}$. The displacements of both bodies in time, caused by the initial displacement of the second body, are shown in Figure 6.23. It is seen that after the second body is released, the two of them move synchronously (first mode) while their amplitudes are decreasing exponentially. The fact that only the low frequency is visible indicates that the high frequency (second mode) is damped out. This is a general feature of vibrating systems, namely, that *high frequency components decline faster.*

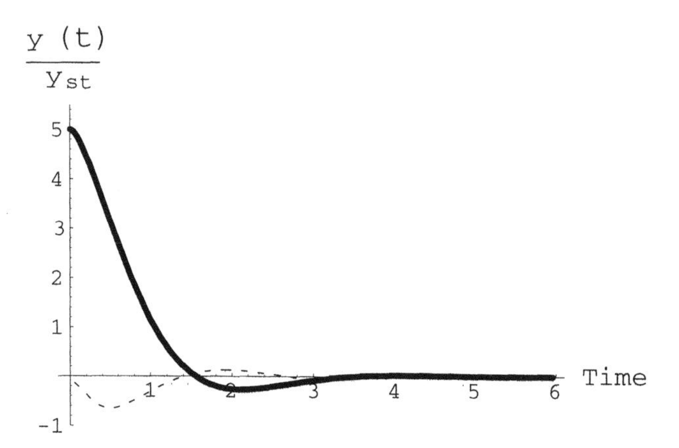

FIGURE 6.22 Normalized displacements of m_2 (solid line) and m_1 (dashed line) caused by initial displacement $y_2(0)/y_{st} = 5$ of the second mass ($\xi_2^{cr} < \xi = 1.0 < \xi_1^{cr}$).

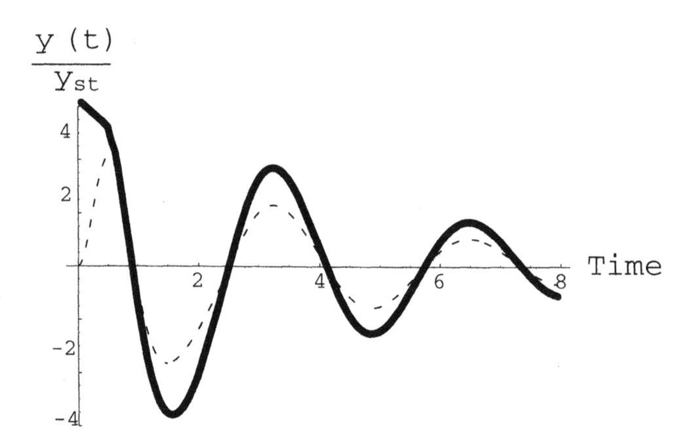

FIGURE 6.23 Normalized displacements of m_2 (solid line) and m_1 (dashed line) caused by initial displacement $y_2(0)/y_{st} = 5$ of the second mass ($\xi = 0.2 < \xi_2^{cr}$).

6.13.3 Forced Vibrations of a 2DOF System with Damping

The matrix form of the equation is

$$\mathbf{M}\ddot{\mathbf{y}} + \mathbf{C}\dot{\mathbf{y}} + \mathbf{K}\mathbf{y} = \mathbf{P} \tag{6.201}$$

where for a 2DOF system shown in Figure 6.18, the matrices **M**, **C**, and **K** are the same as in Equation 6.186, and the vector **P** is as follows

$$\mathbf{P} = \begin{vmatrix} p_1 \\ p_2 \end{vmatrix} e^{i\omega t} \tag{6.202}$$

The steady-state (particular) solutions are of the form (see Section 6.8)

$$\mathbf{Y} = \begin{vmatrix} D_1 \\ D_2 \end{vmatrix} e^{i\omega t} \tag{6.203}$$

After substitution of Equation 6.203 into Equation 6.201, and taking into account Equation 6.202, one obtains

$$\left[-\begin{vmatrix} m_1 & \\ & m_2 \end{vmatrix} \omega^2 + i \begin{vmatrix} c_1 + c_2 & -c_2 \\ -c_2 & c_2 \end{vmatrix} \omega + \begin{vmatrix} k_1 + k_2 & -k_2 \\ -k_2 & k_2 \end{vmatrix} \right] \begin{vmatrix} D_1 \\ D_2 \end{vmatrix} = \begin{vmatrix} p_1 \\ p_2 \end{vmatrix} \tag{6.204}$$

or, after summing up matrices on the left,

$$\begin{vmatrix} k_1 + k_2 - m_1\omega^2 + i\omega(c_1 + c_2) & -k_2 - i\omega c_2 \\ -k_2 - i\omega c_2 & k_2 - m_2\omega^2 + i\omega c_2 \end{vmatrix} \begin{vmatrix} D_1 \\ D_2 \end{vmatrix} = \begin{vmatrix} p_1 \\ p_2 \end{vmatrix} \tag{6.205}$$

The matrix on the left is a *complex dynamic stiffness matrix*, which is also called an *impedance matrix*. Denote the elements of the latter as $z_{ks}(i\omega)$, $k,s = 1,2$. Then the impedance matrix has the form

$$\mathbf{Z}(i\omega) = \begin{vmatrix} z_{11}(i\omega) & z_{12}(i\omega) \\ z_{21}(i\omega) & z_{22}(i\omega) \end{vmatrix} \tag{6.206}$$

where

$$z_{11}(i\omega) = k_1 + k_2 - m_1\omega^2 + i\omega(c_1 + c_2) \tag{6.207}$$

$$z_{12}(i\omega) = z_{21}(i\omega) = -k_2 - i\omega c_2 \tag{6.208}$$

$$z_{22}(i\omega) = k_2 - m_2\omega^2 + i\omega c_2 \tag{6.209}$$

The solution of Equation 6.205 is

$$\begin{vmatrix} D_1 \\ D_2 \end{vmatrix} = \mathbf{Z}^{-1}(i\omega) \begin{vmatrix} p_1 \\ p_2 \end{vmatrix} \tag{6.210}$$

where

$$\mathbf{Z}^{-1}(i\omega) = \frac{1}{z_{11}(i\omega)z_{22}(i\omega) - z_{12}(i\omega)^2} \begin{vmatrix} z_{22}(i\omega) & -z_{12}(i\omega) \\ -z_{12}(i\omega) & z_{11}(i\omega) \end{vmatrix} \tag{6.211}$$

where it was taken into account that $z_{12}(i\omega) = z_{21}(i\omega)$.

Thus, the inverse of the impedance matrix is the *dynamic frequency response matrix.* The expressions for dynamic responses of individual masses in Figure 6.18a follow from Equations 6.210 and 6.211

$$D_1 = \frac{z_{22}(i\omega)p_1 - z_{12}(i\omega)p_2}{z_{11}(i\omega)z_{22}(i\omega) - z_{12}(i\omega)^2} \tag{6.212}$$

and

$$D_2 = \frac{-z_{12}(i\omega)p_1 + z_{11}(i\omega)p_2}{z_{11}(i\omega)z_{22}(i\omega) - z_{12}(i\omega)^2} \tag{6.213}$$

Consider the same numerical example as in Section 6.13.2, namely, when $m_1 = m_2 = m$ and $k_1 = k_2 = k$. In addition, assume that $c_1 = c_2 = c$ and $p_2 = 0$. Then the vibrations amplitudes are as follows:

$$D_1 = \frac{-z_{12}(i\omega)p_2}{z_{11}(i\omega)z_{22}(i\omega) - z_{12}(i\omega)^2} \tag{6.214}$$

and

$$D_2 = \frac{z_{11}(i\omega)p_2}{z_{11}(i\omega)z_{22}(i\omega) - z_{12}(i\omega)^2} \tag{6.215}$$

where

$$z_{11}(i\omega) = 2k - m\omega^2 + 2i\omega c \tag{6.216}$$

$$z_{12}(i\omega) = z_{21}(i\omega) = -k - i\omega c \tag{6.217}$$

$$z_{22}(i\omega) = k - m\omega^2 + i\omega c \tag{6.218}$$

The expressions for the amplitudes and phase angles, obtained with the help of Mathematica, are as follows:

$$|D_1| = \left(\frac{k^2 + c^2\omega^2}{k^4 - 6k^3m\omega^2 - 6km\omega^4(c^2 + m^2\omega^2) + k^2(2c^2\omega^2 + 11m^2\omega^4) + \omega^4(c^4 + 7c^2m^2\omega^2 + m^4\omega^4)}\right)^{1/2} \tag{6.219}$$

$$|D_2| = \left(\frac{4c^2 + (-2k + m\omega^2)^2}{k^4 - 6k^3m\omega^2 - 6km\omega^4(c^2 + m^2\omega^2) + k^2(2c^2\omega^2 + 11m^2\omega^4) + \omega^4(c^4 + 7c^2m^2\omega^2 + m^4\omega^4)}\right)^{1/2} \tag{6.220}$$

$$\tan\phi_1 = \frac{-k\omega(2ck - 3cm\omega^2) + c\omega(k^2 - c^2\omega^2 - 3km\omega^2 + m^2\omega^4)}{c\omega^2(2ck - 3cm\omega^2) + k(k^2 - c^2\omega^2 - 3km\omega^2 + m^2\omega^4)} \quad (6.221)$$

and

$$\tan\phi_2 = \frac{-(-2k + m\omega^2)\omega(2ck - 3cm\omega^2) - 2c(k^2 - c^2\omega^2 - 3km\omega^2 + m^2\omega^4)}{-2c\omega(2ck - 3cm\omega^2) + (-2k + m\omega^2)(k^2 - c^2\omega^2 - 3km\omega^2 + m^2\omega^4)} \quad (6.222)$$

In Figures 6.24 through 6.26 the amplitude–frequency and phase angle diagrams for the first and second bodies are shown in the case when $k/m = 10/\text{sec}^2$ and $\xi = 0.05$ (so that $c/m = 0.0316/\text{sec}$).

One can see from Figures 6.24 and 6.25 that in the case of damping almost 10 times smaller than the lowest critical, the second resonance cannot be practically excited. This is also seen in Figure 6.23 in the case of free vibrations, whereas when the damping is absent then both frequencies are present (see Figures 6.19 and 6.20). The comparison of phase angles in Figure 6.26 shows that damping does not change qualitatively the phase angle function, except that it varies more smoothly near resonances. One should recall that the phase angle characterizes the shift in time between the maximum of the force and that of the displacement.

6.14 ROTORDYNAMICS

Rotating machine components (shafts, rotors) are not perfectly manufactured and assembled and this causes machine vibrations. The result of a nonideal manufacturing of, for example, a rotor with a disk mounted on it, is that the center mass of the disk may not lie on the axis of rotor rotation. This, in turn, leads to unbalanced forces during rotation and, thus, to vibrations. There may also be other causes of vibrations, such as a nonideal geometry of supports and bearings, misalignment of multistage rotors, periodic forces generated by the meshing gears.

The common factor for all of the above dynamic forces is that their periodicity is proportional to the speed of rotation. If the periodicity of the dynamic forces is that of the speed of rotation and they cause the system to resonate, then this speed of rotation is called the *critical speed*. Below, some simple examples of rotors with one mounted disk illustrating the dynamic phenomena in rotating machinery are presented.

6.14.1 Rigid Rotor on Flexible Supports

The rotor is considered to be rigid if its critical speed is higher than its operating speed of rotation. However, such a system may still vibrate due to the flexibility of the supports (journal or roller bearings). A schematic diagram of such a system is shown in Figure 6.27, where e is the eccentricity of the mass center of the disk, k is the support stiffness, and m is the mass of the disk (it is assumed that the mass of the rotor is small compared with that of the disk).

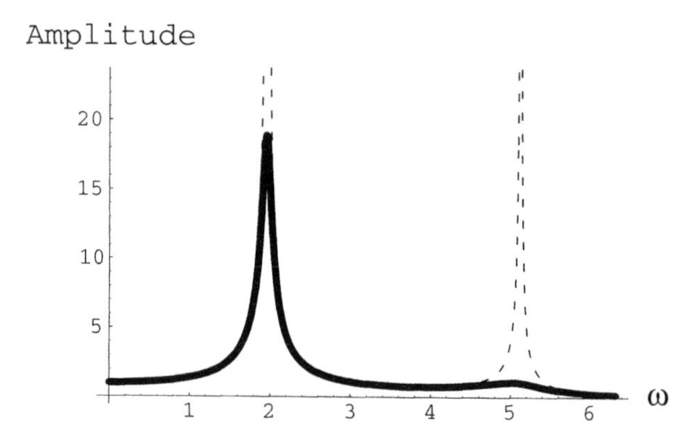

FIGURE 6.24 Normalized amplitude of the first body vs. frequency. Solid line = $\xi = 0.05$; dashed line = $\xi = 0$.

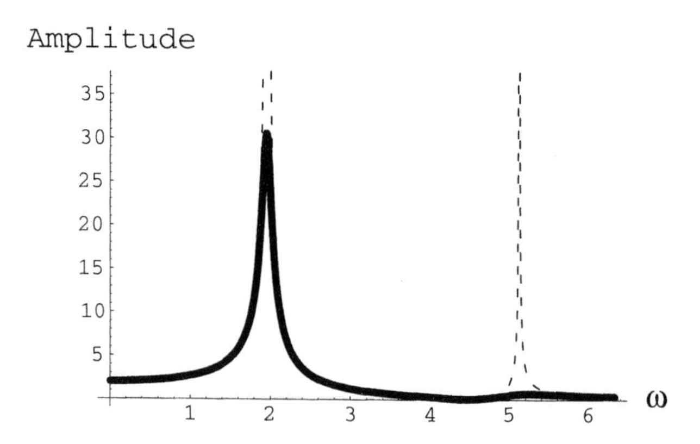

FIGURE 6.25 Normalized amplitude of the second body vs. frequency. Solid line = $\xi = 0.05$; dashed line = $\xi = 0$.

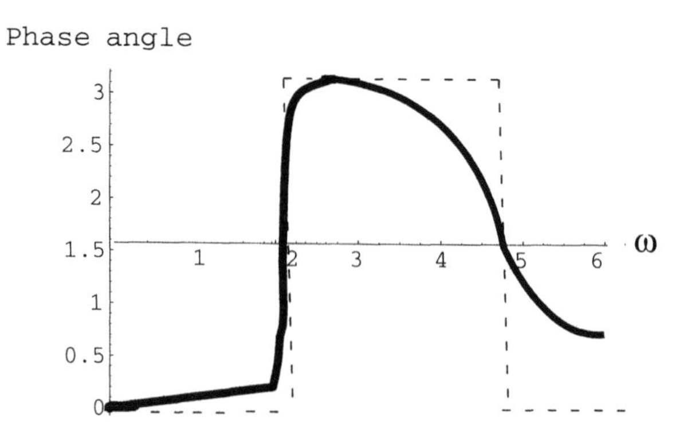

FIGURE 6.26 Phase angle of the first body vs. frequency. Solid line = $\xi = 0.05$; dashed line = $\xi = 0$.

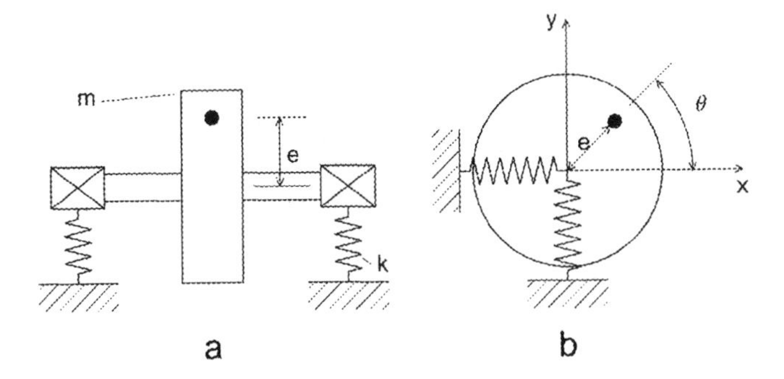

FIGURE 6.27 Rigid rotor on flexible supports.

If the angular speed of the rotor is ω, then an eccentrically located (with respect to the rotor axis) mass will generate a centrifugal force:

$$F_c = me\omega^2 \tag{6.223}$$

This force rotates with the angular velocity ω in the (x,y)-coordinate system. Thus, at any angle θ the x- and y-projections of this force are

$$F_{cx} = F_c \cos\theta \text{ and } F_{cy} = F_c \sin\theta \tag{6.224}$$

Taking into account that the angle of rotation is $\theta = \omega t$, the latter expressions can be written in terms of the rotor angular velocity and time, t.

$$F_{cx} = F_c \cos\omega t \text{ and } F_{cy} = F_c \sin\omega t \tag{6.225}$$

As one can see, the centrifugal force, Equation 6.223, is not periodic, whereas its x- and y-components are. The centrifugal force is in fact a vector and can be written as

$$\mathbf{F_c} = \mathbf{i}F_{cx} + \mathbf{j}F_{cy} \tag{6.226}$$

where **i** and **j** are the unit vectors directed along the x- and y-coordinates, respectively.

Now one can use the *superposition principle* and consider the effect of each force component separately and then sum up the results to obtain the effect of the centrifugal force on the rotor dynamics. If the lateral deflections of the springs are small so that their lateral stiffness can be neglected, then the x- and y-motions of the rotor are uncoupled. In this case the corresponding motion equations for each direction are similar to those of the mass on a spring under the action of a periodic force (see Equation 6.12). Namely,

$$m\ddot{x} + kx = me\omega^2 \cos\omega t \tag{6.227}$$

and

$$m\ddot{y} + ky = me\omega^2 \sin \omega t \tag{6.228}$$

The solutions for the steady-state vibrations are given by Equation 6.64, in which it should be taken that $p_0 = e\omega^2$. Thus, the solutions are

$$x(t) = \frac{e\omega^2}{\omega_n^2 - \omega^2} \cos \omega t \tag{6.229}$$

and

$$y(t) = \frac{e\omega^2}{\omega_n^2 - \omega^2} \sin \omega t \tag{6.230}$$

The axis of the rotor is displaced from the static position of equilibrium by the radial distance r, which is equal to

$$r = \sqrt{x^2(t) + y^2(t)} = \frac{e\omega^2}{\omega_n^2 - \omega^2} \tag{6.231}$$

As one can see from the latter formula, for any given frequency ω the rotor axis displacement remains constant. The rotor axis moves along the constant radius orbit. Such motion is called *whirling*. If the x- and y-springs have the same stiffness, then the orbit is circular (as is the case considered above); otherwise, the orbit becomes elliptical.

From Equation 6.231 it is seen that when the speed of rotation ω equals that of natural frequency ω_n, then the displacement of the rotor is infinite; i.e., it is in resonance. The speed ω_n is called the *critical speed*. In Figure 6.28 the rotor displacement normalized with respect to the eccentricity e is shown as a function of a nondimensional frequency. It is seen from Equation 6.231 that when $\omega \to \infty$, then $|r| \to e$ (in Figure 6.28 this limit is equal to 1 and it is shown by a dashed line).

The position of the mass center during the rotation is the sum of the rotor axis displacement and the eccentricity (Figure 6.29)

$$r_c = r + e \tag{6.232}$$

The latter, using Equation 6.231 for r, is equal to

$$r_c = \frac{e\omega^2}{\omega_n^2 - \omega^2} + e \tag{6.233}$$

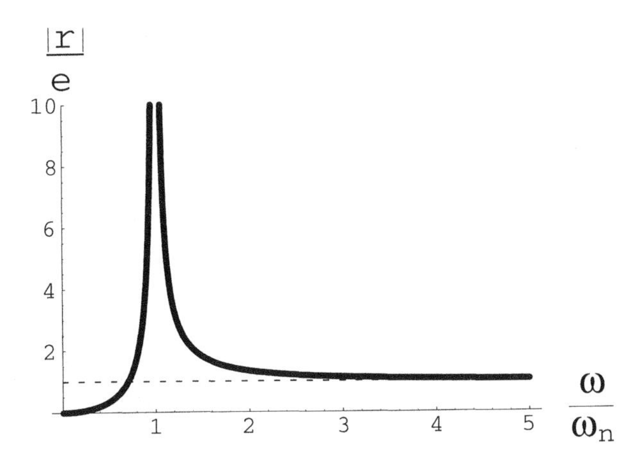

FIGURE 6.28 Normalized radius of rotor whirling vs. nondimensional angular velocity.

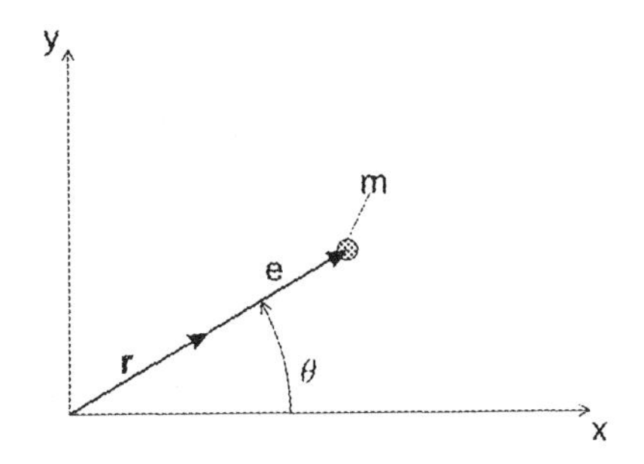

FIGURE 6.29 Position of mass center of a disk on a rigid rotor when $\omega < \omega_n$.

One can see that if the speed of the rotor is less than critical, $\omega < \omega_n$, then $r_c > e$, if $\omega > \omega_n$, then $r_c < e$, and when $\omega \rightarrow \infty$, then $r_c \rightarrow 0$. In other words, in unbalanced rotors the mass center changes its position with respect to a whirling rotor axis at the critical speed, and then with the increase in speed this mass center tends to the axis of the undisturbed rotor.

It is important to understand that the rotation of the mass center is the result of summation of two rotations: one is the rotation of the rotor itself with respect to its center (the latter coincides with the tip of the vector **r** in Figure 6.29), and the other is the rotation of the rotor axis (rotation of the vector **r**) around the coordinate center. Since both these angular velocities are equal to ω, the whirl in this case is called *synchronous*. The two vectors, **r** and **e**, are always collinear.

6.14.2 Flexible Rotor on Rigid Supports

Such a system is shown in Figure 6.30, where for simplicity the disk is placed in the middle of the shaft (rotor). The rotating unbalanced disk generates the centrifugal forces given by Equation 6.223. Then, the motion equations are the same as in the

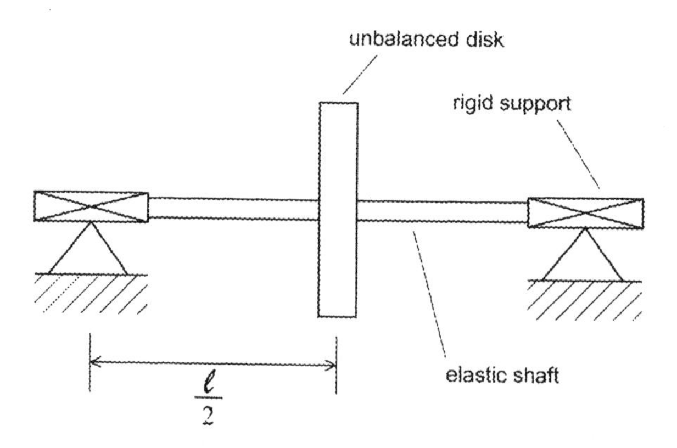

FIGURE 6.30 Unbalanced disk on a flexible shaft.

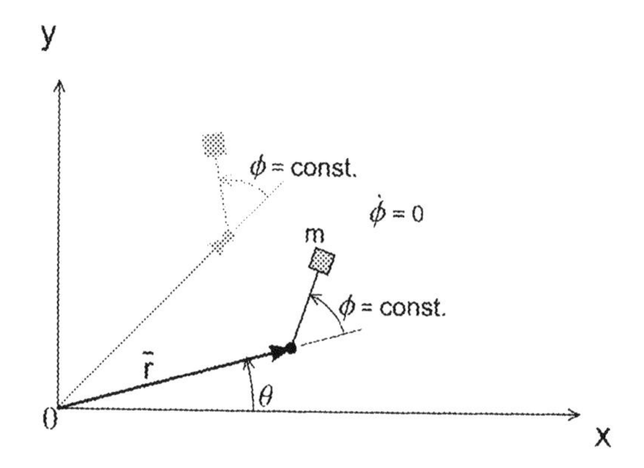

FIGURE 6.31 Positions of mass center of a disk on a flexible rotor with damping.

case of a rigid rotor (Equations 6.227 and 6.228), except that instead of spring stiffness the shaft stiffness at the disk location should be used. Thus, all the results obtained in Section 6.14.1 are applicable in this case as well.

6.14.3 Flexible Rotor with Damping on Rigid Supports

The damping in the disk–rotor system may be caused by the material damping properties of the rotor itself, by the friction at rotor supports (bearings), or by the air resistance to a rotor motion. The effect of damping is that the vectors **r** and **e** in Figure 6.29 are no longer collinear, because **e** is associated with the direction of the external force $me\omega^2$ and **r** with the reaction of the system to this force. The angle ϕ between the vectors **r** and **e** is the phase angle (Figure 6.31). When the angular velocity ω of the rotor is constant, the angle ϕ also remains constant. Thus, the rotor imbalance vector **e** leads (or lags behind) the rotor whirl vector **r** by a constant phase angle ϕ. If $\omega = \dot{\theta}$, then the corresponding whirl having constant velocity $\dot{\theta}$ is called *synchronous*.

In the case of a *nonsynchronous* whirl the time rate of change of the angle ϕ is the spin velocity of the rotor relative to the rotating whirl vector **r**. Thus, the rotor angular velocity in this case is $\omega = \dot{\theta} + \dot{\phi}$. The synchronous whirl takes place during a steady-state rotation, while nonsteady rotation leads to a nonsynchronous whirl.

The case of a synchronous whirl ($\dot{\phi} = 0$) will be considered. The equations of motion in this case are the same as Equations 6.227 and 6.228, but, in addition, they contain terms associated with the damping forces. Thus,

$$m\ddot{x} + c_e\dot{x} + kx = me\omega^2\cos\omega t \tag{6.234}$$

and

$$m\ddot{y} + c_e\dot{y} + ky = me\omega^2\sin\omega t \tag{6.235}$$

where, as before in Equations 6.227 and 6.228, x and y are the coordinates of the rotor axis, so that the coordinates of the center mass are

$$x^m = x + e\cos(\theta + \phi) \text{ and } y^m = y + e\sin(\theta + \phi) \tag{6.236}$$

Note that in Equations 6.234 and 6.235 the damping coefficient has a subscript e. It indicates that in the case of a synchronous whirl the rotation of a bent rotor does not cause internal damping forces, and so the damping may be only of external nature (bearings, air).

Instead of solving Equations 6.234 and 6.235 for x and y, one can transform this system into a single equation with respect to a complex variable $z = x + iy$, which is another form of the vector **r**. To achieve such transformation, multiply Equation 6.235 by an imaginary unit i, add the two equations together, and take into account Euler's formula ($e^{i\omega t} = \cos\omega t + i\sin\omega t$). The result is

$$m\ddot{z} = c_e\dot{z} + kz = me\omega^2 e^{i\omega t} \tag{6.237}$$

For a steady-state regime, the solution of the above equation has the form:

$$z = z_0 e^{i\omega t} \tag{6.238}$$

By substituting the latter into Equation 6.237 and canceling the exponential terms, the following expression for the complex amplitude of rotor axis displacement is obtained:

$$z_0 = \frac{me\omega^2}{(k - m\omega^2) + ic_e\omega} \tag{6.239}$$

By denoting, as before (see Section 6.14.1), the critical speed of the undamped system by $\omega_n = (k/m)^{0.5}$, and expressing the coefficient of damping as $c_e = 2m\omega_n\xi_e$

(see Equation 6.17), Equation 6.239 can be written in nondimensional parameters as follows:

$$z_0 = \frac{e\omega^2}{(\omega_n^2 - \omega^2) + 2i\omega\omega_n\xi_e} \tag{6.240}$$

By multiplying the denominator in the latter by a complex conjugate number, it is transformed into the following form:

$$z_0 = \frac{e\omega^2[(\omega_n^2 - \omega^2) - 2i\omega\omega_n\xi_e]}{(\omega_n^2 - \omega^2)^2 + (2\omega\omega_n\xi_e)^2} \tag{6.241}$$

Thus, the complex amplitude of the axis position of the rotor has the form:

$$z_0 = x_0 + iy_0 \tag{6.242}$$

where x_0 and y_0 are the amplitudes associated with solutions of Equations 6.234 and 6.235.

The complex number in Equation 6.242 can also be written in the form:

$$z_0 = |z_0|e^{i\phi} = \sqrt{x_0^2 + y_0^2}e^{i\phi} \tag{6.243}$$

where $|z_0|$ is the absolute value of the amplitude and ϕ is the argument of the complex number, which in this case is the phase angle.

The expressions for the amplitude $|z_0|$ and the phase angle ϕ are as follows

$$|z_0| = \frac{e\omega^2}{\sqrt{(\omega_n^2 - \omega^2)^2 + (2\omega\omega_n\xi_e)^2}} \tag{6.244}$$

and

$$\phi = \arctan\left[\frac{2\omega\omega_n\xi_e}{\omega_n^2 - \omega^2}\right] \tag{6.245}$$

It is seen that the above equations are identical to Equations 6.89 and 6.90, if it is taken into account that in this case $p_0 = e\omega^2$. In this case, however, the critical speed is

$$\omega_c = \frac{\omega_n}{\sqrt{1 - 2\xi_e^2}} \tag{6.246}$$

and the corresponding maximum amplitude is

$$|z_0|_{\max} = \frac{e}{2\xi_e\sqrt{1 - \xi_e^2}} \tag{6.247}$$

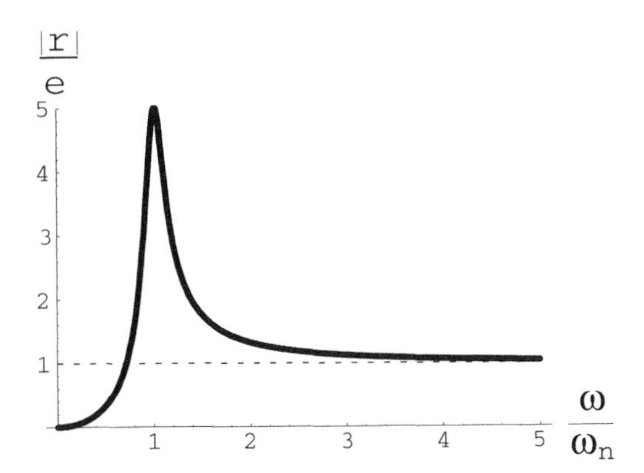

FIGURE 6.32 Normalized radius of rotor whirling vs. nondimensional angular velocity.

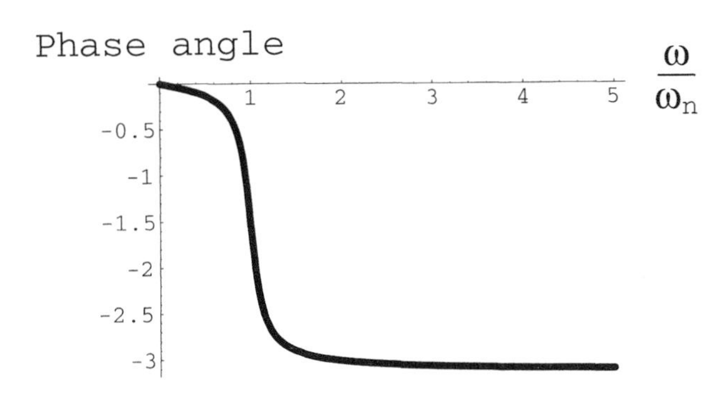

FIGURE 6.33 Phase angle vs. normalized angular rotor speed.

In Figure 6.32 the normalized radius vs. the normalized angular speed is shown when the nondimensional coefficient of damping equals $\xi_e = 0.1$. The effect of damping on the system dynamics is a limiting of the maximum amplitude at the critical speed (compare with Figure 6.28). For the chosen value of ξ_e the maximum normalized amplitude is 5.025. Note that if $\xi_e \ll 1$, then the amplitude can be estimated by the formula:

$$|z_0|_{max} = \frac{e}{2\xi_e} \tag{6.248}$$

The above formula gives the value of the normalized amplitude equal to 5.

The change of the phase angle as a function of normalized rotor speed is shown in Figure 6.33. First of all, it is seen that the phase angle changes with the speed of rotation and, second, since this angle is negative the vector **r** lags behind the vector **e** (see Figure 6.32). Recall that the phase angle is the angle between the direction

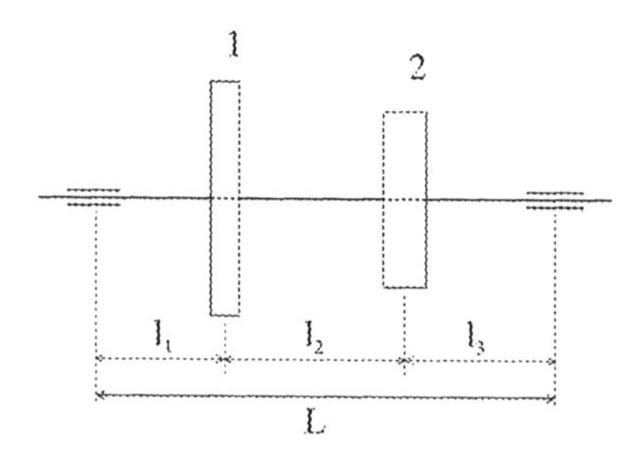

FIGURE 6.34 A two-disk rotor.

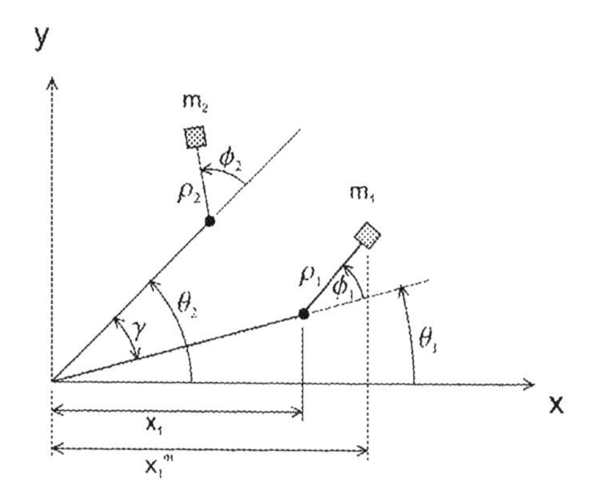

FIGURE 6.35 Displaced positions of mass centers.

of the force (in this case it is $me\omega^2$ directed along the vector **e**) and the direction of the system response (in this case it is the bending of the rotor represented by the vector **r**).

6.14.4 Two-Disk Flexible Rotor with Damping

The two-disk rotor is shown schematically in Figure 6.34. Here a case of synchronous motion is considered. Accordingly, the eccentricity vector identifying the center mass of each disk will have an angle with the plane of rotor bending. These phase angles are shown in Figure 6.35 and they are constant.

The coordinates of mass centers in this case are given by

$$x_1^m = x_1 + e_1\cos(\theta_1 + \phi_1) \tag{6.249}$$

$$y_1^m = y_1 + e_1\sin(\theta_1 + \phi_1) \tag{6.250}$$

$$x_2^m = x_2 + e_2\cos(\theta_2 + \phi_2) \tag{6.251}$$

$$y_2^m = y_2 + e_2\sin(\theta_2 + \phi_2) \tag{6.252}$$

where $x_i = x_i(t)$, $y_i = y_i(t)$ (i = 1,2) are the displacements of the rotor at disk locations, and $\theta_i = \theta_i(t)$ (i = 1,2) identifies a plane in which the bent rotor lies in the vicinity of the disk i.

The planes in which the bending of rotors at each disk location takes place depend on the direction of unbalanced forces. Denote the angle between the latter by γ (see Figure 6.35), and assume that this angle is known. Thus, $\theta_2 = \theta_1 + \gamma$. For a constant angular velocity of the rotor ω, $\theta_1 = \omega t$, and one can take that $\phi_1 = \phi_2 = 0$.

The motion equation in a matrix form in, for example, the x-plane is

$$\mathbf{M}\mathbf{x}^{m} + \mathbf{C}\dot{\mathbf{x}} + \mathbf{K}\mathbf{x} = \mathbf{0} \tag{6.253}$$

where the mass matrix **M**, the damping matrix **C**, and the stiffness matrix **K** are equal to

$$\mathbf{M} = \begin{vmatrix} m_1 & \\ & m_2 \end{vmatrix} \tag{6.254}$$

$$\mathbf{C} = \begin{vmatrix} c_{11} & c_{12} \\ c_{21} & c_{22} \end{vmatrix} \tag{6.255}$$

$$\mathbf{K} = \begin{vmatrix} k_{11} & k_{12} \\ k_{21} & k_{22} \end{vmatrix} \tag{6.256}$$

$$\mathbf{x}^{\mathrm{m}} = \begin{vmatrix} x_1^{m} \\ x_2^{m} \end{vmatrix} \tag{6.257}$$

and

$$\mathbf{x} = \begin{vmatrix} x_1 \\ x_2 \end{vmatrix} \tag{6.258}$$

In Equation 6.255, c_{ij} $(i,j = 1,2)$ are the damping coefficients with the following meaning: c_{11} is the force at position 1 corresponding to a unit velocity at this position $\dot{x}_1 = 1$, c_{12} is the force at position 1 corresponding to a unit velocity at position 2 $\dot{x}_2 = 1$, c_{21} is the force at position 2 corresponding to a unit velocity at position 1 $\dot{x}_1 = 1$, c_{22} is the force at position 2 corresponding to a unit velocity at this position $\dot{x}_2 = 1$. Note that $c_{12} = c_{21}$.

In Equation 6.256, k_{ij} $(i,j = 1,2)$ are the stiffness coefficients with the following meaning: k_{11} is the force at position 1 corresponding to a unit displacement at this position $x_1 = 1$, k_{12} is the force at position 1 corresponding to a unit displacement at position 2 $x_2 = 1$, k_{21} is the force at position 2 corresponding to a unit displacement at position 1 $x_1 = 1$, k_{22} is the force at position 2 corresponding to a unit displacement at this position $x_2 = 1$. Note that $k_{12} = k_{21}$.

Substituting x_1^m, x_2^m, x_1, and x_2 into Equation 6.253 yields the following differential equation:

$$\mathbf{M\ddot{x}} + \mathbf{C\dot{x}} + \mathbf{Kx} = -\mathbf{M\ddot{e}} \tag{6.259}$$

where

$$\mathbf{e} = \begin{vmatrix} e_1 \cos(\omega t) \\ e_2 \cos(\omega t + \gamma) \end{vmatrix} \tag{6.260}$$

The term on the right in Equation 6.259 after differentiation of **e** becomes

$$\mathbf{M\ddot{x}} + \mathbf{C\dot{x}} + \mathbf{Kx} = \mathbf{P} \tag{6.261}$$

where

$$\mathbf{P} = \begin{vmatrix} m_1 e_1 \omega^2 \cos(\omega t) \\ m_2 e_2 \omega^2 \cos(\omega t + \gamma) \end{vmatrix} \tag{6.262}$$

Recall that the cosine functions can be represented as the real parts of a complex exponential function:

$$e^{i\omega t} = \cos \omega t + i \sin \omega t \tag{6.263}$$

and

$$e^{i(\omega t + \gamma)} = e^{i\omega t} e^{i\gamma} = e^{i\gamma}(\cos \omega t + i \sin \omega t) \tag{6.264}$$

Now one can represent the force on the right-hand side $\mathbf{P} = \mathbf{P}^* \cos \omega t$ as a real part of a vector $\mathbf{P}^* e^{i\omega t}$, where $\mathbf{P}^*$ is equal to

$$\mathbf{P}^* = \begin{vmatrix} m_1 e_1 \omega^2 \\ m_2 e_2 \omega^2 e^{i\gamma} \end{vmatrix} \tag{6.265}$$

Instead of solving Equation 6.261, one can solve the following equation:

$$\mathbf{M\ddot{x}} + \mathbf{C\dot{x}} + \mathbf{Kx} = \mathbf{P}^* e^{i\omega t} \tag{6.266}$$

The advantage of this substitution of the original equation by Equation 6.266 is that one can use already derived solutions for Equation 6.201, in which one has to take into account that the damping and stiffness matrices and the force vector are different in this case. As a result, instead of Equations 6.207 through 6.209, one has the following:

$$z_{11}(i\omega) = k_{11} - m_1\omega^2 + i\omega c_{11} \quad (6.267)$$

$$z_{12}(i\omega) = z_{21}(i\omega) = k_{12} + i\omega c_{12} \quad (6.268)$$

$$z_{22}(i\omega) = k_{22} - m_2\omega^2 + i\omega c_{22} \quad (6.269)$$

In Equations 6.212 and 6.213 take into account that

$$p_1 = m_1 e_1 \omega^2 \quad (6.270)$$

and

$$p_2 = m_2 e_2 \omega^2 e^{i\gamma} \quad (6.271)$$

The stiffness matrix **K** can be found as an inverse of the flexibility matrix **A**

$$\mathbf{A} = \begin{vmatrix} \alpha_{11} & \alpha_{12} \\ \alpha_{21} & \alpha_{22} \end{vmatrix} \quad (6.272)$$

The elements of the flexibility matrix are called the *influence coefficients*. Their meaning is as follows: α_{ij} is the displacement at the position i from the unit force applied at the position j. The influence coefficients can thus be found by analyzing the displacements in a beam. For the case of a beam on simple supports shown in Figure 6.34 the influence coefficients are as follows

$$\alpha_{11} = \frac{l_1^2(l_2 + l_3)^2}{3EIL} \quad (6.273)$$

$$\alpha_{22} = \frac{l_3^2(l_1 + l_2)^2}{3EIL} \quad (6.274)$$

$$\alpha_{12} = \alpha_{21} = \frac{l_1 l_3}{6EIL}(L^2 - l_1^2 - l_3^2) \quad (6.275)$$

where E is the modulus of elasticity of the rotor material, and $I = \pi d^4/64$ is the moment of inertia of the cross section of the rotor.

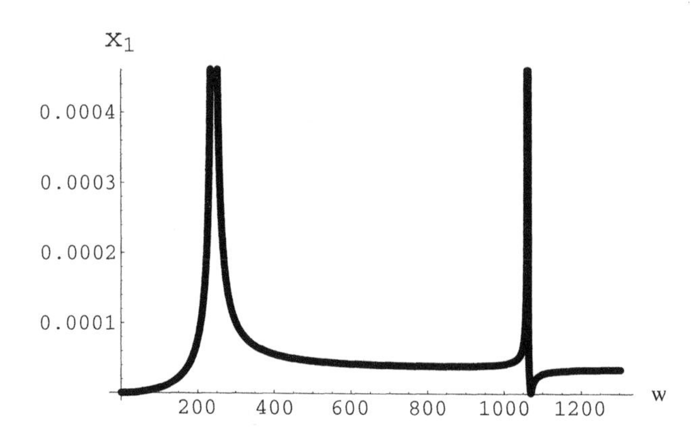

FIGURE 6.36 Amplitude–angular velocity diagram.

The corresponding stiffness coefficients are

$$k_{11} = \frac{\alpha_{22}}{\Delta},\ k_{22} = \frac{\alpha_{11}}{\Delta},\ k_{12} = k_{21} = -\frac{\alpha_{12}}{\Delta},\ \text{where } \Delta = \alpha_{11}\alpha_{22} - \alpha_{12}^2. \quad (6.276)$$

The *characteristic equation* in this case is found as the determinant of the following matrix:

$$\det\left|\mathbf{K} - \omega^2\mathbf{M}\right| = 0 \quad (6.277)$$

By using Equations 6.254 and 6.256,

$$m_1m_2\omega^4 - (m_1k_{22} + m_2k_{11})\omega^2 + k_{11}k_{22} - k_{12}^2 = 0 \quad (6.278)$$

The above equation is quadratic with respect to ω^2.

Consider a specific example: l_1 = 0.325 m, l_2 = 0.35 m, l_3 = 0.525 m, D_1 = D_2 = 0.25 m (disk diameters), b = 0.05 m (disk thickness), E = 200 10^9 Pa, ρ = 7860 kg/m^3, e_1 = 0.035 10^{-3} (first disk eccentricity), e_2 = 0.046 10^{-3} (second disk eccentricity), d = 0.05 m (rotor diameter). For the given system the critical speeds (natural frequencies) are ω_1 = 241.533 rad/s and ω_2 = 1061.81 rad/s.

The determination of the damping coefficients is not straightforward because they depend on the masses and natural frequencies of the system. Use a simplified assumption that all damping coefficients are equal, and that their value is $\delta\, m_1\omega_1/\pi$, where δ is the logarithmic decrement, and ω_1 is the first natural frequency. Take that δ = 0.02 (steel).

The amplitude–angular velocity diagram for the first disk for the given data is shown in Figure 6.36.

PROBLEMS AND EXERCISES

PROBLEMS

1. Show that a mass on an ideal spring will have a periodic motion if subjected to initial displacement.
2. Show that a mass on an ideal spring will have a periodic motion if subjected to initial velocity.
3. For an SDOF, what is the effect of (a) mass and (b) spring stiffness on the resonance frequency?
4. If the SDOF system is not ideal, i.e., its damping properties are taken into account, would it affect the period of response when subjected to initial disturbance?
5. What is critical damping for an SDOF?
6. Why in systems with damping subjected to periodic forces can the initial conditions be neglected when finding the system response?
7. Find the solution of the equation $\ddot{y} + 2\xi\omega_n\dot{y} + \omega_n^2 y = p_0 \sin(\omega t)$ in a steady-state regime. What are the effects of damping on an SDOF system in terms of its response to a periodic force?
8. Show that if in the equation $\ddot{y} + 2\xi\omega_n\dot{y} + \omega_n^2 y = p_0 \sin\omega t$ the forcing function is substituted by $p_0 \cos\omega t$, the expressions for the amplitude and phase angle remain the same.
9. State the initial conditions and find the two constants in the following general solution of a motion equation for an SDOF system:

$$y_c(t) = y_g(t) + y_p(t) = c_1 e^{\lambda_1 t} + c_2 e^{\lambda_2 t} + d_1 \cos\omega t + d_2 \sin\omega t$$

10. What is the definition of a logarithmic decrement? How it is related to the coefficient of damping for an SDOF system?
11. What is called the kinematic excitation? Write the motion equation for an SDOF system. How does it differ from the case of forced excitation?
12. What is the principle of superposition? Explain, using the example of a polyharmonic force acting on an SDOF system.
13. What causes torsional vibrations? Write the motion equation for an SDOF system and compare it with the equation for the mass–spring system. Is there any similarity between the two in terms of the system response to a periodic excitation?
14. How many natural frequencies characterize a 2DOF system without damping? What is called a mode of vibration? Explain.
15. Derive the characteristic equation for a 2DOF system without damping.

16. The algebraic system corresponding to free vibrations of a 2DOF system is as follows:

$$\{-m_1\omega^2 + (k_1 + k_2)\}X_1 - k_2X_2 = 0$$
$$-k_2X_1 + \{-m_2\omega^2 + (k_2 + k_3)\}X_2 = 0$$

   a. Why does a meaningful solution exist only for specific frequencies?
   b. For specific frequencies found in part (a), is it possible to find explicitly the amplitudes X_1 and X_2?

17. Does damping affect the modes of vibration in a 2DOF system? Explain.
18. Show that if in Figure 6.27 the stiffness in the x-direction is not equal to that in the y-direction, then the shaft will whirl on an elliptical trajectory.
19. What is the effect of damping on a synchronous whirl of a single-disk rotor?
20. How are the vibrations caused by kinematic excitation and by unbalanced forces in a rotating disk similar?

Exercises (Projects) with Mathematica

1. An automobile moving over a rough road will experience vibrations, the level of which will depend on the type of roughness and the speed of motion (in addition to the vehicle properties). Consider the following one-dimensional model of the vehicle (Figure P6.1): m_1 is the passenger mass, m_2 is the main body mass, m_3 is the axle mass, k_1 is the seat stiffness, k_2 is the suspension stiffness (main body on axles), k_3 is the stiffness of tires, c_1 is the damping coefficient of the seat, c_2 is the damping coefficient of suspension, and c_3 is the damping coefficient of tires. Assume that the roughness of the road can be described by a periodic function $y = A\sin(2\pi x/L)$, where A is the amplitude and L is the period of roughness. As is clear, y is a function of time since it depends on the velocity of motion ($x = Vt$, where V is velocity, $0 \le x \le L$). Choose system parameters and the road roughness constants and analyze the level of passenger vibrations for vehicle velocities from 10 to 110 km/h.

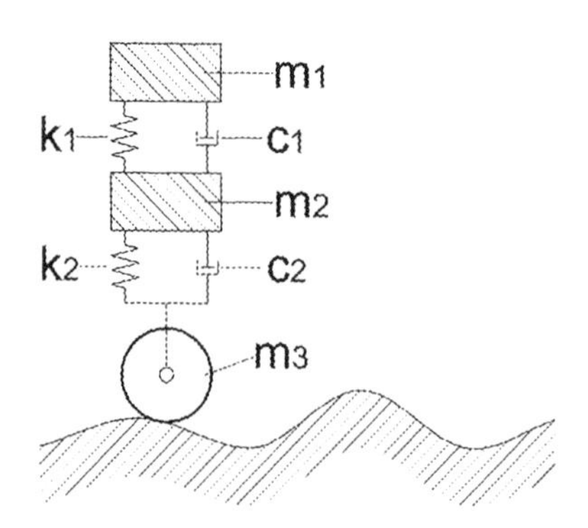

FIGURE P6.1

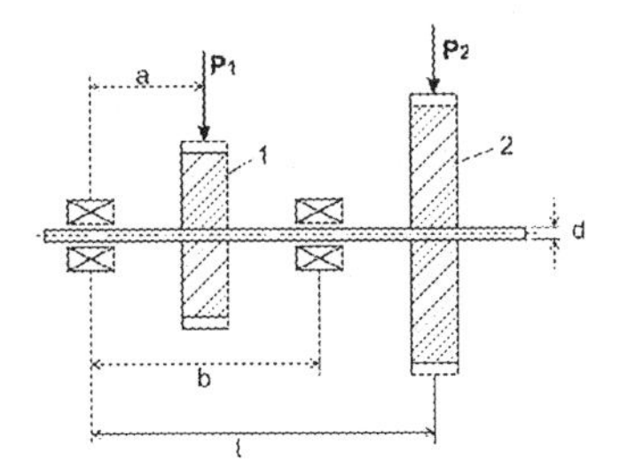

FIGURE P6.2

2. A steel shaft on two supports carries two spur gears (Figure P6.2). One of the causes of shaft vibrations is the impact of teeth, which, in turn, is due to the imperfection of gear geometry. The frequency of these vibrations is called the tooth frequency, and it is equal to the product of the rotational frequency and the number of teeth. For the following data, $l = 165$ mm, $a = 40$ mm, $b = 130$ mm, $N_1 = 25$, $N_2 = 30$, $m_1 = 1.3$ kg, $m_2 = 0.7$ kg, find the diameter of the shaft d such that there will be no resonances within the speed range of 2000 to 3000 rpm of the given shaft. Note, N_1 and N_2 are the teeth numbers of gears 1 and 2, respectively.

Bibliography

Chironis, N.P., *Mechanisms and Mechanical Devices Sourcebook*, McGraw-Hill, New York, 1991.

Erdman, A.G. and Sandor, G.N., *Mechanical Design: Analysis and Synthesis*, Vol. 1, Prentice-Hall, Englewood Cliffs, NJ, 1991.

Kimbrell, J.T., *Kinematic Analysis and Synthesis*, McGraw-Hill, New York, 1991.

Mabie, H.H. and Reinholtz, C.F., *Mechanisms and Dynamics of Machinery*, John Wiley & Sons, New York, 1987.

Malik, A.K., Ghosh, A., and Dittrich, G., *Kinematic Analysis and Synthesis of Mechanisms*, CRC Press, Boca Raton, FL, 1994.

Martin, G.H., *Kinematics and Dynamics of Machines*, McGraw-Hill, New York, 1982.

Myszka, D.H., *Machines and Mechanisms: Applied Kinematic Analysis*, Prentice-Hall, Englewood Cliffs, NJ, 1999.

Nikravesh, P.E., *Computer-Aided Analysis of Mechanical Systems*, Prentice-Hall, Englewood Cliffs, NJ, 1988.

Norton, R.L., *Design of Machinery: An Introduction to the Synthesis and Analysis of Mechanisms and Machines*, McGraw-Hill, New York, 1992.

Rao, S.S., *Mechanical Vibrations*, Addison-Wesley, Reading, MA, 1990.

Shigley, J.E. and Uicker, J.J., Jr., *Theory of Machines and Mechanisms*, McGraw-Hill, New York, 1995.

Waldron, K.J. and Kinzel, G.L., *Kinematics, Dynamics, and Design of Machinery*, John Wiley & Sons, New York, 1999.

Wolfram, S., *Mathematica*, Cambridge University, Cambridge, U.K., 1999.

Appendix: Use of Mathematica as a Tool

A.1 INTRODUCTION TO MATHEMATICA

To start Mathematica, double-click the Mathematica icon and wait until an empty notebook appears. Type a command, e.g., a numerical command; then press SHIFT + ENTER and wait until the kernel is loaded (since this is the first evaluation).

In[1]:= 2 + 3

Out[1]:= 5

To the left of the command, 2 + 3, a label In[1] appears, and the result of evaluation is labeled as Out[1]. Both input and output are identified on the right by brackets, which are called *cells*. By moving the cursor below the Out[1] cell and then clicking, a straight line appears, which opens a new cell ready for an input. Thus, the *notebook* consists of a series of cells.

Numerical Calculations

Mathematica can be used as a calculator, in which case numbers and corresponding arithmetic operators are typed in and then executed by pressing SHIFT + ENTER, as in the example above. Mathematica has many built-in functions (see *Built-in Functions* in *Help*), the numerical values of which can be found for any given argument (1) by placing the argument in square brackets, (2) then by placing the function itself in square brackets, (3) then by preceding the latter with capital N, and (4) then by executing the hierarchy of two commands.

In[2]:= N[Cos[π/3]]

Out[2]:= 0.5

In the above input In[2], N[ ] is a function that gives a numerical approximation of a real number. Note also that all commands and all built-in functions are written with the first letter capitalized.

Symbolic Calculations

Consider a quadratic equation $ax^2 + bx + c = 0$. We find the roots by using the command Solve[ ].

In[3]:= quadraticEq = $ax^2 + bx + c == 0$

Out[3]:= $c + bx + ax^2 == 0$

In[4]:= solution = Solve [quadraticEq, x]

Out[4]:= $\left\{\left\{x \to \frac{-b-\sqrt{b^2-4ac}}{2a}\right\}, \left\{x \to \frac{-b+\sqrt{b^2-4ac}}{2a}\right\}\right\}$

The quadratic equation is defined by the user-defined name *quadraticEq*. If we type *quadraticEq* and execute, the equation will be the output.

In[5]:= quadraticEq

Out[5]:= $c + bx + ax^2 == 0$

The solution of the equation is given as a *List of Replacement Rules*, which means that the expressions for the roots cannot be used explicitly in any other symbolic or numerical operation. We can obtain explicit expressions for the roots by extracting corresponding information and assigning it to specific names.

In[6]:= rootOne = x/. solution[[1]]

Out[6]:= $\frac{-b-\sqrt{b^2-4ac}}{2a}$

The meaning of the above is "replace x by the expression in the first braces in the list *solution* (/. denotes a *Replacement Rule*) and give it the name *rootOne*." Now, rootOne can be used in any numerical or symbolic operation. Consider, for example, rootOne squared:

In[7]:= rootOne^2

Out[7]:= $\frac{(-b-\sqrt{b^2-4ac})^2}{4a^2}$

In[8]:= rootOneSpecific = rootOne /. {a → 2, b → 5, c → 10}

Out[8]:= $\frac{1}{4}(-5 - \mathrm{I}\sqrt{55})$

We found a specific value of the root rootOne by replacing symbols with numbers. It turns out that for these numbers the root is a complex number (I denotes an imaginary unit).

Plotting Results

To plot a two-dimensional function, use the command *Plot*. Let us say that the function to be plotted is x – Sin[x] and it is to be plotted from $x = 0$ to $x = 2\pi$. We name this plot *plotOne*.

In[9]:= plotOne = Plot[x – Sin[x], {x, 0, 2 Pi}]

Out[9]:=

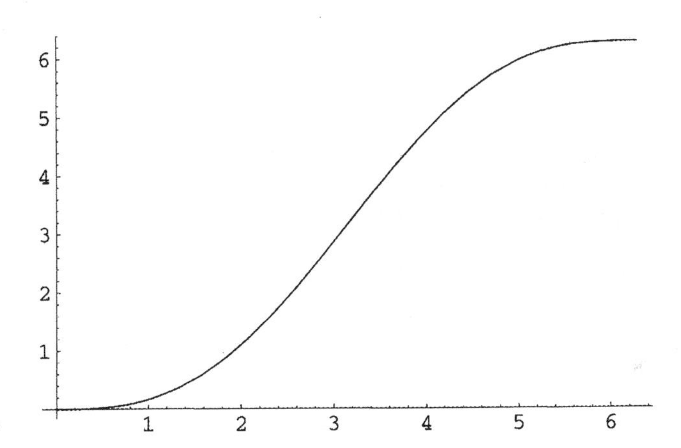

Animation

We would like to animate the motion of a wheel moving forth along a straight line. Mathematica has a *Circle* primitive, which is a command.

In[10]:= circl = Circle[{0, rad}, rad]

Out[10]:= Circle[{0, rad}, rad]

In the brackets 0,rad gives the coordinates of the circle center, and rad is its radius. This circle can be plotted with the Show and Graphics commands. Let us assign a specific value to the radius:

In[11]:= rad = 1.

Out[11]:= 1.

In[12]:= circleGraphics = Show[Graphics[circl], AspectRatio → Automatic, Axes → True]

Out[12]:=

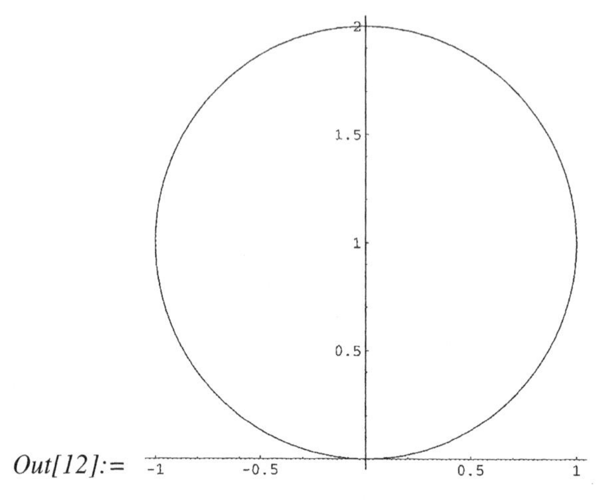

The option in Graphics AspectRatio → Automatic gives the same scale for x- and y-coordinates. Now we can use another Mathematica primitive graphics, *Line*, to draw a line passing through the circle center.

In[13]:= diameterY = Line[{{0, 0}, {0, 2 rad}}]

Out[13]:= Line[{{0, 0}, {0, 2.}}]

The two primitives, Circle and Line, can be shown together as one object, wheelGraphics.

In[14]:= wheelGraphics = Show[Graphics[circl], Graphics[diameterY], AspectRatio → Automatic]

Out[14]:=

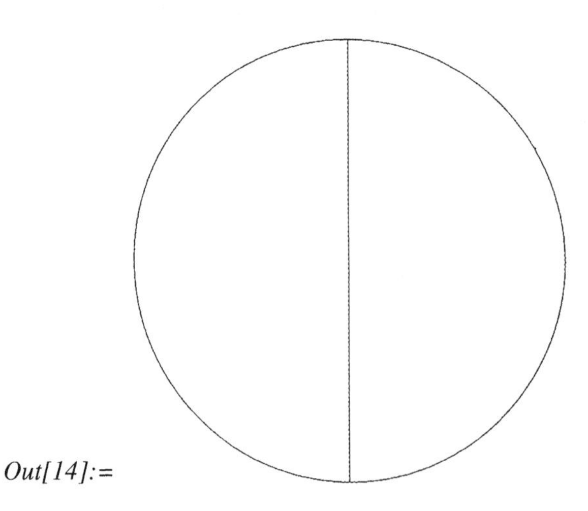

Now we introduce a function that describes the circle and the diameter at any position along the line in terms of the rotation *angle* and the number *j* characterizing its discrete position.

```
In[15]:= wheelFunction[angle_, j_] :=
    (
      xCentre = rad angle[[j]]; yCentre = rad;
      xOne = rad(angle[[j]] – Sin[angle[[j]]]); yOne = rad (1 – Cos[angle[[j]]]);
      xTwo = rad(angle[[j]] + Sin[angle[[j]]]); yTwo = rad (1 + Cos[angle[[j]]]);
      {
      Thickness[0.01],
      {
        {RGBColor[0, 0, 1], Circle[{xCentre, yCentre}, rad]},
        {RGBColor[1, 0, 0], Line[{{xOne, yOne}, {xTwo, yTwo}}]}
      }
     }
    )
    Clear[j]
```

The angle angle is expressed in radians; xCentre and yCentre are the coordinates of the circle center; *xOne*, *yOne* and *xTwo*, *yTwo* are the coordinates of two points on the circle through which the line passes; RGBColor[0, 0, 1] is an option for the color of the circle; and RGBColor[1, 0, 0] for the color of the line (these colors are not shown).

Animation is achieved by displaying the wheel at various positions sequentially. The positions are given as functions of the angle of wheel rotation, i.e., *angle* in our notation. The discrete set of values for the variable angle is introduced in the form of an array, and this array is generated with the help of the *Table* command.

```
In[16]:=   angleDiscrete = Table[t, {t, 0., N[2 Pi], N[Pi/10]}]

Out[16]:= {0., 0.314159, 0.628319, 0.942478, 1.25664, 1.5708, 1.88496, 2.19911,
           2.51327, 2.82743, 3.14159, 3.45575, 3.76991, 4.08407, 4.39823,
           4.71239, 5.02655, 5.34071, 5.65487, 5.96903, 6.28319}
```

Let us find the number of elements in the array angleDiscrete:

```
In[17]:=   nPositions = Length[angleDiscrete]

Out[17]:= 21
```

Now we create an array of all wheel positions by again using the command Table:

```
In[18]:=   plots = Table[Graphics[wheelFunction[angleDiscrete, j],
              AspectRatio → Automatic,
              Axes → True, PlotRange → {{–rad, N[2 Pi rad]}, {0, 2 rad}}],
              {j, 1, nPositions}
           ]
```

Out[18]:= {– Graphics –, – Graphics –}

Now we can plot the wheel at all nPositions positions:

In[19]:= Show[plots]

Out[19]:=

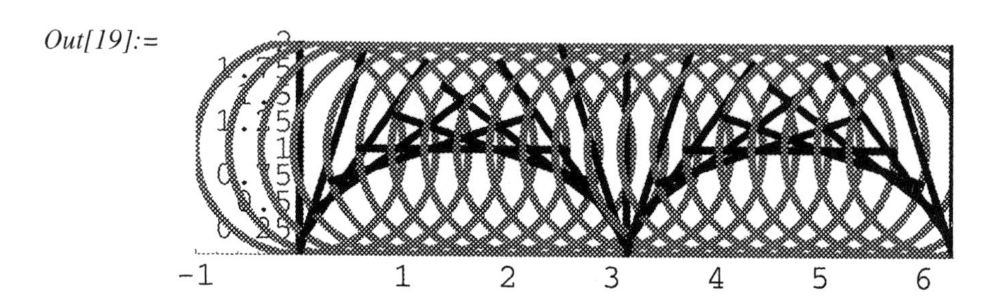

To animate the sequence of wheel positions Mathematica uses a special code, i.e., a standard package. The animation package can be loaded by the following command:

In[20]:= Needs["Graphics`Animation`"]

The following command displays all wheel positions. To animate the motion, double-click on any of the positions. To stop animation, click anywhere within the cell once. Of course, the animation cannot be shown here.

In[21]:= ShowAnimation[plots]

A.2 VECTOR ALGEBRA

A vector is represented as a *list*.

In[1]:= vectorA = {ax, ay}

Out[1]:= {ax, ay}

In[2]:= vectorB = {bx, by}

Out[2]:= {bx, by}

The sum of two vectors **vectorA** and **vectorB** is **vectorC**.

In[3]:= vectorC = vectorA + vectorB

Out[3]:= {ax + bx, ay + by}

The scalar product of two vectors is obtained by placing a *dot* between the vectors:

In[4]:= vectorD = vectorA . vectorB

Out[4]:= ax bx + ay by

To obtain a cross-product, **vectorA** and **vectorB** should be represented as three-dimensional (3D) vectors:

In[5]:= vectorA3D = {ax, ay, 0}

Out[5]:= {ax, ay, 0}

In[6]:= vectorB3D = {bx, by, 0}

Out[6]:= {bx, by, 0}

The cross-product of two vectors is obtained by using a *Cross* command:

In[7]:= vectorU3D = Cross[vectorA3D, vector B3D]

Out[7]:= {0, 0, –ay bx + ax by}

Now we will extract the nonzero component and call it *zComponent*:

In[8]:= zComponent = vectorU3D[[3]]

Out[8]:= –ay bx + ax by

We will repeat it for the vectors with the trigonometric representation of coordinates:

In[9]:= vectorAtrig = a{Cos[α], Sin[α]}

Out[9]:= {a Cos[α], a Sin[α]}

In[10]:= vectorBtrig = b{Cos[β], Sin[β]}

Out[10]:= {b Cos[β], b Sin[β]}

In[11]:= vectorDtrig = vectorAtrig . vectorBtrig

Out[11]:= ab Cos[α] Cos[β] + ab Sin[α] Sin[β]

This expression can be simplified by using a *TrigReduce* command:

In[12]:= vectorDsimplified = TrigReduce[vectorDtrig]

Out[12]:= ab Cos[α – β]

A.3 VECTOR ANALYSIS

If vector parameters are functions of time t, then a vector is written as follows:

In[13]:= vectorFunction = am[t]{Cos[γ[t]], Sin[γ[t]]}

Out[13]:= {am[t] Cos[γ[t]], am[t] Sin[γ[t]]}

The time-derivative is

In[14]:= vectorDerivative = D[vectorFunction, t]

Out[14]:= {Cos[γ[t]] am′[t] – am[t] Sin[γ[t]] γ′[t],
Sin[γ[t]] am′[t] + am[t] Cos[γ[t]] γ′[t]}

The second time-derivative is

In[15]:= vectorDerivativeSec = D[vectorFunction, {t, 2}]

Out[15]:= {–2 Sin[γ[t]] am′[t]γ′[t] – am[t] Cos[γ[t]] γ′[t]^2 +
Cos[γ[t]] am″[t] – am[t] Sin[γ[t]] γ″[t],
2 Cos[γ[t]] am′[t]γ′[t] – am[t] Sin[γ[t]] γ′[t]^2 +
Sin[γ[t]] am″[t] + am[t] Cos[γ[t]] γ″[t]}

A.4 KINEMATIC AND FORCE ANALYSIS OF MECHANISMS

A.4.1 Slider-Crank Mechanism

Position Analysis

Let us consider the case when the crank is the driver. Then, the distance from the piston to the crankshaft center (r_1) and the position of the connecting rod (θ_3) are the two unknowns. This case falls into the second category (see text). In this case $\alpha = \theta_2 - \pi$, $\theta_1 = \pi$, $b = r_2$, and r_2 and r_3 are constants.

We put a semicolon at the end of the input if we do not want the output to be displayed.

In[1]:= $\alpha = \theta_2 + \pi$;

In[2]:= $b = r_2$;

In[3]:= $\theta_1 = \pi$;

In[4]:= $r_1 = r_2$ Cos[$\alpha - \theta_1$] + Sqrt[$r_3^2 - r_2^2$ Sin[$\alpha - \theta_1$]^2]

Out[4]:= r_2 Cos[θ_2] + $\sqrt{r_3^2 - r_2^2 \mathrm{Sin}[\theta_2]^2}$

Now we find Sin[$\theta_3 - \theta_1$] and Cos[$\theta_3 - \theta_1$] from Equations 2.29 and 2.30:

In[5]:= $\quad \text{Sin}\theta_3\theta_1 = \dfrac{\text{bSin}[\alpha - \theta_1]}{r_3}$

Out[5]:= $\quad \dfrac{r_2\text{Sin}[\theta_2]}{r_3}$

In[6]:= $\quad \text{Cos}\theta_3\theta_1 = \dfrac{\text{bCos}[\alpha - \theta_1] - r_1}{r_3}$

Out[6]:= $\quad -\dfrac{\sqrt{r_3^2 - r_2^2\text{Sin}[\theta_2]^2}}{r_3}$

The correct angle $\theta_3 - \theta_1 = \theta_3 - \pi$ is found using the command *ArcTan[Cos$\theta_3\theta_1$, Sin$\theta_3\theta_1$]*:

In[7]:= $\quad \theta_3 = \pi + \text{ArcTan}[\text{Cos}\theta_3\theta_1, \text{Sin}\theta_3\theta_1]$

Out[7]:= $\quad \pi + \text{ArcTan}\left[-\dfrac{\sqrt{r_3^2 - r_2^2\text{Sin}[\theta_2]^2}}{r_3}, \dfrac{r_2\text{Sin}[\theta_2]}{r_3}\right]$

At this point, we should assign some values to r_2 and r_3:

In[8]:= $\quad r_2 = 1.;\ r_3 = 4.;$

In[9]:= $\quad \theta_3\text{Specific} = \theta_3$

Out[9]:= $\quad \pi + \text{ArcTan}[-0.25\sqrt{16.-1.\text{Sin}[\theta_2]^2}, 0.25\text{Sin}[\theta_2]]$

In[10]:= $\quad r_1\text{Specific} = r_1$

Out[10]:= $\quad 1.\text{Cos}[\theta_2] + \sqrt{16.-1.\text{Sin}[\theta_2]^2}$

Now we can plot both functions.

In[11]:= $\quad \text{Plot}[\theta_3\text{Specific}, \{\theta_2, 0, 2\pi\}]$

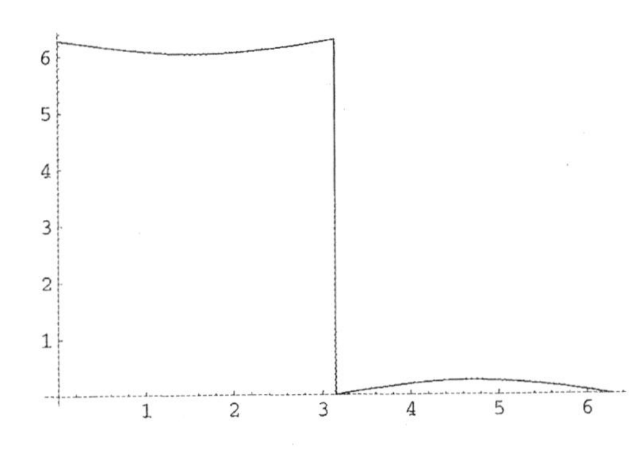

Out[11]:= – Graphics –

We can see that θ_3 is described by the positive values of the angle. It is more convenient to look at this angle as being within $\pm\theta_3$. We can transform the above function:

In[12]:= θ_3Transformed = If[$\theta_2 \leq \pi$, θ_3Specific – 2π, θ_3Specific]

Out[12]:= If[$\theta_2 \leq \pi$, θ_3Specific – 2π, θ_3Specific]

In[13]:= Plot[θ_3Transformed, {θ_2, 0, 2π},
AxesLabel → {"θ_2", "θ_3"}, PlotStyle → {AbsoluteThickness[2.4]}]

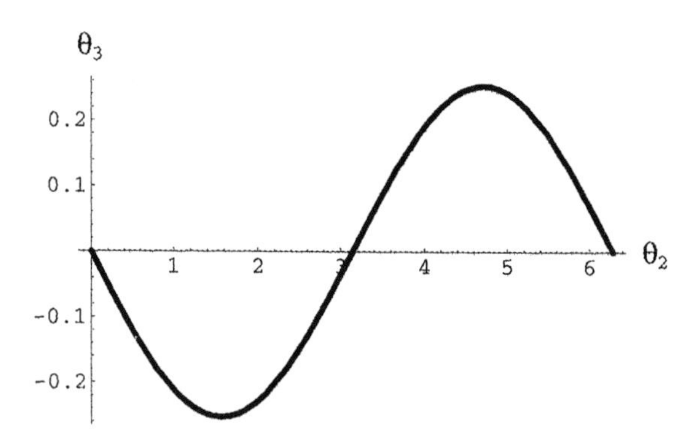

Out[13]:= – Graphics –

In[14]:= Plot[r_1Specific, {θ_2, 0, 2π},
AxesLabel → {"θ_2", "r_1"}, PlotStyle → {AbsoluteThickness[2.4]}]

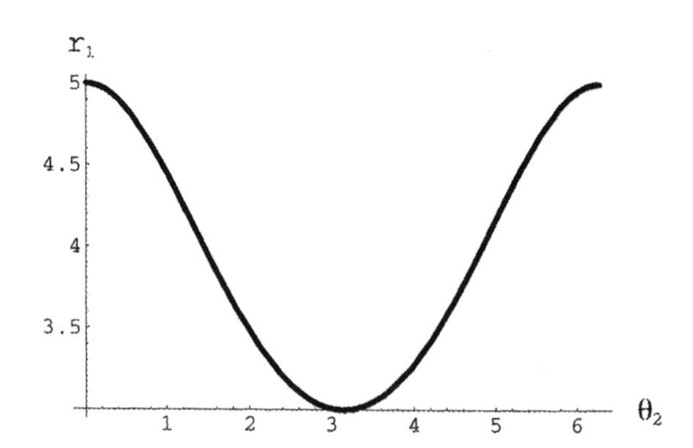

Out[14]:= – Graphics –

To animate the mechanism, we discretize the input angle θ_2:

In[15]:= theta2 = Table[iπ/10, {i, 0, 20}]

Out[15]:= $\left\{0, \frac{\pi}{10}, \frac{\pi}{5}, \frac{3\pi}{10}, \frac{2\pi}{5}, \frac{\pi}{2}, \frac{3\pi}{5}, \frac{7\pi}{10}, \frac{4\pi}{5}, \frac{9\pi}{10}, \pi,\right.$

$\left.\frac{11\pi}{10}, \frac{6\pi}{5}, \frac{13\pi}{10}, \frac{7\pi}{5}, \frac{3\pi}{2}, \frac{8\pi}{5}, \frac{17\pi}{10}, \frac{9\pi}{5}, \frac{19\pi}{10}, 2\pi\right\}$

By replacing the symbolic variable θ_2 by its numerical counterpart theta2 in all expressions we automatically discretize these functions:

In[16]:= r_1Discr = r_1 /. $\theta_2 \to$ theta2

Out[16]:= {5., 4.9391, 4.76559, 4.50512, 4.19431, 3.87298, 3.57627, 3.32955, 3.14756, 3.03699, 3., 3.03699, 3.14756, 3.32955, 3.57627, 3.87298, 4.19431, 4.50512, 4.76559, 4.9391, 5.}

In[17]:= θ_3Discr = θ_3 /. $\theta_2 \to$ theta2

Out[17]:= {2π, 6.20585, 6.1357, 6.07953, 6.04312, 6.03051, 6.04312, 6.07953, 6.1357, 6.20585, 2π, 0.0773313, 0.14748, 0.203659, 0.240063, 0.25268, 0.240063, 0.203659, 0.14748, 0.0773313, 2π}

Animation is achieved by drawing the mechanism in Length[theta2] positions, and then displaying them in sequence. To accomplish this, we have to create discrete data for all variables (which we already did), and then make the drawings as functions of these variables. We define a function that describes the mechanism geometry at any angle θ_2:

In[18]:= mechanismLines[theta2_, θ_3Discr_, j_] :=

(

$x_2 = r_2$ Cos[theta2[[j]]]; $y_2 = r_2$ Sin[theta2[[j]]];

$x_3 = x_2 + r_3$ Cos[θ_3Discr[[j]]]; $y_3 = y_2 + r_3$ Sin[θ_3Discr[[j]]];

r_1Mag = Sqrt[$x_3^2 + y_3^2$];

(*Coordinates of the piston in the global system*)

$$x_1P = \left(r_1\text{Mag} - \frac{1}{8}r_2\right)\text{Cos}[\pi - \theta_1] - \frac{1}{4}r_2\text{Sin}[\pi - \theta_1];$$

$$y_1P = -\left(r_1\text{Mag} - \frac{1}{8}r_2\right)\text{Sin}[\pi - \theta_1] - \frac{1}{4}r_2\text{Cos}[\pi - \theta_1];$$

$$x_2P = \left(r_1\text{Mag} - \frac{1}{8}r_2\right)\text{Cos}[\pi - \theta_1] + \frac{1}{4}r_2\text{Sin}[\pi - \theta_1];$$

$$y_2P = -\left(r_1\text{Mag} - \frac{1}{8}r_2\right)\text{Sin}[\pi - \theta_1] + \frac{1}{4}r_2\text{Cos}[\pi - \theta_1];$$

$$x_3P = \left(r_1 Mag + \frac{1}{8}r_2\right)Cos[\pi - \theta_1] + \frac{1}{4}r_2 Sin[\pi - \theta_1];$$

$$y_3P = -\left(r_1 Mag + \frac{1}{8}r_2\right)Sin[\pi - \theta_1] + \frac{1}{4}r_2 Cos[\pi - \theta_1];$$

$$x_4P = \left(r_1 Mag + \frac{1}{8}r_2\right)Cos[\pi - \theta_1] - \frac{1}{4}r_2 Sin[\pi - \theta_1];$$

$$y_4P = -\left(r_1 Mag + \frac{1}{8}r_2\right)Sin[\pi - \theta_1] - \frac{1}{4}r_2 Cos[\pi - \theta_1];$$

(*Coordinates of the cylinder in the global system*)

$$x_1C = \left(-r_2 + r_3 - \frac{1}{6}r_2\right)Cos[\pi - \theta_1] - \left(\frac{1}{4} + \frac{1}{10}\right)r_2 Sin[\pi - \theta_1];$$

$$y_1C = -\left(-r_2 + r_3 - \frac{1}{6}r_2\right)Sin[\pi - \theta_1] - \left(\frac{1}{4} + \frac{1}{10}\right)r_2 Cos[\pi - \theta_1];$$

$$x_2C = \left(-r_2 + r_3 - \frac{1}{6}r_2\right)Cos[\pi - \theta_1] + \left(\frac{1}{4} + \frac{1}{10}\right)r_2 Sin[\pi - \theta_1];$$

$$y_2C = -\left(-r_2 + r_3 - \frac{1}{6}r_2\right)Sin[\pi - \theta_1] + \left(\frac{1}{4} + \frac{1}{10}\right)r_2 Cos[\pi - \theta_1];$$

$$x_3C = \left(r_2 + r_3 + \frac{1}{6}r_2\right)Cos[\pi - \theta_1] + \left(\frac{1}{4} + \frac{1}{10}\right)r_2 Sin[\pi - \theta_1];$$

$$y_3C = -\left(r_2 + r_3 + \frac{1}{6}r_2\right)Sin[\pi - \theta_1] + \left(\frac{1}{4} + \frac{1}{10}\right)r_2 Cos[\pi - \theta_1];$$

$$x_4C = \left(r_2 + r_3 + \frac{1}{6}r_2\right)Cos[\pi - \theta_1] - \left(\frac{1}{4} + \frac{1}{10}\right)r_2 Sin[\pi - \theta_1];$$

$$y_4C = -\left(r_2 + r_3 + \frac{1}{6}r_2\right)Sin[\pi - \theta_1] - \left(\frac{1}{4} + \frac{1}{10}\right)r_2 Cos[\pi - \theta_1];$$

```
{
  Thickness[0.015],
  {
    {RGBColor[1, 0, 0], Line[{{0, 0}, {x2, y2}}]},
    {RGBColor[0, 1, 0], Line[{{x2, y2}, {x3, y3}}]},
    (*Drawing piston*)
      {RGBColor[0, 0, 1], Line[{{x1P, y1P}, {x2P, y2P}}]},
      {RGBColor[0, 0, 1], Line[{{x2P, y2P}, {x3P, y3P}}]},
      {RGBColor[0, 0, 1], Line[{{x3P, y3P}, {x4P, y4P}}]},
      {RGBColor[0, 0, 1], Line[{{x1P, y1P}, {x4P, y4P}}]},
    (*Drawing cylinder*)
      {RGBColor[1, 0, 1], Line[{{x2C, y2C}, {x3C, y3C}}]},
      {RGBColor[1, 0, 1], Line[{{x3C, y3C}, {x4C, y4C}}]},
      {RGBColor[1, 0, 1], Line[{{x4C, y4C}, {x1C, y1C}}]},
```

```
        (*Drawing joints*)
          {RGBColor[0, 0, 1], Circle[{0, 0}, 1/10 r2]},
          {RGBColor[1, 0, 0], Circle[{x3, y3}, 1/20 r2]},
          {RGBColor[0, 0, 1], Circle[{x2, y2}, 1/20 r2]}
      }
   }
)
In[19]:=Clear[j]
```

Now we plot the mechanism at Length[theta2] positions:

```
In[20]:=   Mechplots = Table[Graphics[mechanismLines[theta2, θ3Discr, j],
             AspectRatio → Automatic, Axes → True,
             PlotRange → {{-2, 6}, {-2., 2.}}], {j, 1, Length[theta2]}
           ]
Out[20]:= {- Graphics -, - Graphics -, - Graphics -, - Graphics -,
            - Graphics -, - Graphics -, - Graphics -, - Graphics -,
            - Graphics -, - Graphics -, - Graphics -, - Graphics -,
            - Graphics -, - Graphics -, - Graphics -, - Graphics -,
            - Graphics -, - Graphics -, - Graphics -, - Graphics -, - Graphics -}
```

To animate the Length[theta2] plots we call upon a special package Graphics\ Animation\:

```
In[21]:=   Needs["Graphics`Animation`"]
```

Instead of ShowAnimation[Mechplots], below we use the *GraphicsArray* command to show the slider-crank mechanism in the first four positions:

```
In[22]:=   Show[GraphicsArray[{{Mechplots[[1]], Mechplots[[2]]},
             {Mechplots[[3]], Mechplots[[4]]}}, AspectRatio → Automatic]]
```

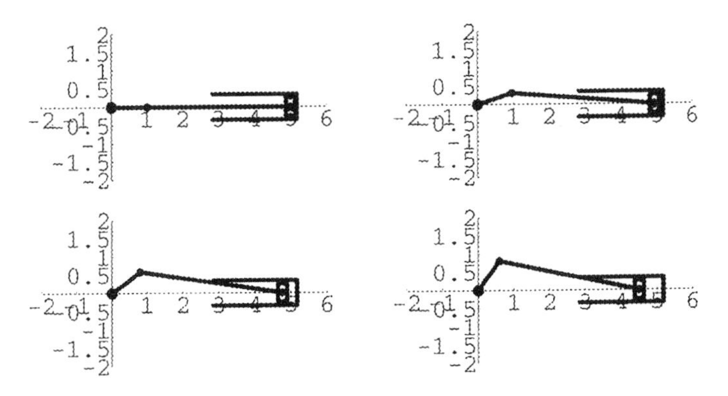

```
Out[22]:= - GraphicsArray -
```

Velocity Analysis

We use the solutions given in the text for the angular velocity of the connecting rod and for the translational velocity of the slider:

In[22]:= $\text{omega3} = -\text{omega2}\dfrac{r_2\text{Cos}[\theta_2]}{r_3\text{Cos}[\theta_3]}$

Out[22]:= $0.25\text{omega2}(\text{Cos}[\theta_2]\text{Sec}[\text{ArcTan}[-0.25\sqrt{16.-1.\text{Sin}[\theta_2]^2}, 0.25\text{Sin}[\theta_2]]])$

In[23]:= $\text{sliderVel} = -r_2\text{omega2}\dfrac{\text{Sin}[\theta_2-\theta_3]}{\text{Cos}[\theta_3]}$

Out[23]:= $-1.\text{omega2}(\text{Sec}[\text{ArcTan}[-0.25\sqrt{16.-1.\text{Sin}[\theta_2]^2}, 0.25\text{Sin}[\theta_2]]]$
$\text{Sin}[\theta_2-\text{ArcTan}[-0.25\sqrt{16.-1.\text{Sin}[\theta_2]^2}, 0.25\text{Sin}[\theta_2]]])$

Let us asume that the input angular velocity is constant and give it a value:

In[24]:= omega2 = 10

Out[24]:= 10

In[25]:= plotω3 = Plot[omega3, {θ_2, 0, 2π}, AxesLabel → {"θ_2", "ω_3"}, PlotStyle → (AbsoluteThickness[2.4]}]

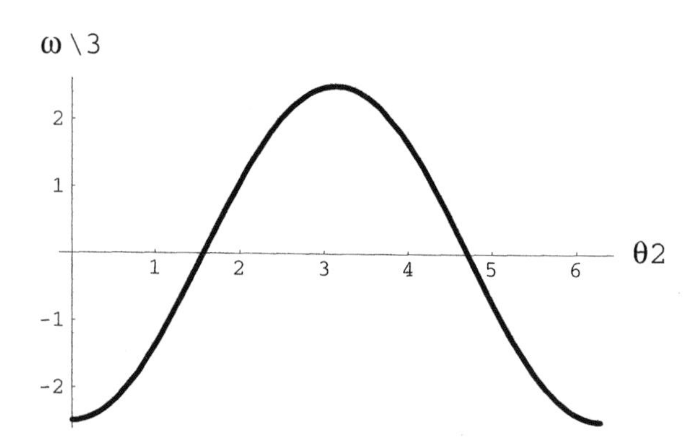

Out[25]:= – Graphics –

In[26]:= plotSliderVel = Plot[sliderVel, {θ_2, 0, 2π},
AxesLabel → {"θ_2", "Slider Velocity"},
PlotStyle → {AbsoluteThickness[2.4]}]

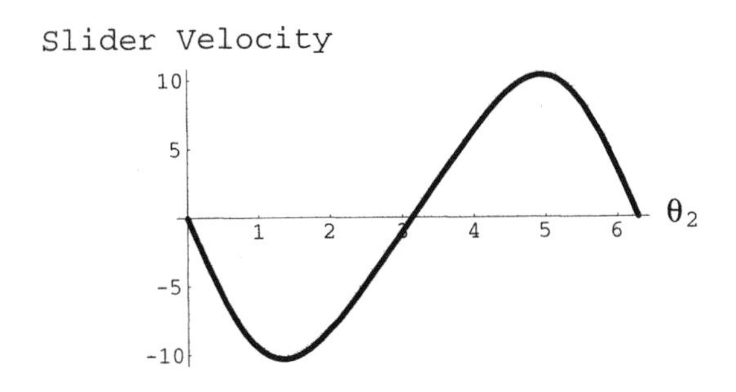

Out[26]:= – Graphics –

Acceleration Analysis

We use the solutions given in the text for the angular acceleration of the connecting rod and for the translational acceleration of the slider:

In[27]:= $\text{angularAccel3} = \dfrac{r_2\text{omega2}^2\text{Sin}[\theta_2] + r_3\text{omega3}^2\text{Sin}[\theta_3]}{r_3\text{Cos}[\theta_3]}$

Out[27]:=

$-0.25(\text{Sec}[\text{ArcTan}[-0.25\sqrt{16. - 1.\text{Sin}[\theta_2]^2}, 0.25\text{Sin}[\theta_2]]]$
$(100.\text{Sin}[\theta_2] - 25.(\text{Cos}[\theta_2]^2(\text{Sec}[\text{ArcTan}[-0.25\sqrt{16. - 1.\text{Sin}[\theta_2]^2}, 0.25\text{Sin}[\theta_2]]]$
$\text{Tan}[\text{ArcTan}[-0.25\sqrt{16. - 1.\text{Sin}[\theta_2]^2}, 0.25\text{Sin}[\theta_2]]]))))$

In[28]:= $\text{sliderAccel} = \dfrac{\text{sliderVel omega3Sin}[\theta_3] - r_2\text{omega2}(\text{omega2} - \text{omega3})}{\text{Cos}[\theta_3]}$

Out[28]:= $-\text{Sec}[\text{ArcTan}[-0.25\sqrt{16. - 1.\text{Sin}[\theta_2]^2}, 0.25\,\text{Sin}[\theta_2]]]$
$(-10.(10 - 2.5\,\text{Cos}[\theta_2]\text{Sec}[\text{ArcTan}[-0.25\sqrt{16. - 1.\text{Sin}[\theta_2]^2}, 0.25\,\text{Sin}[\theta_2]])+$
$25.\text{Cos}[\theta_2]\text{Sec}[\text{ArcTan}[-0.25\sqrt{16. - 1.\text{Sin}[\theta_2]^2}, 0.25\,\text{Sin}[\theta_2]]]$
$\text{Sin}[\theta_2 - \text{ArcTan}[-0.25\sqrt{16. - 1.\text{Sin}[\theta_2]^2}, 0.25\,\text{Sin}[\theta_2]]]$
$\text{Tan}[\text{ArcTan}[-0.25\sqrt{16. - 1.\text{Sin}[\theta_2]^2}, 0.25\,\text{Sin}[\theta_2]]])$

In[29]:= plotAngularAccel3 = Plot[angularAccel3, {θ_2, 0, 2π},
AxesLabel → {"θ_2", "α_3"}, PlotStyle → (AbsoluteThickness[2.4]}]

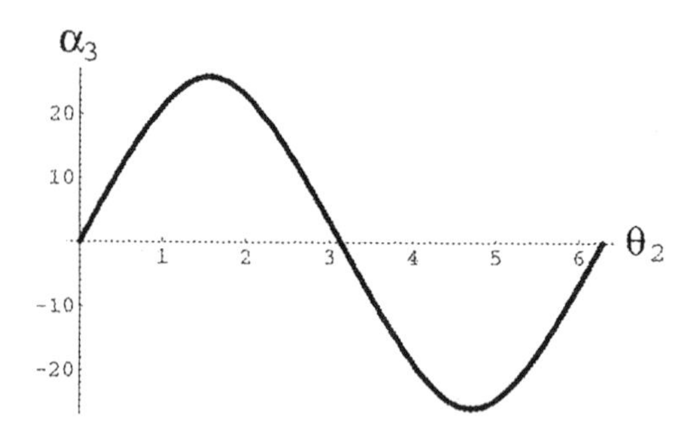

Out[29]:= – Graphics –

In[30]:=
```
plotSliderAccel = Plot[sliderAccel, {θ2, 0, 2π},
   AxesLabel → {"θ2", "Slider acceleration"},
PlotStyle → {AbsoluteThickness[2.4]}]
```

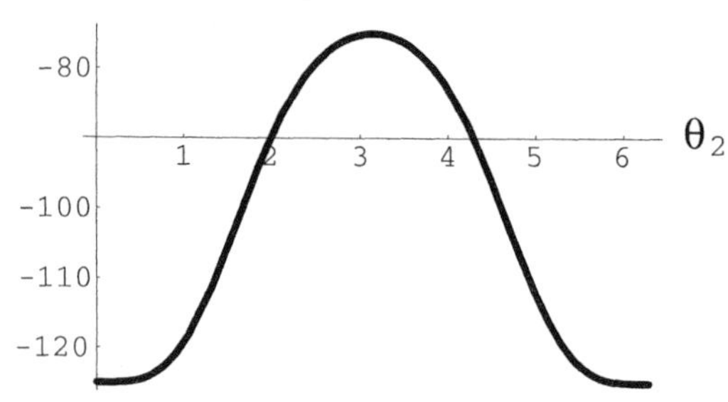

Out[30]:= – Graphics –

Force Analysis

We assume that the moment *moment2* is applied to the crank, and the resistance force P_4 is applied to the piston. We also assume that there are two concentrated masses: one at point *A*, which is the revolute joint connecting the crank and the connecting rod, and the other at point *B*, identifying the revolute joint on the slider. Note that we have to express all vectors as 3D objects.

In[31]:= $r_2IN3D = r_2\{Cos[\theta_2], Sin[\theta_2], 0\}$

Out[31]:= $\{1.Cos[\theta_2], 1.Sin[\theta_2], 0\}$

In[32]:= $r_3IN3D = r_3\{Cos[\theta_3], Sin[\theta_3], 0\}$

Out[32]:= $\{-4.Cos[ArcTan[-0.25\sqrt{16. - 1.Sin[\theta_2]^2}, 0.25\,Sin[\theta_2]]],$
$-4.Sin[ArcTan[-0.25\sqrt{16. - 1.Sin[\theta_2]^2}, 0.25\,Sin[\theta_2]]], 0\}$

Forces and moments as 3D vectors:

In[33]:= sliderForce = P4{–1, 0, 0};

In[34]:= crankMoment = moment2{0, 0, 1};

Force in the crank-connecting rod joint acting on the crank:

In[35]:= jointForce32 = {f32x, f32y, 0};

Force in the connecting rod–slider joint acting on the rod:

In[36]:= jointForce13 = f13{0, 1, 0};

Force in the crank–frame joint acting on the crank:

In[37]:= jointForce12 = {f12x, f12y, 0};

Inertial forces

Acceleration of point *A*:

In[38]:= accelA = r_2 D[{Cos[θ_2[t]], Sin[θ_2[t]], 0}, {t, 2}]

Out[38]:= {1.(–Cos[θ_2[t]]θ_2'[t]^2 – Sin[θ_2[t]]θ_2''[t]),
1.(–Sin[θ_2[t]]θ_2'[t]^2 + Cos[θ_2[t]]θ_2''[t]), 0}

Assume that omega2 is constant. Also, let us replace $\theta_2[t]$ by θ_2, the notation used above.

In[39]:= accelAMod = accelA /. {θ_2[t] → θ_2, θ_2'[t] → omega2, θ_2''[t] → 0}

Out[39]:= {–100.Cos[θ_2], –100.Sin[θ_2], 0}

Inertial force at joint *A*:

In[40]:= inertForceA = –massA accelAMod;

Acceleration of point *B*. But first we have to replace θ_2 by $\theta_2[t]$ in the expression for θ_3.

In[41]:= θ_3T = θ_3 /. θ_2 → θ_2[t]

Out[41]:= π + ArcTan[–0.25 $\sqrt{16. - 1.\text{Sin}[\theta_2[t]]^2}$, 0.25 Sin[θ_2[t]]]

In[42]:= accelB = D[r_2{Cos[θ_2[t]], Sin[θ_2[t]], 0} + r_3{Cos[θ_3T], Sin[θ_3T], 0}, {t, 2}];

Again, we take into account that omega2 is constant, and also replace $\theta_2[t]$ with θ_2.

In[43]:= accelBMod = accelB /. {θ_2[t] → θ_2, θ_2'[t] → omega2, θ_2''[t] → 0};

Equilibrium equations

Forces in link 2, crank:

In[44]:= eqOne = jointForce12 + jointForce32 − inertForceA == 0

Out[44]:= {f12x + f32x − 100.massA Cos[θ_2], f12y + f32y − 100.massA Sin[θ_2], 0} == 0

Moments in link 2 about point 0:

In[45]:= eqTwo = Cross[jointForce32 − inertForceA, r_2IN3D] − crankMoment == 0

Out[45]:= {0, 0, −moment2 − 1.f32y Cos[θ_2] + 1.f32x Sin[θ_2] +
0.massA Cos[θ_2]Sin[θ_2]} == 0

Forces in link 3, connecting rod:

In[46]:= eqThree = −jointForce32 + jointForce13 +
sliderForce − massB accelBMod == 0;

Moments in link 3 about point *B*:

In[47]:= eqFour = Cross[r_3IN3D, −jointForce32] == 0

Out[47]:= {0, 0, 4.f32y Cos[ArcTan[$-0.25\sqrt{16. - 1.\text{Sin}[\theta_2]^2}$, 0.25 Sin[$\theta_2$]]] −
4.f32x Sin[ArcTan[$-0.25\sqrt{16. - 1.\text{Sin}[\theta_2]^2}$, 0.25 Sin[$\theta_2$]]]} == 0

Solution of equations

There are six scalar equations with six unknowns: $f_{12}x$, $f_{12}y$, $f_{32}x$, $f_{32}y$, f_{13}, and P_4. We do not display the solutions since they are long. At this point, we assign specific input data:

In[48]:= massA = 5; massB = 20; moment2 = 10;

In[49]:= solution = Solve[{eqOne, eqTwo, eqThree, eqFour},
{f12x, f12y, f32x, f32y, f13, P4}];

The solutions are given in the form of the replacement rules. We transform replacement rules into explicit functions:

In[50]:= f12xFunct = f12x /. solution

Out[50]:= {0.+500.Cos[θ2]+1.(0.+(1.Cot[ArcTan[-0.25 $\sqrt{16.-1.\text{Sin}[\theta_2]^2}$,0.25Sin[θ2]]]
(0. Sin[θ2] – 4. (–10. + 0. Cos[θ2]Sin[θ2])
Sin[ArcTan[–0.25 $\sqrt{16.-1.\text{Sin}[\theta_2]^2}$, 0.25 Sin[θ2]]])) /
(–4. Cos[ArcTan[–0.25 $\sqrt{16.-1.\text{Sin}[\theta_2]^2}$, 0.25 Sin[θ2]]] Sin[θ2] +
4. Cos[θ2]Sin[ArcTan[–0.25 $\sqrt{16.-1.\text{Sin}[\theta_2]^2}$, 0.25 Sin[θ2]]]))}

In[51]:= f12yFunct = f12y /. solution

Out[51]:= {0. + 500. Sin[θ2] + (1. (0. Sin[θ2] – 4. (–10. + 0. Cos[θ2]Sin[θ2])
Sin[ArcTan[–0.25 $\sqrt{16.-1.\text{Sin}[\theta_2]^2}$, 0.25 Sin[θ2]]])) /
(–4. Cos[ArcTan[–0.25 $\sqrt{16.-1.\text{Sin}[\theta_2]^2}$, 0.25 Sin[θ2]]] Sin[θ2] +
4. Cos[θ2]Sin[ArcTan[–0.25 $\sqrt{16.-1.\text{Sin}[\theta_2]^2}$, 0.25 Sin[θ2]]])}

In[52]:= f32xFunct = f32x /. solution

Out[52]:= {0. – (1. Cot[ArcTan[–0.25 $\sqrt{16.-1.\text{Sin}[\theta_2]^2}$, 0.25 Sin[θ2]]]
(0. Sin[θ2] – 4. (–10. + 0. Cos[θ2]Sin[θ2])
Sin[ArcTan[–0.25 $\sqrt{16.-1.\text{Sin}[\theta_2]^2}$, 0.25 Sin[θ2]]])) /
(–4. Cos[ArcTan[–0.25 $\sqrt{16.-1.\text{Sin}[\theta_2]^2}$, 0.25 Sin[θ2]]] Sin[θ2] +
4. Cos[θ2]Sin[ArcTan[–0.25 $\sqrt{16.-1.\text{Sin}[\theta_2]^2}$, 0.25 Sin[θ2]]])}

In[53]:= f32yFunct = f32y /. solution

Out[53]:= {–(1. (0. Sin[θ2] – 4. (–10. + 0. Cos[θ2]Sin[θ2])
Sin[ArcTan[–0.25 $\sqrt{16.-1.\text{Sin}[\theta_2]^2}$, 0.25 Sin[θ2]]])) /
(–4. Cos[ArcTan[–0.25 $\sqrt{16.-1.\text{Sin}[\theta_2]^2}$, 0.25 Sin[θ2]]] Sin[θ2] +
4. Cos[θ2]Sin[ArcTan[–0.25 $\sqrt{16.-1.\text{Sin}[\theta_2]^2}$, 0.25 Sin[θ2]]])}

In[57]:= f13Funct = f13 /. solution;

In[58]:= P4Funct = P4 /. solution;

All of the above expressions are functions of the independent variable θ_2. Now we can plot any of these functions. For example, we plot P_4:

In[59]:= Plot[P4Funct, {θ2, 0, 2π}, AxesLabel → {"θ2", "Slider force"},
PlotStyle → (AbsoluteThickness[2.4]}]

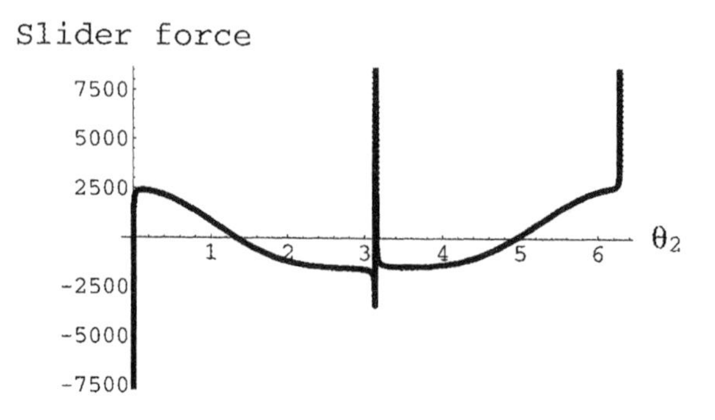

Out[59]:= – Graphics –

As we can see, there are three spikes in the resistance force at the 0, π, and 2π positions of the crank. These spikes correspond to the dead point positions of the slider (piston), since at these positions an infinite force is needed to maintain the equilibrium. In other words, the system becomes singular.

A.4.2 Four-Bar Linkage

The $\mathbf{r}_1$ vector describes the distance between the support joints, and then $\mathbf{r}_2$, $\mathbf{r}_3$, and $\mathbf{r}_4$ follow $\mathbf{r}_1$ in a clockwise direction. If θ_2 is an independent variable, then the two vectors, $\mathbf{r}_1$ and $\mathbf{r}_2$, are known. In this case $\mathbf{b} = -(\mathbf{r}_1 + \mathbf{r}_2)$.

In[1]:= $b = -(r_1\{Cos[\theta_1], Sin[\theta_1]\} + r_2\{Cos[\theta_2], Sin[\theta_2]\})$

Out[1]:= $\{-r_1Cos[\theta_1] - r_2Cos[\theta_2], -r_1Sin[\theta_1] - r_2Sin[\theta_2]\}$

The magnitude of vector **b** is

In[2]:= $bMagn = Sqrt[b[[1]]^2 + b[[2]]^2]$

Out[2]:= $\sqrt{(-r_1Cos[\theta_1] - r_2Cos[\theta_2])^2 + (-r_1Sin[\theta_1] - r_2Sin[\theta_2])^2}$

The Sin α and Cos α components of the vector **b** are

In[3]:= SinAlpha = b[[2]] / bMagn

Out[3]:= $$\frac{-r_1Sin[\theta_1] - r_2Sin[\theta_2]}{\sqrt{(-r_1Cos[\theta_1] - r_2Cos[\theta_2])^2 + (-r_1Sin[\theta_1] - r_2Sin[\theta_2])^2}}$$

In[4]:= CosAlpha = b[[1]] / bMagn

Out[4]:= $$\frac{-r_1\mathrm{Cos}[\theta_1]-r_2\mathrm{Cos}[\theta_2]}{\sqrt{(-r_1\mathrm{Cos}[\theta_1]-r_2\mathrm{Cos}[\theta_2])^2+(-r_1\mathrm{Sin}[\theta_1]-r_2\mathrm{Sin}[\theta_2])^2}}$$

At this point we should assign some values to $\mathbf{r}_1$, $\mathbf{r}_2$, $\mathbf{r}_3$, $\mathbf{r}_4$, and θ_1. Assume in this example such values that $\mathbf{r}_2$ is a crank.

In[5]:= $\mathbf{r}_1 = 4.;\ \mathbf{r}_2 = 1.;\ \mathbf{r}_3 = 6.;\ \mathbf{r}_4 = 5.;\ \theta_1 = \pi;$

The angle α is determined explicitly using the ArcTan command

In[6]:= α = ArcTan[CosAlpha, SinAlpha]

Out[6]:= $$\mathrm{arctan}\left[\frac{4.-1.\mathrm{Cos}[\theta_2]}{\sqrt{(4.-1.\mathrm{Cos}[\theta_2])^2+1.\mathrm{Sin}[\theta_2]^2}}, -\frac{1.\mathrm{Sin}[\theta_2]}{\sqrt{(4.-1.\mathrm{Cos}[\theta_2])^2+1.\mathrm{Sin}[\theta_2]^2}}\right]$$

In[7]:= Plot[α, {θ_2, 0, 2π}]

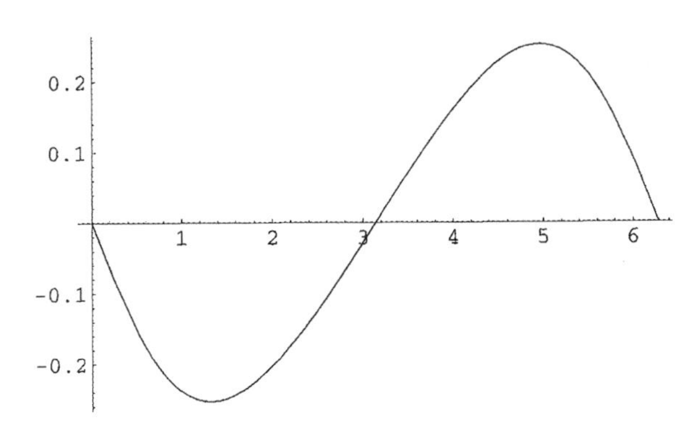

Out[7]:= – Graphics –

Solution of Loop-Closure Equation

From the formulas in the text, in this case $i = 3$, $j = 4$, we have

In[8]:= $$A = \frac{\mathrm{bMagn}^2 - r_3^2 + r_4^2}{2\mathrm{bMagn}\ r_4}$$

Out[8]:= $$\frac{0.1(-11.+(4.-1.\mathrm{Cos}[\theta_2])^2+1.\mathrm{Sin}[\theta_2]^2)}{\sqrt{(4.-1.\mathrm{Cos}[\theta_2])^2+1.\mathrm{Sin}[\theta_2]^2}}$$

In[9]:= $B = \frac{\text{bMagn} - r_4\ A}{r_3}$

Out[9]:= $0.166667\left(-\frac{0.5(-11. + (4. - 1.\text{Cos}[\theta_2])^2 + 1.\text{Sin}[\theta_2]^2)}{\sqrt{(4. - 1.\text{Cos}[\theta_2])^2 + 1.\text{Sin}[\theta_2]^2}} + \sqrt{(4. - 1.\text{Cos}[\theta_2])^2 + 1.\text{Sin}[\theta_2]^2}\right)$

The sign of *C* is either positive or negative. This corresponds to two possible solutions, which is to say, to two possible configurations. Let us take a positive sign:

In[10]:= $\text{Cconst} = \frac{r_4\sqrt{1 - A^2}}{r_3}$

Out[10]:= $0.833333\sqrt{1 - \frac{0.01(-11. + (4. - 1.\text{Cos}[\theta_2])^2 + 1.\text{Sin}[\theta_2]^2)^2}{(4. - 1.\text{Cos}[\theta_2])^2 + 1.\text{Sin}[\theta_2]^2}}$

Now we can find $\alpha - \theta_3$, since $\text{Sin}(\alpha - \theta_3) = C$ and $\text{Cos}(\alpha - \theta_3) = B$.

In[11]:= $\alpha\theta_3 = \text{ArcTan[B, Cconst]}$

Out[11]:= $\text{ArcTan}\left[0.166667\left(-\frac{0.5(-11. + (4. - 1.\text{Cos}[\theta_2])^2 + 1.\text{Sin}[\theta_2]^2)}{\sqrt{(4. - 1.\text{Cos}[\theta_2])^2 + 1.\text{Sin}[\theta_2]^2}} + \sqrt{(4. - 1.\text{Cos}[\theta_2])^2 + 1.\text{Sin}[\theta_2]^2}\right), 0.833333\sqrt{1 - \frac{0.01(-11. + (4. - 1.\text{Cos}[\theta_2])^2 + 1.\text{Sin}[\theta_2]^2)^2}{(4. - 1.\text{Cos}[\theta_2])^2 + 1.\text{Sin}[\theta_2]^2}}\right]$

In[19]:= $\text{Plot}[\alpha\theta_3, \{\theta_2, 0, 2\pi\}]$

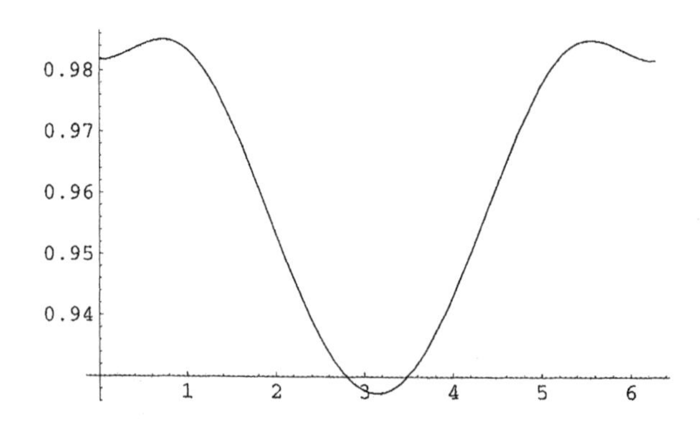

Out[19]:= – Graphics –

If the above attempt to plot gives an error message indicating that it encounters complex numbers, then link 2 is not a crank; i.e., θ_2 varies within limits smaller than 0 to 2π. Let us denote these limits: min θ_2 and max θ_2. These limits can be found by using the *FindRoot* command, and the starting values for finding the root can be approximately taken from the above plot.

If the above attempt to plot does not give error messages, then min $\theta_2 = 0$ and max $\theta_2 = 2\pi$. We assume below that link 2 is a revolving link. However, we describe the case of a nonrevolving link 2 in the form of comments (* ... *).

```
          (*minθ2 = Chop[FindRoot[αθ3 == 0, {θ2, minθ2Appr}]]*)
In[21]:=  (*minθ2Expl = θ2 /. minθ2*)

In[22]:=  (*maxθ2 = Chop[FindRoot[αθ3 == 0, {θ2, maxθ2Appr}]]*)

In[23]:=  (*maxθ2Expl = θ2 /. maxθ2*)

In[24]:=  (*Plot[αθ3, {θ2, minθ2Expl, maxθ2Expl}]*)
```

Now we can find θ_3 as a function of θ_2:

```
In[25]:=  θ3 = α – αθ3;

In[26]:=  Plot[θ3, {θ2, 0, 2π},
             AxesLabel → {"θ2", "θ3"},
          PlotStyle → {AbsoluteThickness[2.4]}]
```

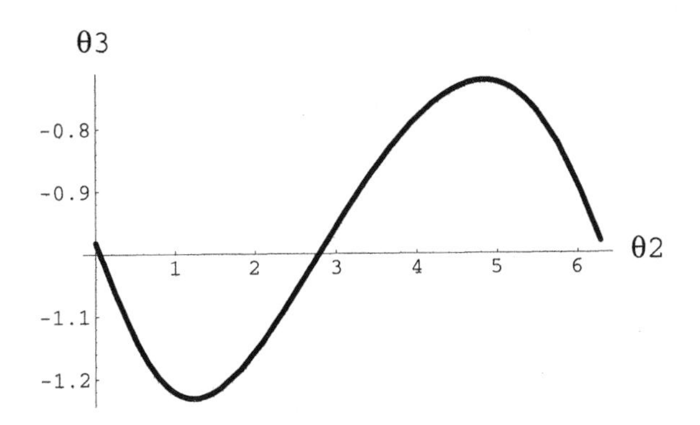

Out[26]:= – Graphics –

Recall that $\mathrm{Cos}(\alpha - \theta_4) = A$ and $\mathrm{Sin}(\alpha - \theta_4) = -r_3C/r_4$. Thus, we can find $\alpha - \theta_4$:

In[27]:= $\alpha\theta_4$ = ArcTan[A, $-r_3$ Cconst / r_4];

In[28]:= $\theta_4 = \alpha - \alpha\theta_4$;

In[29]:= Plot[θ_4, {θ_2, 0, 2π},
AxesLabel → {"θ_2", "θ_4"}, PlotStyle → {AbsoluteThickness[2.4]}]

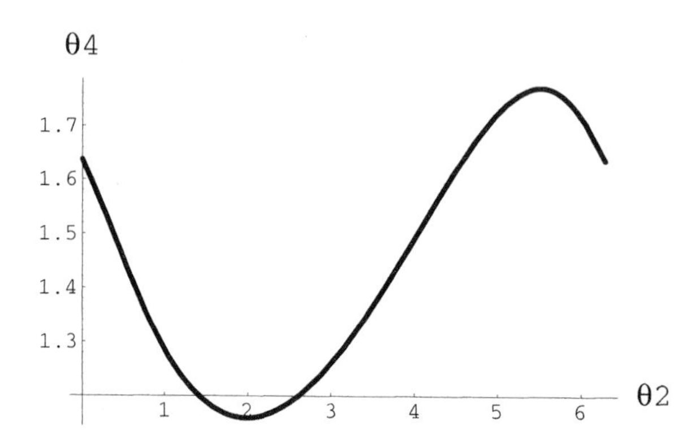

Out[29]:= – Graphics –

Let us now discretize the input angle θ_2, needed for animation, within the limits min θ_2÷max θ_2.

In[30]:= θ_2Interval = 2π;

In[32]:= theta2 = Table[i θ_2Interval/20, {i, 0, 19}]

Out[32]:= $\left\{0, \frac{\pi}{10}, \frac{\pi}{5}, \frac{3\pi}{10}, \frac{2\pi}{5}, \frac{\pi}{2}, \frac{3\pi}{5}, \frac{7\pi}{10}, \frac{4\pi}{5}, \frac{9\pi}{10}, \pi, \frac{11\pi}{10}, \frac{6\pi}{5}, \frac{13\pi}{10}, \frac{7\pi}{5}, \frac{3\pi}{2}, \frac{8\pi}{5}, \frac{17\pi}{10}, \frac{9\pi}{5}, \frac{19\pi}{10}\right\}$

We discretize all found functions by replacing θ_2 → theta2.

In[33]:= θ_3Discr = θ_3 /. θ_2 → theta2

Out[33]:= {–0.981765, –1.08404, –1.16711, –1.21687, –1.23086, –1.21423, –1.17486, –1.12048, –1.05771, –0.991906, –0.927295, –0.867239, –0.814469, –0.771389, –0.7404, –0.724273, –0.726486, –0.751283, –0.802791, –0.882029}

In[34]:= θ_4Discr = θ_4 /. θ_2 → theta2

Out[34]:= {1.63751, 1.5223, 1.40319, 1.30099, 1.22592, 1.17978, 1.16037, 1.16445, 1.18881, 1.23057, 1.287, 1.35524, 1.43205, 1.51354, 1.59483, 1.66974, 1.73029, 1.76658, 1.76751, 1.72431}

Animation

Now we introduce a function that will draw the mechanism for any set of angles $\theta_2(j)$, $\theta_3(j)$, $\theta_4(j)$, where j identifies the position number.

```
In[35]:= mechanismLines[θ2_, θ3_, θ4_, j_] :=
   (
     x2 = r2 Cos[θ2[[j]]];
     y2 = r2 Sin[θ2[[j]]];
     x3 = x2 + r3 Cos[θ3[[j]]];
     y3 = y2 + r3 Sin[θ3[[j]]];
     x4 = x3 + r4 Cos[θ4[[j]]];
     y4 = y3 + r4 Sin[θ4[[j]]];
     {
       Thickness[0.02],
       {
         {RGBColor[1, 0, 0], Line[{{0, 0}, {x2, y2}}]},
         {RGBColor[0, 1, 0], Line[{{x2, y2}, {x3, y3}}]},
         {RGBColor[0, 1, 0], Line[{{x3, y3}, {x4, y4}}]},
         {RGBColor[0, 0, 1], Circle[{0, 0}, 1/5 r2]},
         {RGBColor[0, 0, 1], Circle[{x2, y2}, 1/10 r2]},
           {RGBColor[0, 0, 1], Circle[{x3, y3}, 1/10 r2]},
           {RGBColor[0, 0, 1], Circle[{x4, y4}, 1/5 r2]}
       }
     }
   )
In[36]:= Clear[j]
```

Now we store all plots under the name *plotsOne*.

```
In[37]:=  plotsOne = Table[Graphics[mechanismLines[theta2, θ3Discr, θ4Discr, j],
            AspectRatio → Automatic,
            Axes → True, PlotRange → {{-r4, r2 + r3}, {-r4, r4}}], {j, Length[theta2]}]

Out[37]:= {- Graphics -, - Graphics -, - Graphics -, - Graphics -, - Graphics -,
           - Graphics -, - Graphics -, - Graphics -, - Graphics -, - Graphics -,
           - Graphics -, - Graphics -, - Graphics -, - Graphics -, - Graphics -,
           - Graphics -, - Graphics -, - Graphics -, - Graphics -, - Graphics -}
```

Below we show an array of positions:

In[39]:= Show[GraphicsArray[{{plotsOne[[1]], plotsOne[[3]]}, {plotsOne[[5]], plotsOne[[7]]}, {plotsOne[[9]], plotsOne[[11]]}}, AspectRatio → Automatic]]

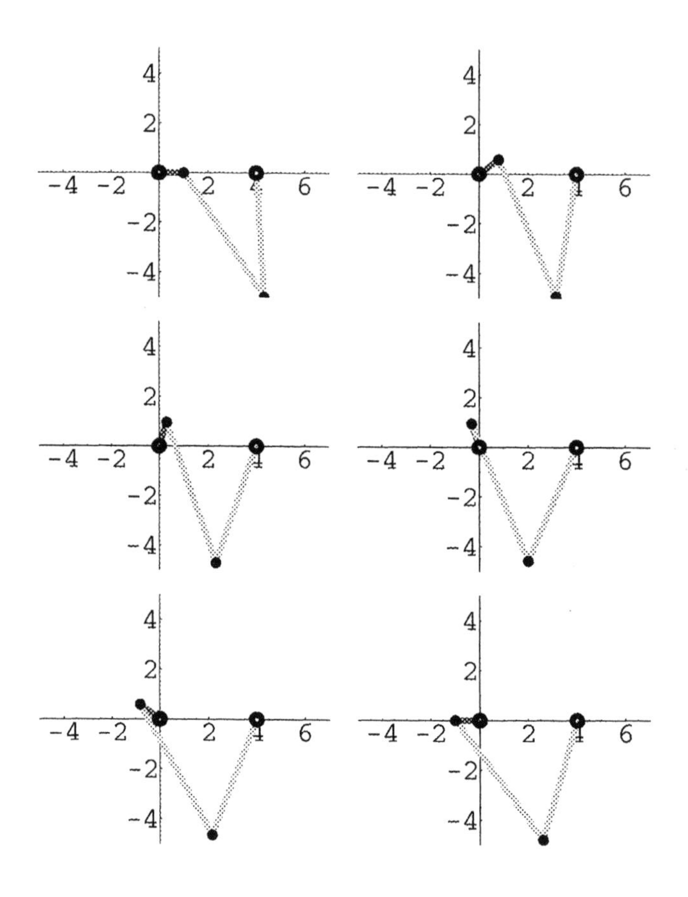

Out[39]:= – GraphicsArray –

Now we can see that by choosing a plus sign for *C*, we have chosen the cross-configuration for the four-bar linkage.

Velocity Analysis

We use formulas derived in the text. Denote: $d\theta_3/dt$ = omega3, and $d\theta_4/dt$ = omega4.

In[76]:= $\text{omega3} = \text{omega2}\frac{r_2\text{Sin}[\theta_2 - \theta_4]}{r_3\text{Sin}[\theta_4 - \theta_3]};$

In[77]:= $\text{omega4} = \text{omega2}\frac{r_2\text{Sin}[\theta_2 - \theta_3]}{r_4\text{Sin}[\theta_3 - \theta_4]};$

Now we assign a specific value to omega2:

In[78]:= omega2 = 10;

In[79]:= omega3Specific = omega3;

In[80]:= Plot[omega3, {θ_2, 0, 2π},
AxesLabel → {"θ_2", "ω_3"}, PlotStyle → {AbsoluteThickness[2.4]}]

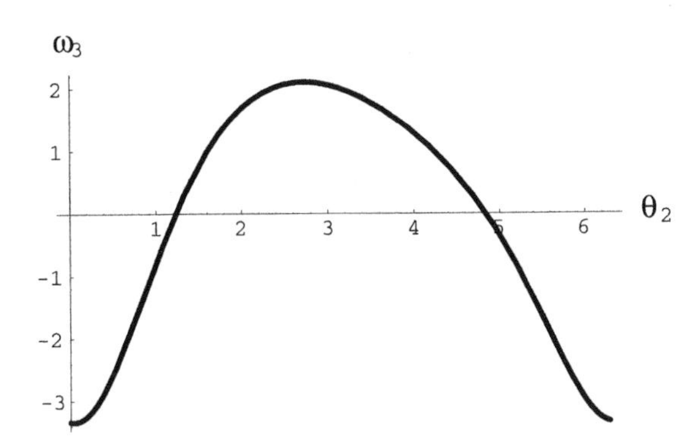

Out[80]:= – Graphics –

In[81]:= omega4Specific == omega4;

In[82]:= Plot[omega4, {θ_2, 0, 2π},
AxesLabel → {"θ_2", "ω_4"}, PlotStyle → {AbsoluteThickness[2.4]}]

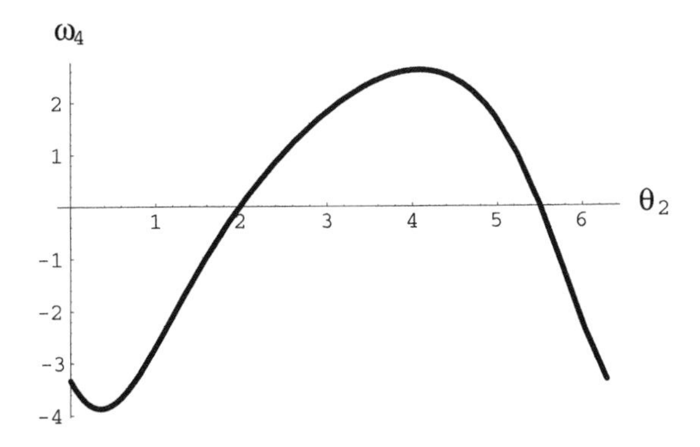

Out[82]:= – Graphics –

Acceleration Analysis

We use formulas derived in the text. Denote: α_3 = alpha3, α_2 = alpha2, and α_4 = alpha4.

In[84]:= alpha3 =

$$\frac{-r_3\omega_3(\omega_4-\omega_3)\text{Cos}[\theta_4-\theta_3]+r_2\text{alpha2Sin}[\theta_2-\theta_4]+r_2\omega_2(\omega_2-\omega_4)\text{Cos}[\theta_2-\theta_4]}{r_3\text{Sin}[\theta_4-\theta_3]};$$

In[85]:= alpha4 =

$$\frac{-r_4\omega_4(\omega_3-\omega_4)\text{Cos}[\theta_3-\theta_4]+r_2\text{alpha2Sin}[\theta_2-\theta_3]+r_2\omega_2(\omega_2-\omega_3)\text{Cos}[\theta_2-\theta_3]}{r_4\text{Sin}[\theta_3-\theta_4]};$$

In[86]:= alpha2 = D[omega2, t]

Out[86]:= 0

Now we replace the angles and angular velocities by their values:

In[93]:= alpha3Specific = alpha3 /. {ω_2 → omega2, ω_3 → omega3, ω_4 → omega4};

In[95]:= alpha4Specific = alpha4 /. {ω_2 → omega2, ω_3 → omega3, ω_4 → omega4};

In[94]:= Plot[alpha3Specific, {θ_2, 0, 2π},
AxesLabel → {"θ_2", "α_3"}, PlotStyle → {AbsoluteThickness[2.4]}]

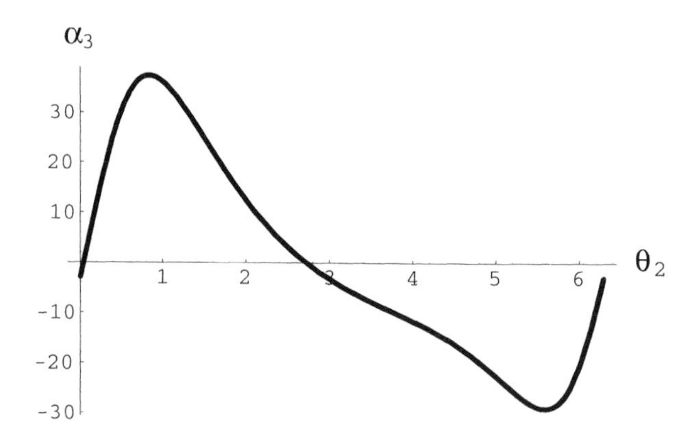

Out[94]:= – Graphics –

In[96]:= Plot[alpha4Specific, {θ_2, 0, 2π},
AxesLabel → {"θ_2", "α_4"}, PlotStyle → {AbsoluteThickness[2.4]}]

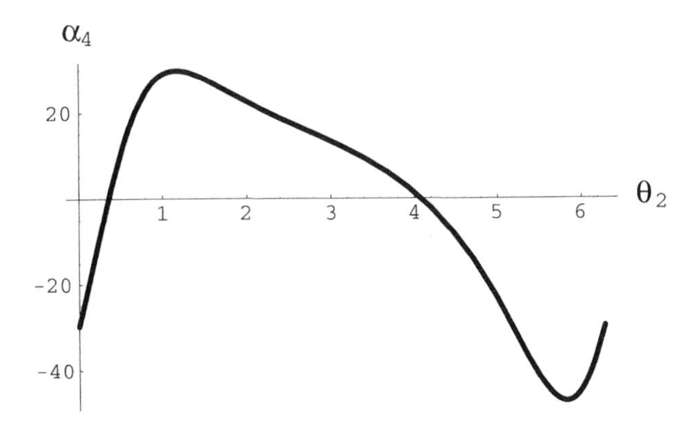

Out[96]:= – Graphics –

A.5 HARMONIC CAM WITH OFFSET RADIAL AND OSCILLATING ROLLER FOLLOWERS

A cam is designed to rise during $0 \le \theta \le \theta\text{One}$ to a level cLift, to dwell during $\theta\text{One} \le \theta \le \theta\text{Two}$ on this level, to return during $\theta\text{Two} \le \theta \le \theta\text{Three}$ to the zero level, and to dwell at the zero level during $\theta\text{Three} \le \theta \le 2\pi$. Two types of followers are considered: an offset roller follower with the distance fOffset from the cam center of rotation and an oscillating roller follower.

The harmonic function describing the follower motion is $S_h = \text{cLift}\,(1 - \text{Cos}[\alpha])$, where $0 \le \alpha \le \pi/2$. Since the boundaries for the angle θ are arbitrary, in order to satisfy the boundary conditions for the harmonic function the ranges 0÷One and θTwo÷θThree must be mapped on the α-range 0 to π/2. The corresponding transformations of angles for the rise and return parts of the cycle are

$$\alpha\text{One} = \frac{\pi\theta}{2\theta\text{One}},\quad \alpha\text{Three} = \frac{\pi(\theta\text{Three} - \theta)}{2(\theta\text{Three} - \theta\text{Two})}$$

In the following, we first introduce the harmonic functions, then describe the cam profile, and, finally, analyze and animate a cam with a radial roller follower and a cam with an oscillating follower. We plot the translational and angular velocities and accelerations of the two followers during one cycle of rotation.

HARMONIC FUNCTIONS

The two functions for the rise and return of the follower are, respectively,

In[1]:= $\text{harmonicOne}[\theta\_] := \text{cLift}\left(1 - \text{Cos}\left[\frac{\pi\theta}{2\theta\text{One}}\right]\right)$

In[2]:= $\text{harmonicThree}[\theta\_] := \text{cLift}\left(1 - \text{Cos}\left[\frac{\pi(\theta\text{Three} - \theta)}{(\theta\text{Three} - \theta\text{Two})}\right]\right)$

These two functions in the case when the angle θ exceeds 2π become

In[3]:= $\text{harmonicOneOver}2\pi[\theta\_] := \text{cLift}\left(1 - \text{Cos}\left[\frac{\pi(\theta - 2\pi)}{2\theta\text{One}}\right]\right)$

In[4]:= $\text{harmonicThreeOver}2\pi[\theta\_] := \text{cLift}\left(1 - \text{Cos}\left[\frac{\pi(\theta\text{Three} - (\theta - 2\pi))}{2(\theta\text{Three} - \theta\text{Two})}\right]\right)$

Cam Profile

The cam profile is made of four functions corresponding to the four regions on the cam displacement diagram. A point on the cam is characterized by the vector rCam (Cos[θ], Sin[θ]), where rCam is a variable for the rising and returning parts and constant for the dwelling parts of the cycle. We denote the base circle radius rBase. For animation purposes, we introduce another angle, φ, characterizing the cam rotation (the angle φ is, in fact, a coordinate transformation angle). Thus, we describe the cam profile coordinates as functions of the angles θ and φ:

In[5]:= xOneAnim[φ_] := (rBase + harmonicOne[θ]) Cos[θ + φ]

In[6]:= yOneAnim[φ_] := (rBase + harmonicOne[θ]) Sin[θ + φ]

In[7]:= xTwoAnim[φ_] := (rBase + cLift) Cos[θ + φ]

In[8]:= yTwoAnim[φ_] := (rBase + cLift) Sin[θ + φ]

In[9]:= xThreeAnim[φ_] := (rBase + harmonicThree[θ]) Cos[θ + φ]

In[10]:= yThreeAnim[φ_] := (rBase + harmonicThree[θ]) Sin[θ + φ]

In[11]:= xFourAnim[φ_] := rBase Cos[θ + φ]

In[12]:= yFourAnim[φ_] := rBase Sin[θ + φ]

To display the profile we need to assign specific values to rBase, cLift, θOne, θTwo, θThree, and angular velocity ω.

In[13]:= rBase = 2.; cLift = 0.4; θOne = π/3; θTwo = 2π/3; θThree = 3π/2;
θFour = 2π; ω = 10;

Let us plot the displacement diagram first:

In[14]:= followerDisplFun[θ_] := Which[
0 ≤ θ ≤ θOne, harmonicOne[θ],
θOne < θ ≤ θTwo, cLift,
θTwo < θ ≤ θThree, harmonicThree[θ],
θThree < θ ≤ 2π, 0]

In[15]:= displDiagram = Plot[followerDisplFun[θ], {θ, 0, 2π},
AspectRatio → Automatic]

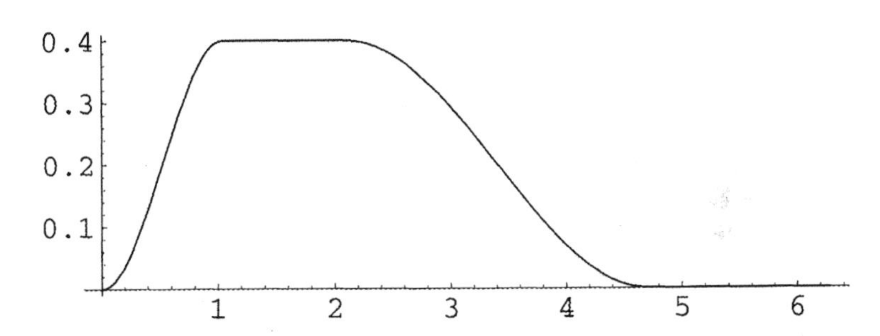

Out[15]:= – Graphics –

In[16]:= velOne = D[harmonicOne[ωt], t]

Out[16]:= 6. Sin[15t]

In[17]:= velThree = D[harmonicThree[ωt], t]

Out[17]:= $-2.4\text{Sin}\left[\frac{3}{5}\left(\frac{3\pi}{2} - 10\text{t}\right)\right]$

In[18]:= angleθ = ωt

Out[18]:= 10 t

In[19]:= followerVel = Which[
0 ≤ ωt ≤ θOne, velOne,
θOne < ωt ≤ θTwo, 0,
θTwo < ωt ≤ θThree, velThree,
θThree < ωt ≤ 2π, 0]

Out[19]:= Which[0 ≤ 10t ≤ $\frac{\pi}{3}$, velOne, θOne < ωt ≤ θTwo,
0, θTwo < ωt ≤ θThree, velThree, θThree < ωt ≤ 2π, 0]

In[20]:= period = 2π / ω

Out[20]:= $\frac{\pi}{5}$

In[21]:= ParametricPlot[{angleθ, followerVel}, {t, 0, period}]

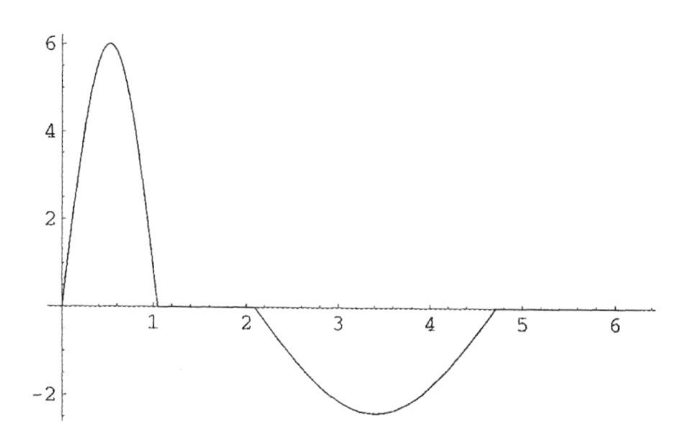

Out[21]:= – Graphics –

In[22]:= accelOne = D[harmonicOne[ωt], {t, 2}]

Out[22]:= 90. Cos[15t]

In[23]:= accelThree = D[harmonicThree[ωt], {t, 2}]

Out[23]:= $14.4\mathrm{Cos}\left[\frac{3}{5}\left(\frac{3\pi}{2}-10t\right)\right]$

In[24]:= followerAccel = Which[
0 ≤ ωt ≤ θOne, accelOne,
θOne < ωt ≤ θTwo, 0,
θTwo < ωt ≤ θThree, accelThree,
θThree < ωt ≤ 2π, 0]

Out[24]:= Which[0 ≤ 10t ≤ $\frac{\pi}{3}$, accelOne, θOne < ωt ≤ θTwo,
0, θTwo < ωt ≤ θThree, accelThree, θThree < ωt ≤ 2π, 0]

In[25]:= ParametricPlot[{angleθ, followerAccel}, {t, 0, period}]

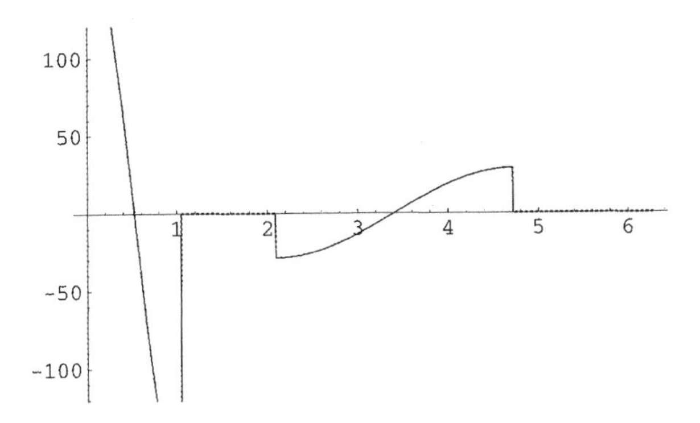

Out[25]:= – Graphics –

Now we introduce a function that will draw the cam profile for any position φ. We describe the profile in a polar coordinate system.

In[26]:=
```
camProfile[φ_] := (
  {  Graphics[Circle[{0, 0}, rBase]],
  ParametricPlot[{xOneAnim[φ], yOneAnim[φ]}, {θ, 0, θOne},
  DisplayFunction → Identity, PlotStyle → {AbsoluteThickness[2]}],
  ParametricPlot[{xTwoAnim[φ], yTwoAnim[φ]}, {θ, θOne, θTwo},
  DisplayFunction → Identity, PlotStyle → {AbsoluteThickness[2]}],
  ParametricPlot[{xThreeAnim[φ], yThreeAnim[φ]}, {θ, θTwo, θThree},
  DisplayFunction → Identity, PlotStyle → {AbsoluteThickness[2]}],
  ParametricPlot[{xFourAnim[φ], yFourAnim[φ]}, {θ, θThree, θFour},
  DisplayFunction → Identity, PlotStyle → {AbsoluteThickness[2]}]}
  )
```

We denote the number of frames in animation by framesN:

In[27]:=
```
framesN = 20;
```

The rotation of the cam is characterized by the angle φ. We introduce a discrete set of the variables φ corresponding to the number framesN. Note, that the negative sign of the angles means that the rotation will be in a clockwise direction in our right coordinate system.

In[28]:=
```
φDiscr = –Table[i 2π / framesN, {i, 0, framesN – 1}]
```

Out[28]:=

$$\left\{0, -\frac{\pi}{10}, -\frac{\pi}{5}, -\frac{3\pi}{10}, -\frac{2\pi}{5}, -\frac{\pi}{2}, -\frac{3\pi}{5}, -\frac{7\pi}{10}, -\frac{4\pi}{5}, -\frac{9\pi}{10},\right.$$

$$\left. -\pi, -\frac{11\pi}{10}, -\frac{6\pi}{5}, -\frac{13\pi}{10}, -\frac{7\pi}{5}, -\frac{3\pi}{2}, -\frac{8\pi}{5}, -\frac{17\pi}{10}, -\frac{9\pi}{5}, -\frac{19\pi}{10}\right\}$$

Now we store the five plots given by the function camProfile in framesN positions.

```
In[29]:=  camPlots = Table[camProfile[φDiscr[[j]]], {j, Length[φDiscr]}]

Out[29]:= {{– Graphics –, – Graphics –, – Graphics –, – Graphics –, – Graphics –},
          {– Graphics –, – Graphics –, – Graphics –, – Graphics –, – Graphics –},
          {– Graphics –, – Graphics –, – Graphics –, – Graphics –, – Graphics –},
          {– Graphics –, – Graphics –, – Graphics –, – Graphics –, – Graphics –},
          {– Graphics –, – Graphics –, – Graphics –, – Graphics –, – Graphics –},
          {– Graphics –, – Graphics –, – Graphics –, – Graphics –, – Graphics –},
          {– Graphics –, – Graphics –, – Graphics –, – Graphics –, – Graphics –},
          {– Graphics –, – Graphics –, – Graphics –, – Graphics –, – Graphics –},
          {– Graphics –, – Graphics –, – Graphics –, – Graphics –, – Graphics –},
          {– Graphics –, – Graphics –, – Graphics –, – Graphics –, – Graphics –},
          {– Graphics –, – Graphics –, – Graphics –, – Graphics –, – Graphics –},
          {– Graphics –, – Graphics –, – Graphics –, – Graphics –, – Graphics –},
          {– Graphics –, – Graphics –, – Graphics –, – Graphics –, – Graphics –},
          {– Graphics –, – Graphics –, – Graphics –, – Graphics –, – Graphics –},
          {– Graphics –, – Graphics –, – Graphics –, – Graphics –, – Graphics –},
          {– Graphics –, – Graphics –, – Graphics –, – Graphics –, – Graphics –},
          {– Graphics –, – Graphics –, – Graphics –, – Graphics –, – Graphics –},
          {– Graphics –, – Graphics –, – Graphics –, – Graphics –, – Graphics –},
          {– Graphics –, – Graphics –, – Graphics –, – Graphics –, – Graphics –},
          {– Graphics –, – Graphics –, – Graphics –, – Graphics –, – Graphics –}}

In[30]:=  plotSizeX = rBase + 1.2 cLift;
```

The following command displays the cam in all φDiscr positions. Here, we show only one position.

```
In[31]:=  Needs["Graphics`Animation`"]

In[32]:=  ShowAnimation[camPlotsDispl]
```

Cam with Radial Roller Follower

Let us consider a roller follower, with the roller having the radius rRoller. Then the center of the roller is located on a curve equidistant from the cam curve. The x-coordinate of the roller center is equal to the fOffset, whereas the y-coordinate is equal to the y-coordinate of the roller center on the equidistant cam profile. Let us define the latter:

```
In[34]:=  rRoller = 0.1 rBase

Out[34]:= 0.2
```

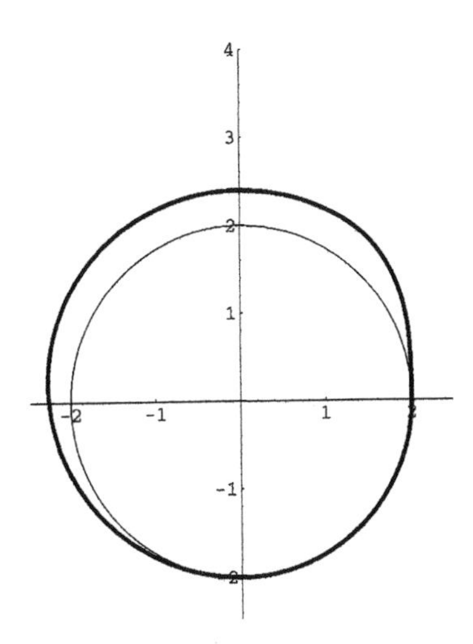

In[35]:= fOffset = 0.2 rBase

Out[35]:= 0.4

We introduce discrete θ- angles corresponding to the number of animation positions. But first we have to find the angle θ corresponding to the intersection of the line x=fOffset at the initial cam position. We will denote this angle θInitial.

In[36]:= eqθInitial = xTwoAnim[0] – fOffset == 0

Out[36]:= –0.4 + 2.4 Cos[θ] == 0

In[37]:= soleqθInitial = Solve[eqθInitial, θ]

Out[37]:= {{θ → –1.40335}, {θ → 1.40335}}

If the follower is placed above the cam, the second solution is correct for the right coordinate system.

In[38]:= θInitial = θ /. soleqθInitial[[2]]

Out[38]:= 1.40335

In[39]:= θDiscrete = –φDiscr + θInitial

Out[39]:= {1.40335, 1.71751, 2.03167, 2.34583, 2.65999, 2.97414, 3.2883, 3.60246, 3.91662, 4.23078, 4.54494, 4.8591, 5.17326, 5.48742, 5.80158, 6.11574, 6.4299, 6.74406, 7.05822, 7.37237}

```
In[40]:=  followerFunctionY[θ_] := Which[
            0 ≤ θ ≤ θOne, harmonicOne[θ],
            θOne < θ ≤ θTwo, cLift,
            θTwo < θ ≤ θThree, harmonicThree[θ],
            θThree < θ ≤ 2π, 0,
            2π < θ ≤ 2π + θOne, harmonicOneOver2π[θ],
            2π + θOne < θ ≤ 2π + θTwo, cLift,
            2π + θTwo < θ ≤ 2π + θThree, harmonicThreeOver2π[θ]
          ]
```

The follower displacements at framesN positions:

```
In[41]:=  followerDispl = Table[followerFunctionY[θDiscrete[[j]]],
            {j, Length[θDiscrete]}]

Out[41]:= {0.4, 0.4, 0.4, 0.339885, 0.266849, 0.198529, 0.137347, 0.0854693,
           0.044734, 0.0165842, 0.0020171, 0, 0, 0, 0, 0, 0.00964685,
           0.0918343, 0.241198, 0.4}
```

We can plot the above displacements to see how closely we approximate the displacement diagram:

```
In[42]:=  discrDisplacements = MapThread[List, {θDiscrete, followerDispl}]

Out[42]:= {{1.40335, 0.4}, {1.71751, 0.4}, {2.03167, 0.4}, {2.34583, 0.339885},
           {2.65999, 0.266849}, {2.97414, 0.198529}, {3.2883, 0.137347},
           {3.60246, 0.0854693}, {3.91662, 0.044734}, {4.23078, 0.0165842},
           {4.54494, 0.0020171}, {4.8591, 0}, {5.17326, 0}, {5.48742, 0},
           {5.80158, 0}, {6.11574, 0}, {6.4299, 0.00964685}, {6.74406, 0.0918343},
           {7.05822, 0.241198}, {7.37237, 0.4}}
```

Now we introduce a function describing the follower at any position, which is synchronized with the cam position:

```
In[43]:= followerLine[followerDispl_, j_] :=
        (
        xLower = fOffset; (*x-coordinate of the follower*)
        yLower = rBase + rRoller + followerDispl[[j]];
          (*y-coordinate of the follower at the lower end*)
        yUpper = yLower + 2 rBase; (*y-coordinate of the follower at the upper end*)
          (*Below are the coordinates of the frame supporting the follower*)
          xLeftSup = xLower – 0.3 fOffset;
          xRightSup = xLower + 0.3 fOffset;
          yLowerSup = 1.5 rBase;
          yUpperSup = 3.0 rBase;
```

```
        {Thickness[0.015],
          {RGBColor[0, 0, 1], Line[{{xLeftSup, yLowerSup},
             {xLeftSup, yUpperSup}}]},
          {RGBColor[0, 0, 1], Line[{{xRightSup, yLowerSup},
             {xRightSup, yUpperSup}}]},
          {RGBColor[1, 0, 0], Line[{{xLower, yLower}, {xLower, yUpper}}]},
          {RGBColor[1, 0, 0], Circle[{xLower, yLower}, rRoller]}
        }
     )
```

In[44]:= followerPositions = Table[Graphics[followerLine[followerDispl, j], DisplayFunction → Identity], {j, Length[φDiscr]}]

Out[44]:= {– Graphics –, – Graphics –, – Graphics –, – Graphics –, – Graphics –, – Graphics –, – Graphics –, – Graphics –, – Graphics –, – Graphics –, – Graphics –, – Graphics –, – Graphics –, – Graphics –, – Graphics –, – Graphics –, – Graphics –, – Graphics –, – Graphics –, – Graphics –}

In[45]:= followerPlots = Table[Show[followerPositions[[j]], AspectRatio → Automatic, Axes → True, Plot Range → {{–plotSizeX, plotSizeX}, {–plotSizeX, 3 rBase}}], {j, Length[φDiscr]}]

Out[45]:= {– Graphics –, – Graphics –, – Graphics –, – Graphics –, – Graphics –, – Graphics –, – Graphics –, – Graphics –, – Graphics –, – Graphics –, – Graphics –, – Graphics –, – Graphics –, – Graphics –, – Graphics –, – Graphics –, – Graphics –, – Graphics –, – Graphics –, – Graphics –}

In[46]:= camSet = MapThread[List, {camPlotsDispl, followerPlots}]

Out[46]:= {{– Graphics –, – Graphics –}, {– Graphics –, – Graphics –}, {– Graphics –, – Graphics –}, {– Graphics –, – Graphics –}, {– Graphics –, – Graphics –}, {– Graphics –, – Graphics –}, {– Graphics –, – Graphics –}, {– Graphics –, – Graphics –}, {– Graphics –, – Graphics –}, {– Graphics –, – Graphics –}, {– Graphics –, – Graphics –}, {– Graphics –, – Graphics –}, {– Graphics –, – Graphics –}, {– Graphics –, – Graphics –}, {– Graphics –, – Graphics –}, {– Graphics –, – Graphics –}, {– Graphics –, – Graphics –}, {– Graphics –, – Graphics –}, {– Graphics –, – Graphics –}, {– Graphics –, – Graphics –}}

We show only one position here:

In[47]:= ShowAnimation[camSet]

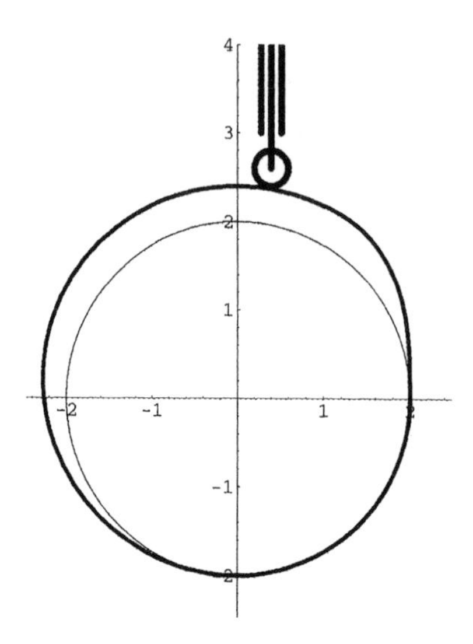

Cam with Oscillating Roller Follower

In[48]:= φDiscr = –φDiscr

Out[48]:= $\left\{0, \frac{\pi}{10}, \frac{\pi}{5}, \frac{3\pi}{10}, \frac{2\pi}{5}, \frac{\pi}{2}, \frac{3\pi}{5}, \frac{7\pi}{10}, \frac{4\pi}{5}, \frac{9\pi}{10}, \pi, \frac{11\pi}{10}, \frac{6\pi}{5}, \frac{13\pi}{10}, \frac{7\pi}{5}, \frac{3\pi}{2}, \frac{8\pi}{5}, \frac{17\pi}{10}, \frac{9\pi}{5}, \frac{19\pi}{10}\right\}$

In[49]:= camProfileDiscr = Table[followerDisplFun[θDiscr[[j]]], {j, Length[θDiscr]}]

Out[49]:= {0, 0.0435974, 0.164886, 0.337426, 0.4, 0.4, 0.4, 0.374884, 0.300524, 0.229688, 0.164886, 0.108413, 0.0622688, 0.0280894, 0.0070851, 0, 0, 0, 0, 0}

In[50]:= followerLength = 2 rBase

Out[50]:= 4.

In[51]:= followerCenterPos = (camProfileDiscr[[1]] + rBase) {Cos[θDiscr[[1]]], Sin[θDiscr[[1]]]} + followerLength{Cos[θDiscr[[1]] – π/2], Sin[θDiscr[[1]] – π/2]}

Out[51]:= {2., –4.}

```
In[52]:=  camPoints = Table[(camProfileDiscr[[j]] + rBase + rRoller)
            {Cos[θDiscr[[j]]], Sin[θDiscr[[j]]]}, {j, 1, Length[θDiscr]}]

Out[52]:= {{2.2, 0}, {2.2436, 0}, {2.36489, 0}, {2.53743, 0}, {2.6, 0}, {2.6, 0},
           {2.6, 0}, {2.57488, 0}, {2.50052, 0}, {2.42969, 0}, {2.36489, 0},
           {2.30841, 0}, {2.26227, 0}, {2.22809, 0}, {2.20709, 0}, {2.2, 0}, {2.2, 0},
           {2.2, 0}, {2.2, 0}, {2.2, 0}}

In[53]:= followerGraphics[camPoints_, j_] :=
        (
        xCam = camPoints[[j, 1]]; yCam = camPoints[[j, 2]];
          xCent = followerCenterPos[[1]]; yCent = followerCenterPos[[2]];
          xCentLeft = xCent - 0.2 rBase; xCentRight = xCent + 0.2 rBase;
          yCentLow = yCent - 0.2 rBase; yCentHigh = yCent +0.2 rBase;
        {Thickness[0.008],
          {RGBColor[1, 0, 0], Line[{{xCam, yCam}, {xCent, yCent}}]},
          {RGBColor[0, 0, 1], Circle[{xCent, yCent}, 0.1 rBase]},
          {RGBColor[1, 0, 0], Circle[{xCam, yCam}, rRoller]},
            { AbsoluteThickness[1],
          Line[{{xCentLeft, yCent}, {xCentRight, yCent}}]},
            { AbsoluteThickness[1],
          Line[{{xCent, yCentLow}, {xCent, yCentHigh}}]}
        }
        )

In[54]:=  flatFolPositions = Table[Graphics[followerGraphics[camPoints, j],
            DisplayFunction → Identity], {j, Length[θDiscr]}
          ]

Out[54]:= {- Graphics -, - Graphics -, - Graphics -, - Graphics -, - Graphics -, - Graphics -,
           - Graphics -, - Graphics -, - Graphics -, - Graphics -, - Graphics -, - Graphics -,
           - Graphics -, - Graphics -, - Graphics -, - Graphics -, - Graphics -, - Graphics -,
           - Graphics -, - Graphics -}

In[55]:=  camFol = Table[Show[flatFolPositions[[j]], AspectRatio → Automatic,
            Axes → True, PlotRange → {{-2 plotSizeX, 2 plotSizeX},
            {-3 plotSizeX, 2 rBase}}], {j, Length[θDiscr]}]

Out[55]:= {- Graphics -, - Graphics -, - Graphics -, - Graphics -, - Graphics -, - Graphics -,
           - Graphics -, - Graphics -, - Graphics -, - Graphics -, - Graphics -, - Graphics -,
           - Graphics -, - Graphics -, - Graphics -, - Graphics -, - Graphics -, - Graphics -,
           - Graphics -, - Graphics -}

In[56]:=  camSetFlat = MapThread[List, {camFol, camPlots}]
```

```
Out[56]:= {{-Graphics-, {-Graphics-, -Graphics-, -Graphics-, -Graphics-, -Graphics-}},
  {-Graphics-, {-Graphics-, -Graphics-, -Graphics-, -Graphics-, -Graphics-}},
  {-Graphics-, {-Graphics-, -Graphics-, -Graphics-, -Graphics-, -Graphics-}},
  {-Graphics-, {-Graphics-, -Graphics-, -Graphics-, -Graphics-, -Graphics-}},
  {-Graphics-, {-Graphics-, -Graphics-, -Graphics-, -Graphics-, -Graphics-}},
  {-Graphics-, {-Graphics-, -Graphics-, -Graphics-, -Graphics-, -Graphics-}},
  {-Graphics-, {-Graphics-, -Graphics-, -Graphics-, -Graphics-, -Graphics-}},
  {-Graphics-, {-Graphics-, -Graphics-, -Graphics-, -Graphics-, -Graphics-}},
  {-Graphics-, {-Graphics-, -Graphics-, -Graphics-, -Graphics-, -Graphics-}},
  {-Graphics-, {-Graphics-, -Graphics-, -Graphics-, -Graphics-, -Graphics-}},
  {-Graphics-, {-Graphics-, -Graphics-, -Graphics-, -Graphics-, -Graphics-}},
  {-Graphics-, {-Graphics-, -Graphics-, -Graphics-, -Graphics-, -Graphics-}},
  {-Graphics-, {-Graphics-, -Graphics-, -Graphics-, -Graphics-, -Graphics-}},
  {-Graphics-, {-Graphics-, -Graphics-, -Graphics-, -Graphics-, -Graphics-}},
  {-Graphics-, {-Graphics-, -Graphics-, -Graphics-, -Graphics-, -Graphics-}},
  {-Graphics-, {-Graphics-, -Graphics-, -Graphics-, -Graphics-, -Graphics-}},
  {-Graphics-, {-Graphics-, -Graphics-, -Graphics-, -Graphics-, -Graphics-}},
  {-Graphics-, {-Graphics-, -Graphics-, -Graphics-, -Graphics-, -Graphics-}},
  {-Graphics-, {-Graphics-, -Graphics-, -Graphics-, -Graphics-, -Graphics-}},
  {-Graphics-, {-Graphics-, -Graphics-, -Graphics-, -Graphics-, -Graphics-}}}
```

We show here only one position in animation:

```
In[57]:= ShowAnimation[camSetFlat]
```

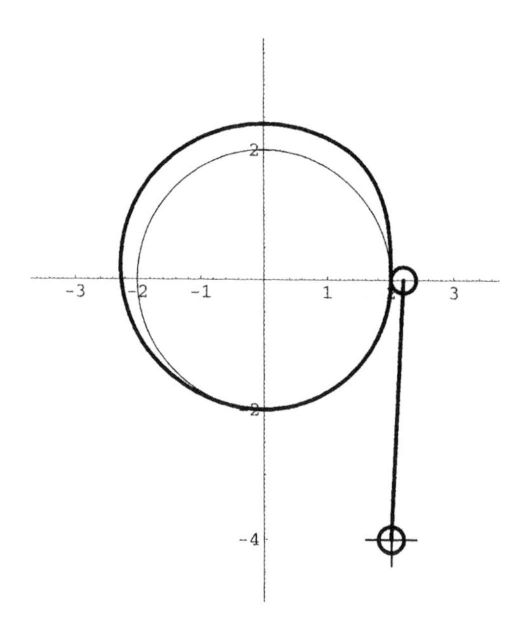

A.6 VIBRATIONS

Free and forced vibrations of a two-degree-of-freedom (2DOF) system are considered below as examples.

A.6.1 Free Vibrations of a 2DOF System

The two-mass system has the following configuration: the m_1-mass is connected to the foundation; the m_2-mass is connected to the mass m_1. The stiffness and damping coefficients of springs and dashpots between the m_1 and the foundation and between the m_1 and m_2 are, correspondingly, k_1, k_2, c_1, c_2. We use the following notations: $u_1[t]$ is the displacement of m_1 and $u_2[t]$ is the displacement of m_2; t is the time.

Dynamic Equations

In[1]:= $eqOne = m_1 u_1''[t] + k_1 u_1[t] + c_1 u_1'[t] + k_2(u_1[t] - u_2[t]) + c_2(u_1'[t] - u_2'[t]) == 0$

Out[1]:= $k_1 u_1[t] + k_2(u_1[t] - u_2[t]) + c_1 u_1'[t] + c_2(u_1'[t] - u_2'[t]) + m_1 u_1''[t] == 0$

In[2]:= $eqTwo = m_2 u_2''[t] + k_2(u_2[t] - u_1[t]) + c_2(u_2'[t] - u_1'[t]) == 0$

Out[2]:= $k_2(-u_1[t] + u_2[t]) + c_2(-u_1'[t] + u_2'[t]) + m_2 u_2''[t] == 0$

Let us substitute solutions in the form: $u_1[t] = d_1\, e^{(\lambda t)}$, $u_2[t] = d_2\, e^{(\lambda t)}$.

In[3]:= $subst = \{u_1[t] \to d_1 E^{(\lambda t)}, u_2[t] \to d_2 E^{(\lambda t)}, u_1'[t] \to D[d_1 E^{(\lambda t)}, t],$
$u_1''[t] \to D[d_1 E^{(\lambda t)}, \{t, 2\}], u_2'[t] \to D[d_2 E^{(\lambda t)}, t], u_2''[t] \to D[d_2 E^{(\lambda t)}, \{t, 2\}]\}$

Out[3]:= $\{u_1[t] \to d_1 e^{t\lambda}, u_2[t] \to d_2 e^{t\lambda}, u_1'[t] \to d_1 e^{t\lambda}\lambda, u_1''[t] \to d_1 e^{t\lambda}\lambda^2,$
$u_2'[t] \to d_2 e^{t\lambda}\lambda, u_2''[t] \to d_2 e^{t\lambda}\lambda^2\}$

In[4]:= eqOneSub = eqOne /. subst

Out[4]:= $d_1 e^{t\lambda} k_1 + (d_1 e^{t\lambda} - d_2 e^{t\lambda}) k_2 + c_1 d_1 e^{t\lambda}\lambda + d_1 e^{t\lambda} m_1 \lambda^2 + c_2(d_1 e^{t\lambda}\lambda - d_2 e^{t\lambda}\lambda) == 0$

In[5]:= eqTwoSub = eqTwo /. subst

Out[5]:= $(-d_1 e^{t\lambda} + d_2 e^{t\lambda})\, k_2 + d_2 e^{t\lambda} m_2 \lambda^2 + c_2(-d_1 e^{t\lambda}\lambda + d_2 e^{t\lambda}\lambda) == 0$

Now let us cancel $E^{(\lambda t)}$ simply by replacing it with 1.

In[6]:= $eqOneAlgebr = Simplify[eqOneSub\ /.\ E^{(\lambda t)} \to 1]$

Out[6]:= $d_1(k_1 + k_2 + \lambda(c_1 + c_2 + m_1\lambda)) == d_2(k_2 + c_2\lambda)$

In[7]:= $eqTwoAlgebr = Simplify[eqTwoSub\ /.\ E^{(\lambda t)} \to 1]$

Out[7]:= $d_2(k_2 + \lambda(c_2 + m_2\lambda)) == d_1(k_2 + c_2\lambda)$

Solve eqOneAlgebr for d_1:

In[8]:= sol = Solve[eqOneAlgebr, d_1]

Out[8]:= $$\left\{\left\{d_1 \to \frac{d_2(k_2 + c_2\lambda)}{k_1 + k_2 + c_1\lambda + c_2\lambda + m_1\lambda^2}\right\}\right\}$$

Make this solution explicit:

In[9]:= d_1Expl = d_1 /. sol[[1]]

Out[9]:= $$\frac{d_2(k_2 + c_2\lambda)}{k_1 + k_2 + c_1\lambda + c_2\lambda + m_1\lambda^2}$$

Let us define the relationship $r_1 = d_1/d_2$, which we will need later,

In[10]:= r_1 = d_1Expl /. $d_2 \to 1$

Out[10]:= $$\frac{k_2 + c_2\lambda}{k_1 + k_2 + c_1\lambda + c_2\lambda + m_1\lambda^2}$$

Substitute d_1Expl into eqTwoAlgebr and replace d_2 with 1 to obtain the characteristic equation:

In[11]:= charactEq = (eqTwoAlgebr /. $d_1 \to d_1$Expl) /. $d_2 \to 1$

Out[11]:= $$k_2 + \lambda(c_2 + m_2\lambda) == \frac{(k_2 + c_2\lambda)^2}{k_1 + k_2 + c_1\lambda + c_2\lambda + m_1\lambda^2}$$

At this point let us choose specific data: m_1=20 kg, m_2 = 10 kg, k_1 = 2000 N/m, k_2 = 3000 N/m, c_1 = 50 N/(m s), c_2 = 50 N/(m s).

In[12]:= $m_1 = 20$; $m_2 = 10$; $k_1 = 2000$; $k_2 = 3000$; $c_1 = 50$; $c_2 = 50$;

Now we solve the characteristic equation to find the four roots:

In[13]:= solForλ = Solve[charactEq, λ];

Now we extract each root in an explicit form. We use the command *N[ ]* to reduce it to a number.

In[14]:= λOne = N[λ /. solForλ[[1]]]

Out[14]:= –4.25552 + 21.6862 i

In[15]:= λTwo = N[λ /. solForλ[[2]]]

Out[15]:= –4.25552 – 21.6862 i

In[16]:= λThree = N[λ /. solForλ[[3]]]

Out[16]:= –0.74448 – 7.80195 i

In[17]:= λFour = N[λ /. solForλ[[4]]]

Out[17]:= –0.74448 + 7.80195 i

The ratio of amplitudes for each of the roots is

In[18]:= r_1One = N[r_1 /. λ → λOne]

Out[18]:= –0.632843 – 0.0269852 i

We can represent this complex number in the form Abs[r_1One] Exp[i Arg[r_1One]], where *i* is the imaginary unit.

In[19]:= Abs[r_1One]

Out[19]:= 0.633418

In[20]:= Arg[r_1One]

Out[20]:= –3.09898

Similarly, for r_1Two, r_1Three, and r_1Four,

In[21]:= r_1Two = N[r_1 /. λ → λTwo]

Out[21]:= –0.632843 + 0.0269852 i

In[22]:= r_1Three = N[r_1 /. λ → λThree]

Out[22]:= 0.794814 + 0.0121931 i

In[23]:= r_1Four = N[r_1 /. λ → λFour]

Out[23]:= 0.794814 – 0.0121931 i

General solutions with eight undetermined coefficients ($a_{11}, a_{12}, b_{11}, b_{12}, a_{21}, a_{22}, b_{21}, b_{22}$) have the following form:

In[24]:= $\mathrm{uOne} = a_{11}E^{(Re[\lambda One]t)}Cos[Im[\lambda One]t] + a_{12}E^{(Re[\lambda One]t)}Sin[Im[\lambda One]t] + b_{11}E^{(Re[\lambda Three]t)}Cos[Im[\lambda Three]t] + b_{12}E^{(Re[\lambda Three]t)}Sin[Im[\lambda Three]t]$

Out[24]:= $b_{11}e^{-0.74448t}Cos[7.80195t] + a_{11}e^{-4.25552t}Cos[21.6862t] - b_{12}e^{-0.74448t}Sin[7.80195t] + a_{12}e^{-4.25552t}Sin[21.6862t]$

Now we take into account that a_{21} = r_1One a_{11} = a_{11} Abs[r_1One] Exp[i Arg[r_1One]], a_{22} = r_1Two a_{12}, and so on.

In[25]:= $\mathrm{uTwo} = a_{11}Abs[r_1One]E^{(Re[\lambda One]t)}Cos[Im[\lambda One]t + Arg[r_1One]] + a_{12}Abs[r_1Two]E^{(Re[\lambda Two]t)}Sin[Im[\lambda Two]t + Arg[r_1One]] + b_{11}Abs[r_1Three]E^{(Re[\lambda Three]t)}Cos[Im[\lambda Three]t + Arg[r_1Three]] + b_{12}Abs[r_1Three]E^{(Re[\lambda Three]t)}Sin[Im[\lambda Three]t + Arg[r_1Three]]$

Out[25]:= $0.633418\ a_{11}e^{-4.25552t}Cos[3.09898 - 21.6862t] + 0.794908\ b_{11}e^{-0.74448t}Cos[0.0153396 - 7.80195t] + 0.794908\ b_{12}e^{-0.74448t}Sin[0.0153396 - 7.80195t] - 0.633418\ a_{12}e^{-4.25552t}Sin[3.09898 + 21.6862t]$

Initial conditions: $u_1[0] = 0$, $u_2[0] = 0$, $u_1'[0] = 1$ m/s, $u_2'[0] = 0$. Now satisfying the first two initial conditions for displacements, we have

In[26]:= eqConOne = uOne == 0 /. t → 0

Out[26]:= $a_{11} + b_{11} == 0$

In[27]:= eqConTwo = uTwo == 0 /. t → 0

Out[27]:= $-0.632843\ a_{11} - 0.0269852\ a_{12} + 0.794814\ b_{11} + 0.0121931\ b_{12} == 0$

Finding the velocities, we have

In[28]:= uOneVel = D[uOne, t]

Out[28]:= $-0.74448\ b_{11}e^{-0.74448t}Cos[7.80195t] - 7.80195\ b_{12}e^{-0.74448t}Cos[7.80195t] - 4.25552\ a_{11}e^{-4.25552t}Cos[21.6862t] + 21.6862\ a_{12}e^{-4.25552t}Cos[21.6862t] - 7.80195\ b_{11}e^{-0.74448t}Sin[7.80195t] + 0.74448\ b_{12}e^{-0.74448t}Sin[7.80195t] - 21.6862\ a_{11}e^{-4.25552t}Sin[21.6862t] - 4.25552\ a_{12}e^{-4.25552t}Sin[21.6862t]$

In[29]:= uTwoVel = D[uTwo, t]

Out[29]:= $-2.69552\ a_{11}e^{-4.25552t}$Cos[3.09898 – 21.6862t] –
$0.591793\ b_{11}e^{-0.74448t}$Cos[0.0153396 – 7.80195t] –
$6.20184\ b_{12}e^{-0.74448t}$Cos[0.0153396 – 7.80195t] –
$13.7365\ a_{12}e^{-4.25552t}$Cos[3.09898 + 21.6862t] +
$13.7365\ a_{11}e^{-4.25552t}$Sin[3.09898 – 21.6862t] +
$6.20184\ b_{11}e^{-0.74448t}$Sin[0.0153396 – 7.80195t] –
$0.591793\ b_{12}e^{-0.74448t}$Sin[0.0153396 – 7.80195t] +
$2.69552\ a_{12}e^{-4.25552t}$Sin[3.09898 + 21.6862t]

Satisfying the initial conditions for the velocities, we have

In[30]:= eqConThree = uOneVel – 1 == 0 /. t → 0

Out[30]:= $-1 - 4.25552\ a_{11} + 21.6862\ a_{12} - 0.74448\ b_{11} - 7.80195\ b_{12} == 0$

In[31]:= eqConFour = uTwoVel == 0 /. t → 0

Out[31]:= $3.27828\ a_{11} + 13.8388\ a_{12} - 0.496593\ b_{11} - 6.21018\ b_{12} == 0$

Solving the four equations for constants a_{11}, a_{12}, b_{11}, and b_{12}, we have

In[32]:= solConst = Solve[{eqConOne, eqConTwo, eqConThree, eqConFour},
{a_{11}, a_{12}, b_{11}, b_{12}}]

Out[32]:= {{a_{11} → 0.0000304662, a_{12} → 0.2326,
b_{11} → –0.0000304662, b_{12} → 0.518345}}

The explicit form of the solutions is

In[33]:= a_{11}Expl = a_{11} /. solConst[[1]]

Out[33]:= 0.0000304662

In[34]:= a_{12}Expl = a_{12} /. solConst[[1]]

Out[34]:= 0.2326

In[35]:= b_{11}Expl = b_{11} /. solConst[[1]]

Out[35]:= –0.0000304662

In[36]:= b_{12}Expl = b_{12} /. solConst[[1]]

Out[36]:= 0.518345

Substituting the constants into the motion equations, we have

In[37]:= motionOne = uOne /. {$a_{11} \rightarrow a_{11}$Expl, $a_{12} \rightarrow a_{12}$Expl, $b_{11} \rightarrow b_{11}$Expl, $b_{12} \rightarrow b_{12}$Expl}

Out[37]:= $-0.0000304662\ e^{-0.74448t}\text{Cos}[7.80195t] + 0.0000304662\ e^{-4.25552t}\text{Cos}[21.6862t] - 0.518345\ e^{-0.74448t}\text{Sin}[7.80195t] + 0.2326\ e^{-4.25552t}\text{Sin}[21.6862t]$

In[38]:= motionTwo = uTwo /. {$a_{11} \rightarrow a_{11}$Expl, $a_{12} \rightarrow a_{12}$Expl, $b_{11} \rightarrow b_{11}$Expl, $b_{12} \rightarrow b_{12}$Expl}

Out[38]:= $0.0000192978\ e^{-4.25552t}\text{Cos}[3.09898 - 21.6862t] - 0.0000242178\ e^{-0.74448t}\text{Cos}[0.0153396 - 7.80195t] + 0.412037\ e^{-0.74448t}\text{Sin}[0.0153396 - 7.80195t] - 0.147333\ e^{-4.25552t}\text{Sin}[3.09898 + 21.6862t]$

Plots of Displacements

In[39]:= p_1One = Plot[motionOne, {t, 0, 5}, PlotStyle → {AbsoluteThickness[2.4]}]

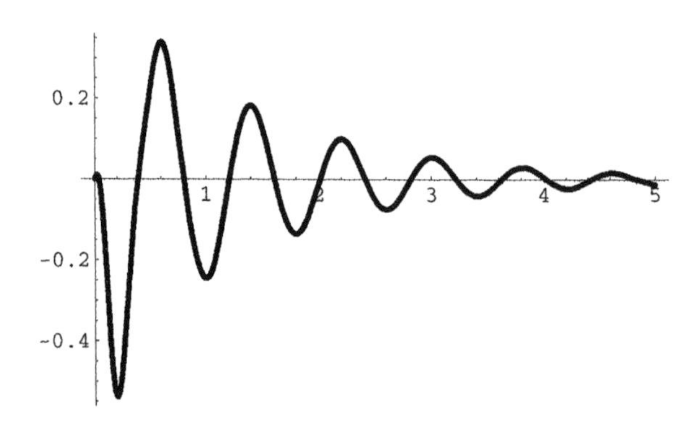

Out[39]:= – Graphics –

In[40]:= plTwo = Plot[motionTwo, {t, 0, 5}, PlotStyle → {Dashing[{0.02}]}]

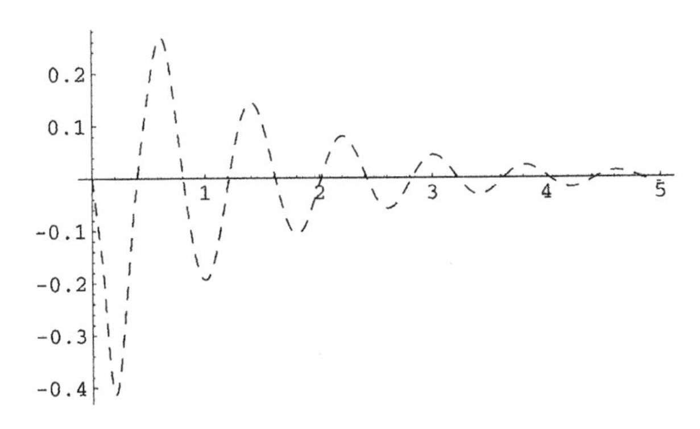

Out[40]:= – Graphics –

In[41]:= Show[plOne, plTwo]

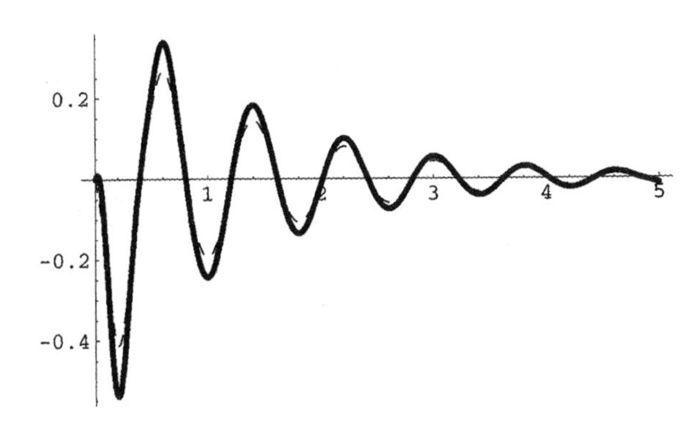

Out[41]:= – Graphics –

Evaluations and Plots of Velocities

In[42]:= velOne = D[motionOne, t]

Out[42]:= $-4.04408\ e^{-0.74448t}$Cos[7.80195t] + $5.04408\ e^{-4.25552t}$Cos[21.6862t] +
$0.386135\ e^{-0.74448t}$Sin[7.80195t] – $0.990493\ e^{-4.25552t}$Sin[21.6862t]

In[43]:= velTwo = D[motionTwo, t]

Out[43]:= $-0.0000821223\ e^{-4.25552t}$Cos[3.09898 – 21.6862 t] –
$3.21467\ e^{-0.74448t}$Cos[0.0153396 – 7.80195t] –
$3.19509\ e^{-4.25552t}$Cos[3.09898 + 21.6862t] +
$0.000418497\ e^{-4.25552t}$Sin[3.09898 – 21.6862t] –
$0.306942\ e^{-0.74448t}$Sin[0.0153396 – 7.80195t] +
$0.626977\ e^{-4.25552t}$Sin[3.09898 + 21.6862t]

In[44]:= plThree = Plot[velOne, {t, 0, 5}, PlotStyle → {AbsoluteThickness[2.4]}]

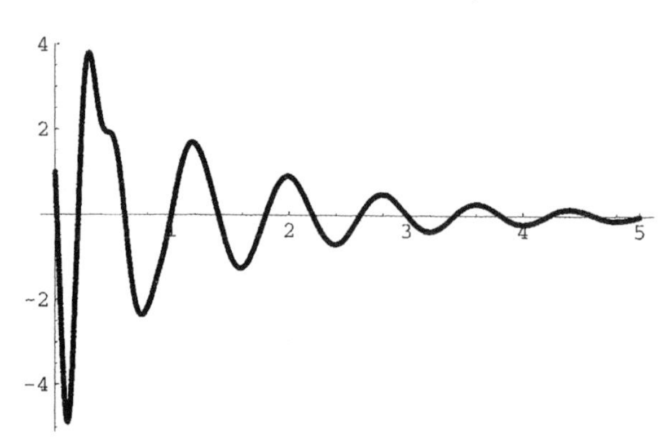

Out[44]:= – Graphics –

In[45]:= plFour = Plot[velTwo, {t, 0, 5}, PlotStyle → {Dashing[{0.02}]}]

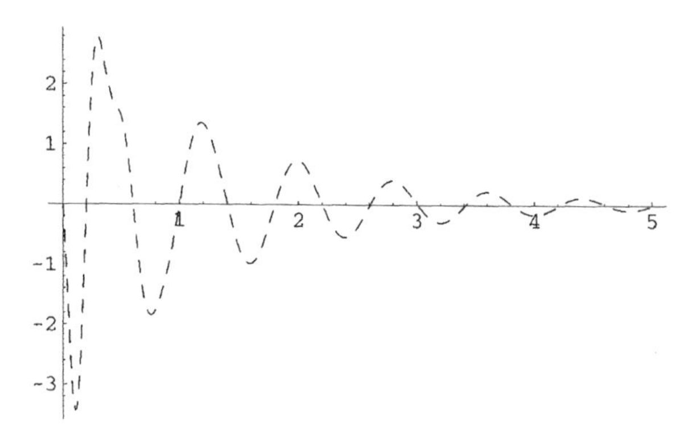

Out[45]:= – Graphics –

In[46]:= Show[plThree, plFour]

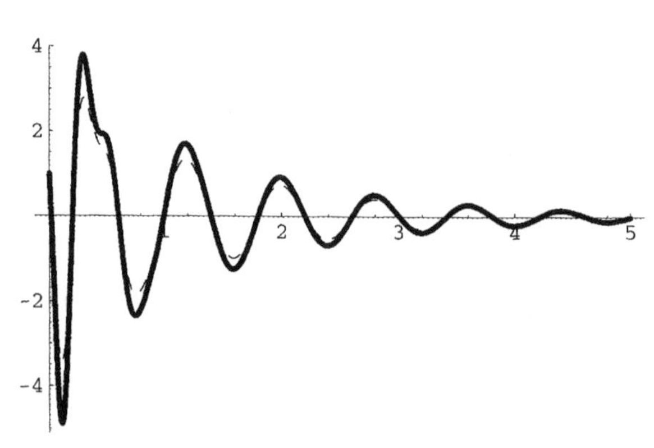

Out[46]:= – Graphics –

A.6.2 Forced Vibrations of a 2DOF System

The system is the same as in the case of free vibrations. In this case, however, a periodic force p2 Exp[iωt] is applied to the second mass.

Dynamic Equations

In[1]:= $\text{eqOne} = m_1 u_1''[t] + k_1 u_1[t] + c_1 u_1'[t] + k_2(u_1[t] - u_2[t]) + c_2(u_1'[t] - u_2'[t]) == 0$

Out[1]:= $k_1 u_1[t] + k_2(u_1[t] - u_2[t]) + c_1 u_1'[t] + c_2(u_1'[t] - u_2'[t]) + m_1 u_1''[t] == 0$

In[2]:= $\text{eqTwo} = m_2 u_2''[t] + k_2(u_2[t] - u_1[t]) + c_2(u_2'[t] - u_1'[t]) == p_2 E^{(\text{I omega t})}$

Out[2]:= $k_2(-u_1[t] + u_2[t]) + c_2(-u_1'[t] + u_2'[t]) + m_2 u_2''[t] == e^{\text{i omega t}} p_2$

Substitute solutions in the form: $u_1[t] = d_1\, e^{(\text{I omega t})}$ and $u_2[t] = d_2\, e^{(\text{I omega t})}$.

In[3]:= $\text{substOne} = \{u_1[t] \to d_1 E^{(\text{I omega t})}, u_2[t] \to d_2 E^{(\text{I omega t})},$
$u_1'[t] \to D[d_1 E^{(\text{I omega t})}, t], u_1''[t] \to D[d_1 E^{(\text{I omega t})}, \{t, 2\}],$
$u_2'[t] \to D[d_2 E^{(\text{I omega t})}, t]\};$

In[4]:= $\text{substTwo} = \{u_1[t] \to d_1 E^{(\text{I omega t})}, u_2[t] \to d_2 E^{(\text{I omega t})},$
$u_1'[t] \to D[d_1 E^{(\text{I omega t})}, t], u_2'[t] \to D[d_2 E^{(\text{I omega t})}, t],$
$u_2''[t] \to D[d_2 E^{(\text{I omega t})}, \{t, 2\}]\};$

In[5]:= eqOneSub = eqOne /. substOne

Out[5]:= $d_1\, e^{\text{i omega t}} k_1 + (d_1\, e^{\text{i omega t}} - d_2\, e^{\text{i omega t}}) k_2 + \text{i}\, c_1\, d_1\, e^{\text{i omega t}} \text{omega} -$
$d_1\, e^{\text{i omega t}} m_1 \text{omega}^2 + c_2(\text{i}\, d_1\, e^{\text{i omega t}} \text{omega} - \text{i}\, d_2\, e^{\text{i omega t}} \text{omega}) == 0$

In[6]:= eqTwoSub = eqTwo /. substTwo

Out[6]:= $(-d_1\, e^{\text{i omega t}} + d_2\, e^{\text{i omega t}}) k_2 - d_2\, e^{\text{i omega t}} m_2 \text{omega}^2 +$
$c_2(-\text{i}\, d_1\, e^{\text{i omega t}} \text{omega} + \text{i}\, d_2\, e^{\text{i omega t}} \text{omega}) == e^{\text{i omega t}} p_2$

Cancel $E^{(\text{I omega t})}$ by replacing it with 1.

In[7]:= $\text{eqOneSub} = \text{eqOneSub} /.\ E^{(\text{I omega t})} \to 1$

Out[7]:= $d_1 k_1 + (d_1 - d_2) k_2 + \text{i}\, c_1\, d_1\, \text{omega} - d_1\, m_1\, \text{omega}^2 +$
$c_2(\text{i}\, d_1\, \text{omega} - \text{i}\, d_2\, \text{omega}) == 0$

In[8]:= $\text{eqTwoSub} = \text{eqTwoSub} /.\ E^{(\text{I omega t})} \to 1$

Out[8]:= $(-d_1 + d_2) k_2 - d_2\, m_2\, \text{omega}^2 + c_2(-\text{i}\, d_1\, \text{omega} + \text{i}\, d_2\, \text{omega}) == p_2$

Solve the above equations for d_1 and d_2:

In[9]:= sol = Solve[{eqOneSub, eqTwoSub}, {d_1, d_2}]

Out[9]:= {{$d_1 \to -((k_2 + i\, c_2\, \text{omega})\, p_2)\, /$
$(-k_1 k_2 - i\, c_2\, k_1\, \text{omega} - i\, c_1\, k_2\, \text{omega} + c_1\, c_2\, \text{omega}^2 + k_2\, m_1\, \text{omega}^2 + k_1\, m_2\, \text{omega}^2 + k_2\, m_2\, \text{omega}^2 + i\, c_2\, m_1\, \text{omega}^3 + i\, c_1\, m_2\, \text{omega}^3 + i\, c_2\, m_2\, \text{omega}^3 - m_1 m_2\, \text{omega}^4)$,
$d_2 \to -((k_1 + k_2 + i\, c_1\, \text{omega} + i\, c_2\, \text{omega} - m_1\, \text{omega}^2) p_2)\, /$
$((-k_2 - i\, c_2\, \text{omega})^2 - (k_1 + k_2 + i\, c_1\, \text{omega} + i\, c_2\, \text{omega} - m_1\, \text{omega}^2)$
$(k_2 + i\, c_2\, \text{omega} - m_2\, \text{omega}^2))$}}

At this point let us choose specific data: m_1 = 20 kg, m_2 = 10 kg, k_1 = 2000 N/m, k_2 = 3000 N/m, c_1 = 50 N/(m s), c_2 = 50 N/(m s):

In[10]:= m_1 = 20; m_2 = 10; k_1 = 2000; k_2 = 3000; c_1 = 50; c_2 = 50;

Find explicitly the amplitudes d_1 and d_2 of the first and second masses, respectively:

In[11]:= displOne = Simplify[d_1 /. sol[[1,1]]]

Out[11]:= $$\frac{(60 + i\, \text{omega}) p_2}{120000 + 5000\, i\, \text{omega} - 2250\, \text{omega}^2 - 40\, i\, \text{omega}^3 + 4\, \text{omega}^4}$$

In[12]:= displTwo = Simplify[d_2 /. sol[[1,2]]]

Out[12]:= $$-\frac{(-250 + 5\, i\, \text{omega} + \text{omega}^2) p_2}{5(60000 + 2500\, i\, \text{omega} - 1125\, \text{omega}^2 - 20\, i\, \text{omega}^3 + 2\, \text{omega}^4)}$$

In[13]:= amplOne = Abs[displOne] /. $p_2 \to 1$

Out[13]:= $$\text{Abs}\left[\frac{60 + i\, \text{omega}}{120000 + 5000\, i\, \text{omega} - 2250\, \text{omega}^2 - 40\, i\, \text{omega}^3 + 4\, \text{omega}^4}\right]$$

In[14]:= amplOneplot = Plot[amplOne, {omega, 0, 50},
PlotStyle → {AbsoluteThickness[2.4]}]

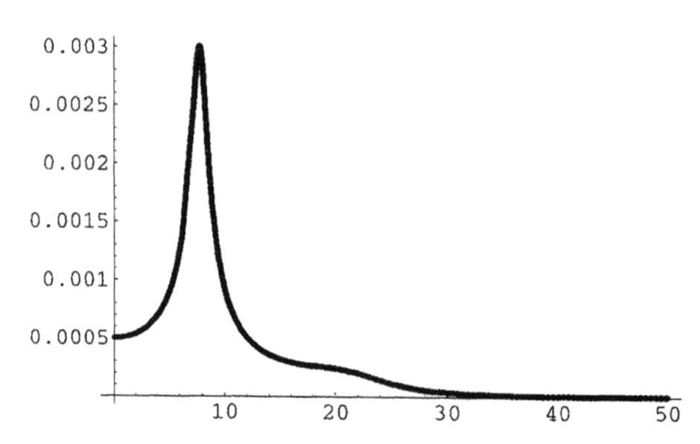

Out[14]:= – Graphics –

In[15]:= amplTwo = Abs[displTwo] /. $p_2 \to 1$

Out[15]:= $\frac{1}{5}\text{Abs}\left[\frac{-250 - 5\ i\ \text{omega} + \text{omega}^2}{60000 + 2500\ i\ \text{omega} - 1125\ \text{omega}^2 - 20\ i\ \text{omega}^3 + 2\ \text{omega}^4}\right]$

In[16]:= amplTwoplot = Plot[amplTwo, {omega, 0, 50},
PlotStyle → {Dashing[{0.02}]}]

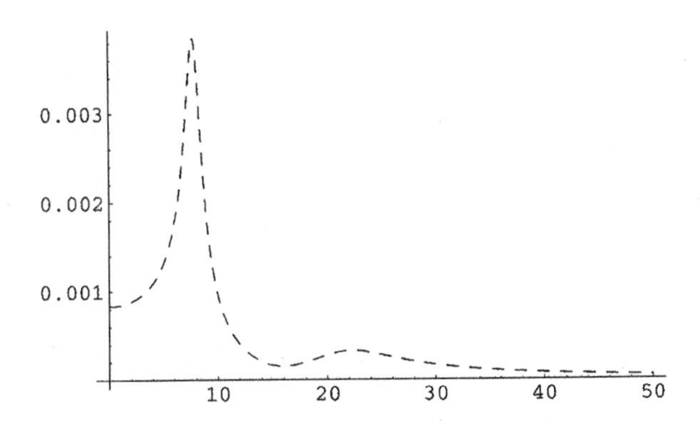

Out[16]:= – Graphics –

In[17]:= Show[amplOneplot, amplTwoplot]

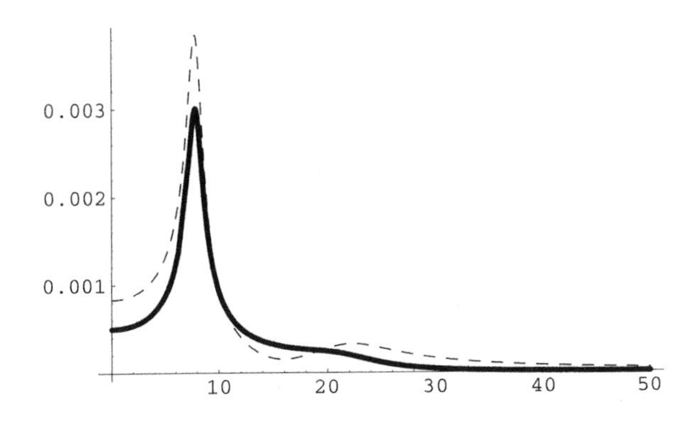

Out[17]:= – Graphics –

In[18]:= phaseAngle = Arg[displOne] /. p2 → 1

Out[18]:= $\text{Arg}\left[\frac{60 + i\ \text{omega}}{120000 + 5000\ i\ \text{omega} - 2250\ \text{omega}^2 - 40\ i\ \text{omega}^3 + 4\ \text{omega}^4}\right]$

In[19]:= phaseOneplot = Plot[phaseAngle, {omega, 0, 50}]

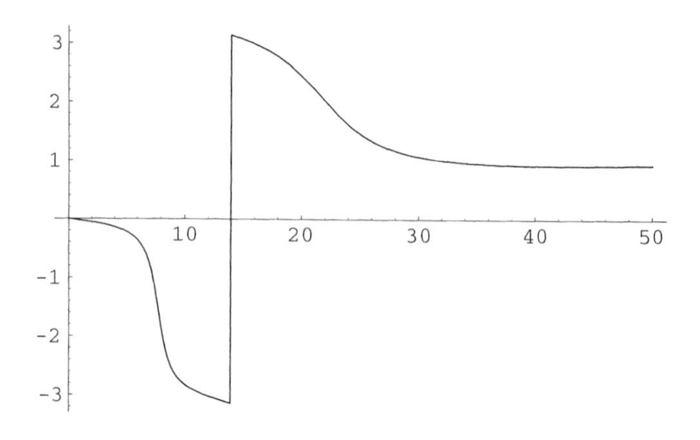

Out[19]:= – Graphics –

We can see that the phase angle makes a 2π jump. Let us adjust it:

In[20]:= phaseOne = If[phaseAngle > 0, phaseAngle – 2π, phaseAngle]

Out[20]:= $\text{If}\left[\text{Arg}\left[\frac{60 + \text{i omega}}{120000 + 5000\ \text{i omega} - 2250\ \text{omega}^2 - 40\ \text{i omega}^3 + 4\ \text{omega}^4}\right] > 0, \text{phaseAngle} - 2\pi, \text{phaseAngle}\right]$

In[21]:= phaseOneplot = Plot[phaseOne, {omega, 0, 50}, PlotStyle → {AbsoluteThickness[2.4]}]

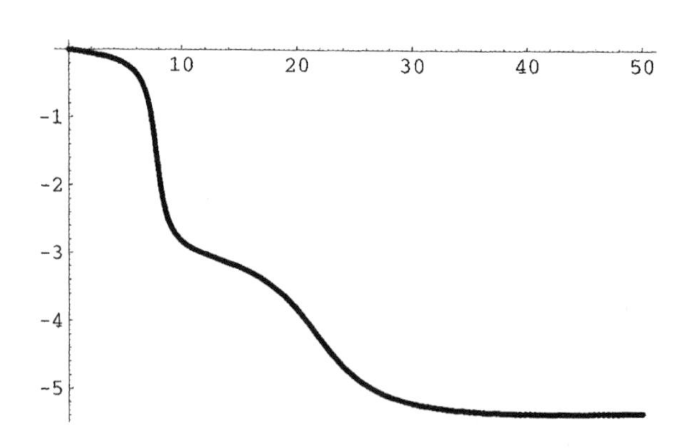

Out[21]:= – Graphics –

In[22]:= phaseTwo = Arg[displTwo] /. p2 → 1

Out[22]:= $\text{Arg}\left[-\frac{-250 - 5\,i\,\text{omega} + \text{omega}^2}{60000 + 2500\,i\,\text{omega} - 1125\,\text{omega}^2 - 20\,i\,\text{omega}^3 + 2\,\text{omega}^4}\right]$

In[23]:= phaseTwoplot = Plot[phaseTwo, {omega, 0, 50},
PlotStyle → {Dashing[{0.02}]}]

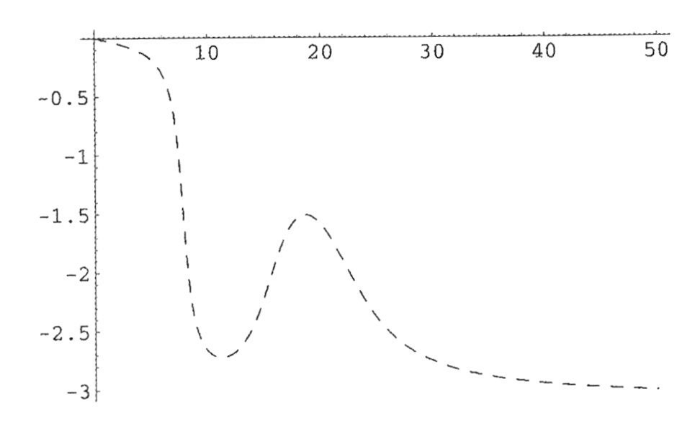

Out[23]:= – Graphics –

In[24]:= Show[phaseOneplot, phaseTwoplot]

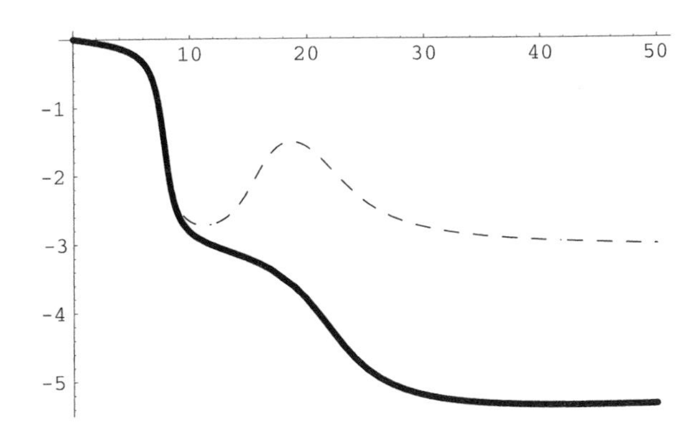

Out[24]:= – Graphics –

We can see that up to the first resonance frequency, 7.8 rad/s, the two masses move in the first mode; i.e., the phase angle is close to 0. Then the phase angle between the two masses increases until it becomes π; i.e., the two masses are in the second mode.

Index

K

L

M

N

P

R

S

T

U

V

W